Urban Agglomeration
http://dx.doi.org/10.5772/intechopen.70025
Edited by Mustafa Ergen

Contributors

Tetsuro Saisho, Kana Ram Godha, Aida Huerta Barrientos, Wolfgang Loibl, Ghazal Etminan, Ernst Gebetsroither, Hans-Martin Neumann, Santiago Sanchez-Guzman, Yu-Hsin Tsai, Jhong-Yun Guan, Yu-Hsin Huang, Rui Alexandre Castanho, Luís Carlos Loures, Ana Vulevic, José Cabezas, José Manuel Naranjo Gómez, Luis Fernández-Pozo, Fred Bidandi, Tamara Streefland, Yoram Krozer, Marcelo Gomes Miguez, Aline Pires Veról, Andréa Queiroz Rêgo, Ianic Bigate Lourenço, Jutta Gutberlet, Jon Burley, Haoxuan Xu, Yuemin Jin, Patricia Machemer, Pat Crawford, Zhen Wu, Luis Loures, Uno Svedin, Hans Liljenström, Miguel Ángel Mejías Vera

Published by InTech
Janeza Trdine 9, 51000 Rijeka, Croatia

Publishing Process Manager Martina Usljebrka
Technical Editor SPi Global
Cover InTech Design team

First published March, 2018
Printed in Croatia
Legal deposit, Croatia: National and University Library in Zagreb

Additional hard copies can be obtained from orders@intechopen.com

Urban Agglomeration, Edited by Mustafa Ergen
p. cm.
Print ISBN 978-953-51-3897-6
Online ISBN 978-953-51-3898-3

For EU product safety concerns:
IN TECH d.o.o., Prolaz Marije Krucifikse Kozulić 3, 51000 Rijeka, Croatia,
info@intechopen.com or visit our website at intechopen.com.

Meet the editor

Mustafa Ergen graduated from Abant İzzet Baysal University's Department of Landscape Architecture in 2000 in Turkey. In 2005, he completed his first master's degree at the Gebze Institute of Technology in Urban and Regional Planning in Turkey and his second master's degree at Anhalt University of Applied Sciences in Landscape Architecture, Germany, in 2006. He studied Geographic Information Systems and Remote Sensing at the Mediterranean Agronomic Institute of Chania between 2007 and 2008 in Greece and was granted a specialization diploma in Environmental Management. He received the Dr.-Ing. degree by the Technical University of Dortmund, Germany, in 2013. Currently, he works in the Department of Architecture at Siirt University in Turkey. He speaks English fluently, he has intermediate level of German language, and his mother tongue is Turkish. His studies focus on cityscapes, urban and landscape planning, design, and geographic information systems and remote sensing.

Contents

Preface

People living in rural areas migrate to urban areas to secure better qualities of life, education, and health facilities and also because they believe that urban settings offer more livable conditions. These appealing features have led to rapid population growth in urban areas, which has resulted in problems that need to be solved through different urban planning and design approaches.

Urbanization has emerged as one of the most important responses to addressing the need of living spaces today. Particularly in recent years, a large portion of the world population has begun to live in cities, the results of which have created a wide variety and ever-increasing number of urban problems. Urban problems can have very different solution approaches; however, one that has been attracting significant attention is urban agglomeration, which uses the appeal of the city to integrate surrounding areas within the city's network. This gathering in of populations to urban areas requires that cities be organized in a well-planned and sustainable manner. In achieving this, more habitable urban areas can be developed.

In addition to the problems related to the characteristic causes of urbanization, the development of which influences the quality of life of the community, urban areas present many challenges, from designing water supply channels and ensuring equal distribution of green spaces to managing land and infrastructure problems. These challenges must be addressed during the urban development stage. To provide healthier and more sustainable development, cities must be planned and designed in such a way as to minimize the problems and adequately address the challenges.

In conjunction with this book, a supplemental resource, which both provides and proposes solutions based on innovative approaches to urbanization problems that emerge from urban agglomeration, has been created. This resource supplement shall also serve as a guide to future urban development efforts. In effect, this book will play an important role in compensating for the limited number of resource books on urbanization. This book is intended to be a reference source for scientists and students interested in the subject.

Dedicated to my daughter (Vesa İlayda)

Assistant Professor Dr.-Ing. Mustafa Ergen
Siirt University, Turkey

Section 1

Urban Development Processes

Urban Agglomeration and Supporting Capacity: The Role of Open Spaces within Urban Drainage Systems as a Structuring Condition for Urban Growth

Marcelo Gomes Miguez, Aline Pires Veról,
Andréa Queiroz da Silva Fonseca Rêgo and
Ianic Bigate Lourenço

Additional information is available at the end of the chapter

http://dx.doi.org/10.5772/intechopen.71658

Abstract

Urbanisation greatly changes the natural environment—city growth may cause urban sprawl, increasing land consumption and infrastructure demands, with consequent built and natural environments degradation. To face this challenge, the supporting capacity of the natural environment needs to be addressed in the urban planning process. This chapter will particularly discuss urban drainage role in the planning context, integrating engineering, urbanism and landscaping in order to set the basic conditions towards a sustainable city development. Urban drainage systems (and the related urban rivers) play a crucial role in city planning, once it intermediates the needs of the built environment, providing safe areas free from flooding, and the demands of the natural environment, giving space and passage to floods. This particular feature gives to the drainage system a spatial structuring characteristic and it provides opportunities to revitalise city areas, improving biodiversity and recovering environmental values. On the other side, a city open spaces system is the main reserve of urban areas for sustainable urban drainage interventions. The adequate land use planning and consequent management of these open spaces shall be in the core discussion to produce integrated and functional solutions for built and natural environments.

Keywords: urban flooding, urban drainage, urban growth, supporting capacity, open spaces, multifunctional landscapes

1. Introduction

The process of city growth and the increase of urban population, during the twentieth century, led to a series of preoccupations with regards to the capacity of sustaining this growing process. There is a real threat of an environmental degradation. In 1968, the Club of Rome published a report where "limits of growth" were questioned [1] and highlighted the preoccupation with the existing economic models. In June 1972, the *Conference of the United Nations on the Human Environment* took place in Stockholm, Sweden, and brought together participants from 110 countries [2]. Starting from this meeting, known as the *Stockholm Conference*, the preoccupation with regards to environmental questions was voiced for the first time on a global scale. In 1987, the *Brundtland Report* called "*Our Common Future*" [3] elaborated by the World Commission on Environment and Development, created by the United Nations (UN), formalised the concept of sustainable development, foreseeing a more intelligent and rational production. The concept of sustainability, necessarily, needs to be seen within three basic pillars: social, economic and environmental. Sustainable urban solutions have to be able to attend the necessities of the society, preserving the environment and being economically viable as a whole and in the long run. Any disequilibrium of one of those pillars, which give support to the concept, can take to non-sustainable situations.

The proposal of a sustainable development was consolidated and widely disseminated after the *Conference of the United Nations on the Environment and Development*, known as *Rio 92* or *Eco 92*, realised in Rio de Janeiro in 1992 [4]. The elaboration of the *Agenda 21* [5] was perhaps the main result of this meeting. It defined objectives to promote sustainable development of human settlements, including: providing adequate housing for all; improving human settlements management; planning a sustainable management of land use; and promoting the provision of an integrated environmental infrastructure, targeting topics such as water, sewage, drainage and solid waste.

According to Lindholm [6], urban planning shows a utopic aspect that has to be complemented with a practical "here and present" perspective. To do this, it is necessary to identify what kind of spatial relations are being created intending to discuss and allow new future possibilities.

Today, we already fully understand that an urban planning cannot be separated from the environmental planning. Cities managed without integrated planning suffer with the loss of ecosystems and of its environmental and cultural values. Inundations, pollution, scarcity of water, shortcomings in public health, among many other grave problems, arise from this lack of integration.

Without intending to create an only definition for a so wide and complex topic, it is here understood that a sustainable city is the one which meets the social functions that are expected from it, through time, in a viable way and with a general positive balance, reaching an adequate quality of life for all its inhabitants, guaranteeing the distributed access to essential services and recourses, respecting the limits imposed by the natural system, being resilient and secure.

The definition of resilience [7–15] is a concept very much present in urban discussion nowadays [16, 17]. The resilience involves, generally (and with various interpretations, depending on the topic of the discussion), three main points: the capacity of a system to continue resisting, even if submitted to stressing conditions beyond those projected; the capacity of the system to recuperate functions in a fast way, continuing to offer its services; and the capacity of the system to recuperate structurally from suffered damages [18]. Meerow et al. [19] stated that resilience is a dynamic concept, recognising the importance of its temporal scale, where a systemic adaptability applies. This perception considers the urban system as a complex and adaptable system composed by socioecological and sociotechnical patterns that go through multiple spatial scales. In this way, the concept of resilience approaches the discussion about sustainability [20–23] as both concepts involve an integration through time, in which the systems must continue to function and provide adequate services today, as well as for future generations, submitted to future challenges.

McHarg [24] points to the necessity of considering landscaping as an ecological system, where the prime concern refers to the integration of social and environmental issues within the planning process. This author proposes a new perspective for a healthy relation between nature and the built environment, understanding the processes that shape landscape and using them as the basis for an efficient planning process. Rogers [25] poses that at the same time, the cities reunite and potentiate physical, intellectual and creative energies, they tend to damage natural environment configuring a threat to the very humanity itself. Today, cities are better known by their conflicts and problems than their potentialities. Harvey [26] highlights that the right to do (and redo) the cities (and ourselves), evolving towards a better place, is one of our most precious and neglected rights.

Various discussions are being motivated by the necessity of making the cities more sustainable [27–29]. Among these discussions, one that has been gaining space refers to the compact and diverse city concept [30–35], in which the networks of the infrastructure become less extensive and more efficient, less land is consumed and the reduced commuting consumes less energy. Burgess [36] defines the compact city as the one that results from a set of politics seeking to increase the constructed area and the demographic densities, to intensify the economic, social and cultural urban activities and manipulate the size, the form and the urban structure as well as the housing systems. This concept benefits from the environmental, social and global sustainability that derives from the concentration of urban activities. It is further believed that in more compact and diverse cities, social relations are intensified, increasing the process of democratisation. For Rogers [37], a city is sustainable when it is fair, pretty, creative, ecological, easy, compact, polycentric and diversified. Nevertheless, if extremely densified, cities that are more compact may lead to greater local negative ecological impacts, when compared with that resulting from less compact cities. There certainly is a limit for compaction, equalising the supporting capacity of the natural environment and the optimisation of the built environment. In this way, a balance shall be established between the built and the open spaces, defining the degree of city compaction affecting its sustainability. The idea is not creating an absolute rule, as all cities are cultural representations of huge magnitudes. However, a point of equilibrium for each city in its diversity must be found.

In this discussion about balancing built and open spaces, urban flooding arise as an element of disruption of urban services, of infrastructure networks, of housing and health systems. When urban rivers and drainage networks fail, leading to flooding, they produce urban negative consequences and, very often, irreparable socioenvironmental damages. However, although extensive flood control structures have been implemented in the recent years (e.g. dikes, dams, canals, pump stations, etc.), cities everywhere remain vulnerable to floods [38]. Ecologically, more sensitive mitigation approaches have appeared in recent years, encouraging flood hazard mitigation to work with nature [39]. Grizzetti et al. [40] remember that a territorial planning for urban water management is needed, and system multifunctionality has to be recognised as well as the benefits of approaching city and nature. Conceptually, urban drainage is an infrastructure system that occupies a key position interfacing natural demands (rainfall-runoff transformation and consequent discharges) and city needs (healthy neighbourhoods and city functions maintenance).

2. The supporting capacity as a limiting factor for urban planning

The capacity of the environment to support a human settlement can be defined as the maximum elasticity to which the environment can be subjected, without any irreversible degradation, to attend the urban necessities that will be developed systematically in this space.

The process of urbanisation, generally, demands multiple and continued efforts to satisfy the necessities of the population. However, this process, if not planned or barely conducted, causes harmful effects on the environment, resulting in a possible degradation of the natural environment as well as of the built environment. As the urban population increases, various problems related with the necessary infrastructure to attend this growing demand also increase. Many times the concept of supporting capacity is reduced to the capacity of the infrastructure to continue meeting the demands of the city development. However, from a systematic point of view, one must not overlook the wider concept of the supporting capacity and the perception that the built environment needs to respect the limits imposed by the natural environment.

As introduced earlier, the urban drainage system occupies a very specific position in this discussion. This system lies between the demands of the natural environment and the necessities of the built environment: the rain that falls on a watershed and needs to flow through the city is able to find its way within the drainage system, which offers, in turn, secure conditions for the city to occupy the natural space [41]. Therefore, the city may coexist with the natural watershed in a healthy combination [42]. Nevertheless, the very urbanisation modifies the natural space and the way in which this space responds to the water cycle, aggravating floods [43, 44]. In this way, environmental issues specifically linked to the urban drainage becomes more evident, once failures of the drainage system affect the functioning of the city, reflecting in vectors of environmental degradation. These are key questions for urban planners and decision makers, since they are structuring factors for planning the space and its sustainable occupation.

However, it is frequent that cities exceed the supporting capacity, consuming more than necessary (depleting natural recourses and expanding over the watershed space, modifying land use) and creating more and more waste (which the environment cannot absorb). The areas of a city which usually are exposed to more extreme conditions, with high densities of population, occupation and construction, with fewer open spaces and higher inundation risks, are also usually related with the more critical shortcomings of infrastructure and higher economic and social losses.

When modifying nature, without considering the supporting capacity of the environment, by means of unrestrictedly building and spreading urban growth, the population suffers with the reduction of environmental quality, which could directly and/or indirectly interfere in the quality of people's lives. Some consequences of this development, such as vegetal cover removal, erosion and/or silting-up of watercourses, lacking of open spaces, air and water pollution, inadequate deposition of garbage and sewage, among others, bear witness to the disequilibrium between the cities growth and the functional aspects of the natural environment.

Several studies have been realised and a series of efforts have been made to establish indicators for sustainable development. In 1996, for example, the Commission of the United Nations for Sustainable Development announced a project to evaluate and compare the degree of sustainable development for each country, generating indicators [45]. Since then, several indicators have been developed to evaluate the results of urban growth. Such indicators generally include social, economic, environmental and institutional categories. Among those, the "environmental" dimension ought to be the prime preoccupation in the pursuit of an environmentally healthy and sustainable development.

Kyushik et al. [46] discussed the urban carrying capacity as a determinant for the developing density and presented some authors who also discussed the supporting capacity concept. In this discussion, they firstly cite Chung [47], who stated that the ecologists generally consider the supporting capacity as the maximum number of individuals who can be supported by the environment in a given area, without compromising future generations living in this area. Planners generally define the supporting capacity as the capacity of a natural or artificial system to absorb the growth of the population or the physical development without considerable degradation or damage [48]. It is also defined as the level of human activities, growth of the population, land use and physical development, which could be sustained by the urban environment without causing serious degradation and irreversible damage [49]. This concept is based on the assumption that a certain environmental thresholds exists, which, when crossed, can cause serious and irreversible damages to the natural environment [50].

Considering the discussion developed and the definitions cited above, one can conclude that the limit of the supporting capacity of the environment is interlinked with the social environmental quality of the urban space. For this reason, the urban planning process and the discussion about socioeconomic, morphologic and functional aspects of the city have to incorporate the supporting capacity concept.

3. Open spaces system

The urban planning integrated with the environmental planning (or, simply, the urban environment planning), can provide an improvement in the quality of life in cities. However, this is only possible when some concepts and relations are understood linking environment, open spaces system and cultural landscapes.

Tuan [51] stated that "environment sustains us as creatures; landscapes display us as cultures". Schlee et al. [52] stressed that "the landscape is a product deeply impregnated with culture, which results in processes of continuous alterations, dictated by biophysical, social, economic and political factors, reflected in the forms of occupation and management of the territory".

Therefore, a landscape, and especially the urban landscape, is not only a cultural representation of a society but also a set of human actions in an environment which establishes itself in an integrated way, but not without conflicts during time. For Meining [53] "any landscape is so dense with evidence and so complex and cryptic that we can never be assured that we have read it all or read it right".

In this way, every urban landscape is distinct, but all of them consist of built spaces and non-built spaces—these last ones are the open spaces free of construction, either being public or private. Macedo and Queiroga [54] understand the open space system as "all the elements and the relationships that organise and structure the set of all open spaces of a determined urban area – from intra-urban to regional scale. It is basic in the existence of the city, because it is fundamental in the performance of the day to day life and in defining the urban form and the image of the city, reflecting its history and memory, participating in the public and private life spheres".

Rego et al. [55] define the open urban spaces system in: (1) environmental open spaces; (2) social interaction open spaces, such as parks, squares or gardens; (3) open spaces of infrastructure character, such as a strip of land protecting transmission lines; and (4) open spaces of mobility character, such as streets, roads and avenues.

Ironically, every city starts to be designed by its streets. These are the ways which connect the various parts and structured elements, leaving the first marks on the land which will define a human settlement. Therefore, the occupation/construction of a city is initiated by an element that afterwards will be named as an open space, which defines not only the border between what is public and private, but also where the largest part of the urban network infrastructure will be implemented.

It has to be pointed out that, at times, the open urban spaces are only apparently free of construction as the substratum is occupied by large structures such as metro, underground roads, garages or even canalised rivers.

In this way, it is understood that the open spaces system of a great number of consolidated cities is not the result of an environmental urban planning but the result of a process of hundreds or thousands of years which today, many times, is insufficient or inadequate to couple with the socioenvironmental demands.

Generally, the segmented organisation of a territory, without the proper preoccupation regarding its recourses, especially referring to open spaces, is one of the facts that characterises the urban sprawl of current metropolis. Urbanisation spreads without borders and without giving attention to the collective necessities, favouring a general disequilibrium [56]. A systematic and integrated vision is necessary, associating environmental questions and urban infrastructures. Public politics should be concerned in re-evaluating land use and the occupation, seeking to adopt more resilient and sustainable urban standards. The open spaces always present a great possibility to suffer and generate transformations (positives or negatives) in the landscape shaping (or re-shaping). Urban spaces are simultaneously the most promising and most fragile spaces, because they are subjected to inadequate or disorderly occupation, when their fundamental importance is not recognised [24].

This fact occurs in many cities in developing countries, from small town to metropolis, where open spaces, generally of environmental character, are ending up being occupied in an irregular manner by substandard dwellings of a population economically and socially less favoured, such as on hill sides, mangroves areas, the margins of rivers and lakes, following an accelerated population growth which is neither accompanied by a socioeconomic development, nor by an adequate infrastructure support.

A strategy to restrain the irregular occupation of those open spaces is to have them recognised as public structures with a social and patrimonial value, that is, spaces for congregating social groups, while appropriate for collective leisure.

Chiesura [57] argues that urban parks and open green spaces are of strategic importance for the quality of life of our increasingly urbanised society. Increasing evidences indicate that the presence of natural assets (i.e. urban parks and forests, green belts) and components (i.e. trees, water) in urban territories contributes to the quality of life in many ways. Besides, important environmental services such as air and water purification, wind and noise filtering, flood control or microclimate stabilisation, natural areas provide social and psychological services, which are of crucial significance for the liveability of modern cities and the well-being of urban dwellers.

Nevertheless, the lack of open spaces in consolidated cities is frequent, as it is frequent that such spaces are not adequately foreseen in still growing cities. To minimise the conflicts in the demand for space, the concept of multifunctional spaces could be the answer to optimise the landscape design, associating the open spaces with the urban occupation demands, without losing any functions already existing today.

A multifunctional space is an urban intervention which foresees more than just one use for the same place, guaranteeing a more rational use of the urban land. Different uses can target infrastructure aspects, such as urban drainage, as well as trying to overcome social privation, such as the absence of leisure areas.

It is necessary to densify cities into more compact arrangements, associating diverse and multiple uses to urban spaces, intending to optimise commuting and the several services offered, without eliminating open spaces. These spaces need to be preserved as a basic support for a city capable of meeting the environment, infrastructural and social needs, corroborating with the very concept of sustainability, also emphasising the well-being and the quality of life of the population.

In this way, the solutions for urban drainage problems should be able to integrate the response to urban demands regarding sanitation purposes with the supporting capacity of the urban watershed, using multifunctional spaces to reorganise the water cycle functions that were modified by urbanisation. Spirn [58] criticises superficial urban projects, created only to beautify the city with an artificial nature. For the author, radical changes are necessary, seeking to recognise and point out a series of damages, mainly in artificialized urban rivers, which do not respect the necessities of nature and do not consider the river as an essential strength that permeate the city. According to Naveh [59], one of the premises for a holistic concept of multifunctional landscapes is that these represent a complex interaction between nature and culture.

In this context, all the aggregated values of landscapes should be encouraged: their environmental aspects; their cultural, social and economical values; defending the environmental legislation as part of the regulation for a proper urban expansion, with the purpose of bringing the concept of supporting capacity to the discussion of urban and environmental planning.

4. Storm waters and cities

Water is probably the most essential natural resource and relates to diverse aspects of human settlements, from the beginning of the history, ranging from water supply, irrigation, transportation, territory defence, among others. From ancient times, man always looked for locations where the environment could give support for his survival. Great part of the first civilisations developed along river banks or near lakes.

Any city, however, tends to introduce modifications in the land use patterns, which induce a series of processes that change the environment, affecting the quality both of the natural and the constructed areas, and many of the supervening consequences are related to urban waters, in its wider sense.

Therefore, a paradox exists in the relationship between water and city [41]. Water is a fundamental resource, which is present in the history and in the origin of cities, but is also a problem in modern cities. Actually, for this reason, it is not unusual that cities turn their back to the rivers, which are degraded, polluted, frequently seen as sewage conveyors or buried and hidden. Most urban rivers are lost as landscape elements, impoverishing the urban biodiversity and degrading also the urban vicinity [41, 42].

The rivers, in general, can be considered as a synthesis of the territory to which they are connected [60]. Or rather, the actions which take place in the watershed reflect in the fluvial corridor. In this way, it can be said that artificial rivers, degraded, suffering from flooding and degrading the city, end up to be the product of the urban environment itself, which have dis-characterised the natural processes of their hydrographic basin.

The urbanisation process strongly alters the natural water cycle and the responses of the fluvial systems on the built environment. The removal of vegetation, the increasing imperviousness that follows this process, the regularisation of the surfaces and the introduction of an artificial drainage system modify significantly the superficial flow patterns, producing larger flow volumes and peak flows, reducing the base flow discharges and the time of the concentration of the basin. A frequent result of this process is the aggravation of flood problems observed in cities and the loss of fluvial ecosystems.

Urban flooding is strongly related with land use questions. Typical urban factors, such as housing shortage, for example, turn up as aggravating agents. The irregular occupation of riverine areas, which should be used as river space, increases flooded areas and flooding consequences. The floods, in turn, are responsible for the degradation of the urban built environment, interfering with various other urban systems. They generate damages to structures and urban equipment, devaluate areas subjected to inundation, induce losses associated to paralysed businesses and services, interrupt the circulation of pedestrians and transport systems. They spread water-borne diseases, affecting and being affected by the improper collection and disposal of urban sewage and solid waste. The system weakens as a whole, risks increase and a degradation cycle emerges [41]. In resume, the impacts of urbanisation are presented in **Figure 1**.

The main gaps of integration between drainage design and urbanisation can be cited in a resumed way, as given in the following:

- When the increasing urban growth is not controlled, the exceeding soil sealing by impervious surfaces generate superficial flows surpassing the capacity of the drainage network, going beyond those discharges foreseen in the project horizon;
- lack of urban growth control in the total project area (mainly upstream), advancing towards areas which were originally natural, also increases discharges in the drainage network;
- unsuitable occupation of the bottom of valleys and riverine areas, which should be preserved as flood plains, directly exposes the communities installed there;
- lack of integration between drainage systems and the other basic sanitation systems, which need to be understood as complementary and not independent systems.

A series of reciprocal interferences can be mapped between the drainage system and the other urban systems, producing faults in cascades, as in a house of cards.

Gusmaroli et al. [61] proposed the adoption of an ecosystem approach to face river problems in urban areas, with the objective to amplify the concept of Waterfront Design, in which one seeks to improve the value of the contact line between city and water surface, reintroducing

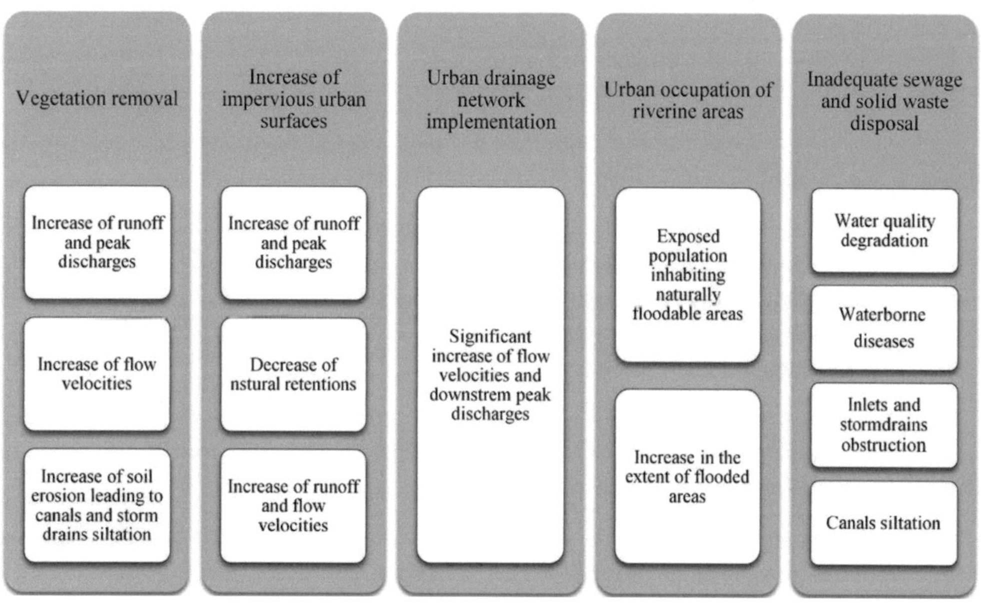

Figure 1. Impacts of urbanisation on inundations [41].

the watercourse in the city landscape. This approach widens the possibility of just using the presence of water as an aesthetic urban value, but also, and mainly, as an environmental asset, an element for connecting the city with nature. This possibility bring an opportunity to pursue the concept of fluvial restoration, from the point of view of an effective environmental improvement, looking at the city as a live organism in constant transformation and, due to this characteristic, capable to remodel and adapt itself (even if only partially, given all the modifications already suffered) to the demands of a more natural watercourse. The perspective to incorporate concepts of environmental sustainability into the process of rethinking the city growth opens a diversified set of opportunities to be explored as integrated solutions in a multidisciplinary context.

5. Urban drainage and open spaces

The traditional practices associated with drainage projects tend to focus on the problem of the resulting discharges in the storm drains and canals, equating them to these discharges. In this concept, the water needs to be rapidly conducted out of the watershed. This concept is very often responsible for transferring of flooding problems to downstream reaches situated in the lower river basin, without really solving the drainage problems.

Based on these questions, along the last decades, this traditional concept is being complemented or replaced by concepts which seek systematic solutions for the basin, with distributed interventions, looking to recover flow patterns similar to those existing before urbanisation. The new approach joins preoccupation of sustainable management of the urban rain waters, integrating those with urban space solutions. Measures for storing waters and incrementing the infiltration appear to be the alternative to treat the principal modifications introduced by the city growth on the water cycle [41]. These are called compensatory techniques [62], because they seek exactly to "compensate" the effects of urbanisation on the water cycle.

A resumed history of the conceptual evolution of urban drainage solutions since the Industrial Revolution is presented as follows:

- The necessity of urgent sanitation appeared as a primary concern in industrial cities. As they suffered from serious problems of epidemics, this fact led to the development of hygienist concepts in drainage projects, reflecting on the necessity to convey as fast as possible rain waters and sewage—storm drains and artificial canals were adopted as frequent solutions and the fluvial and sanitary systems worked together.
- The possibility of environmental degradation, the great rainfall intensities and the difficulty to treat diluted sewage led to the concept of separate sewer network at the beginning of the twentieth century.
- Until approximately the 1970s, the drainage focus was set on the increase of conveyance and the necessity to improve flow conditions to face the increase of run off generated by the cities growth.
- Since 1970, the traditional paradigms started to be broken.

- The vision that the traditional measures of canalisation tended to transfer problems became dominant.
- Compensatory techniques were adopted to control the distributed generated run off.
- On source controls to minimise run off were introduced.
- Quantity and quality of urban waters became integrated concerns.
- The concept of "sustainable drainage" was formalised, seeking to: minimise the transfer of flood problems in space (for other locations) and in time (for future generations); integrate drainage solutions with socioenvironmental aspects; increase the biodiversity in the built environment; revitalise the urban spaces and articulate with urban development plans.
- Rivers came to be seen as possible drivers for restructuring the urban landscape.
- Rain water started to be seen as a usable resource.

In this new approach, the objectives of an urban drainage system, in a wider context, can be defined as:

- The reduction of inundation of a given region of interest and the minimisation of damages to the community installed in the drained basin.
- Integration with the urban developing plan, both in what regards land use questions and the future urban growth.
- Preservation of riverine areas, as well as the integration of drainage solutions within urban landscapes, in multifunctional combinations.
- Integrated evaluation of questions regarding quality and quantity of the urban waters.
- A compromise between the drainage of the interest region and final destination in the receiving water body, without transferring problems downstream.

It is necessary to evaluate the points of conflict between the natural and the built environment, searching for means of convergence, equalising conflicts and potentialising synergies and positive results.

The search for sustainability need to be kept in the conjugation of the social, economic and environmental scales, with lasting results, as sustainability is a concept that directs the vision to the future. Therefore, a solution that intends to be sustainable needs to be able to stand and face future challenges. This characteristic converges to the concept of resilience. In this context, the sustainable development is capable to face future risks supported by the environmental resilience.

A set of premises is suggested here to direct the major drainage projects towards an effective solution for the urban flooding, considering an integrated vision for the river and the city, with a complementary (and not a competitive) approach regarding the natural and built environments. In this sense, the creation of "multifunctional spaces" allows that the open spaces remain in the urban cores already consolidated, associating hydrologic functions to the already existing socioenvironmental and economic functions.

The utilisation of these open spaces can permit:

- to control and reorganise flow discharges avoiding flooding;
- the introduction of fluvial parks, which may be used for leisure, landscaping and flood storage, besides helping the organisation of urban space and creating limits for the urban growth (preventing the city to advance over preservation areas);
- the increment of urban biodiversity;
- the use of green ways with the purpose of mobility and connection of the fragmented environmental areas;
- the valorisation of property in the neighbourhood;
- the eventual waterway transportation in specific cases.

Therefore, the creation of multifunctional spaces can (and should) be linked to a strategy that embodies the concepts of fluvial restoration. Such an action foresees the environmental recovery of the fluvial corridor, as an additional component of the sustainable urban drainage approach, which, in turn, offers a distributed action in the watershed, reorganising flow patterns and restoring water cycle functions, preventing flooding.

6. Case study and discussion

Dona Eugênia River Basin (coordinates 22° 46′ 55″ S, 43° 25′ 44″ O) drains a watershed of 18 km^2 situated in the Rio de Janeiro metropolitan area, crossing two cities of a lowland region called Baixada Fluminense: Nova Iguaçu and Mesquita. Dona Eugênia River is about 10 km long. The first 4 km are located in Nova Iguaçu inside an environmental preservation area called Gericinó/Mendanha. The subsequent 6 km run through the city of Mesquita, crossing a dense urban area until its outfall at Sarapuí River. The preserved upstream area contrasts with the degradation found in the lower urban area, just a few kilometres downstream, and is a reference for restoration goals. The climate is hot and humid with a summer rainy season (Aw in the Koppen climate classification). It has an average annual temperature of 22°C and an average annual rainfall of 1700 mm [63]. Mesquita has approximately 170,000 inhabitants, according to the 2010 Brazilian Census [64] and the Human Development Index calculated for the city and reported by the United Nations Development Programme in 2010, is 0.737, which places Mesquita at the 850th on the list of Brazilian municipalities.

The city has many infrastructure problems. According to Ref. [42], in the most populated areas of the city, there are numerous shantytowns built illegally, without considering environmental risk factors, both in flat flood prone areas and on steep hillsides. Even in areas regularly occupied, urban growth was not controlled, urban occupation sprawled, open areas are few and infrastructure was not adequately provided. The main problems of this city can be summarised as follows:

- Intense and irregular informal settlements on riverine areas;
- Lack of vegetation in the urban reaches of the river;
- Urban settlements in risk areas, including dwellings on river banks, where the houses act as dikes, "canalising" the river;
- Parts of the main river run canalised underground;
- Recurrent problem of flooding, which affects about 80% of the city population;
- Sedimentation at various points of the river;
- Discharge of sewage and solid waste into the river, with visible environmental degradation and health hazards;
- Scarce infrastructure;
- Degradation of the urban environment.

In the recent past, the municipality has canalised Dona Eugênia River towards its upstream reaches in an artificialisation process that is stressing the environmental protection area, while configuring a loss of opportunity to integrate the river as an environmental asset in the urban landscape. The consequences are clear: less biodiversity, fewer open green areas, less leisure opportunities, greater environmental fragmentation (with consequent degradation) and worse floods downstream.

In the context of the discussion developed in the previous sections of this chapter, this case has motivated a research joining the necessity to find a flood control alternative, the aspiration to revitalise the degraded urban areas and the goal to improve environmental conditions. The main part of this research is concerned with the reorganisation of the open spaces system, providing a storage capacity to the urban drainage system, while using the river path as a corridor to integrate environment fragmented areas among them and with the upstream protected forest.

A mathematical hydrodynamic model—MODCEL—developed in the Federal University of Rio de Janeiro (UFRJ) was used to produce flood maps and then evaluate flood conditions. Due to the limited availability of data, the model calibration used the results of a previous study as reference. This study was, carried out by the Rio de Janeiro state government, in the revision of the *Water Resources Master Plan of the Iguaçu-Sarapuí River Basin* [63]. The catchment's outlet was one of the control points adopted for calibration purposes. At this point, the target discharge was 40 m^3/s, and the modelled value reached 41 m^3/s. To ensure a greater reliability, the calibration process also reproduced critical flooding points, based on historic events and population memories.

The diagnosis of the current situation considered as a design storm event with a 25-year return period, as recommended by the Ministry of the Cities, in Brazil. **Figure 2** shows the flood map obtained for this situation, confirming the criticality of the flooding problem in

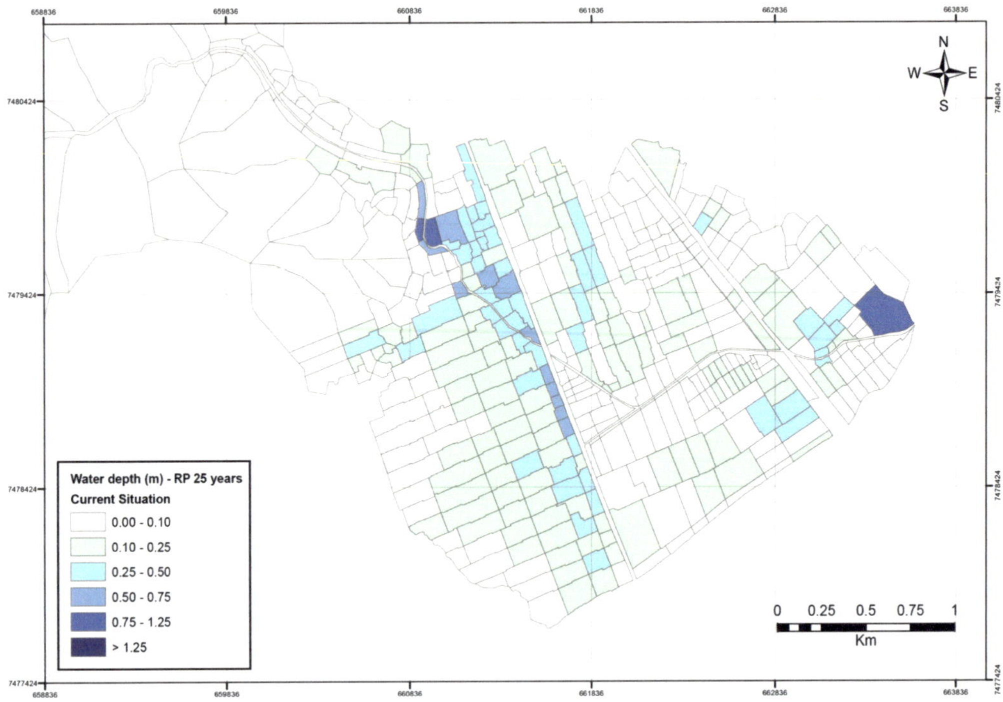

Figure 2. Current situation—flood map—recurrence period 25 years (Source: see Ref. [42]).

Mesquita. The mathematical modelling showed areas with inundation levels reaching more than 0.75 m in some places. The critical points are associated with the city centre region and in riparian areas with informal settlements. This antagonistic situation shows that both formal and informal urban areas suffer from significant flooding. The region located upstream the railway line is also critical, once the railway walls act as a barrier to runoff and urban storm drains are insufficient.

It is important to note that some flooding observed in the current situation is due to the lack of minor drainage in the basin. Thus, floods are retained in the plains and cannot reach the river rapidly. The city centre is greatly affected: both major and the minor drainage system fail. The plains near the river, at the city centre and a natural wetland downstream (where some informal settlements appear) show the highest flooding levels.

An interesting observation related to a past situation helps to illustrate the discussion. In the beginning of 1970 decade, only the area upstream the railway was urbanised (**Figure 3**).

Mesquita population, according to IBGE census for 1970 [65], was 93,678. Considering that soil imperviousness is proportional to the population density, it was possible to simulate floods in 1970. Taking the same rainfall event of 25 years of recurrence period, flood maps simulated for 1970 were compared with the ones obtained for 2010 (**Figure 4**). Two main observations arise from comparing these maps:

- In the downstream areas, where no urbanisation took place, there were no flooding problems.

- In the upstream area, although with fewer people and less imperviousness, floods were already significant (even if less important than today). This means that the space that should be occupied by flood flows was inadvertently occupied by the built environment.

Flood control is a matter of allocating spaces. Flood flows need space and the open spaces system of a city can be used to fulfil this aim.

After completing the flooding diagnosis, an alternative has been studied to solve this problem, with the expectation of avoiding river overflows and organising major drainage demands,

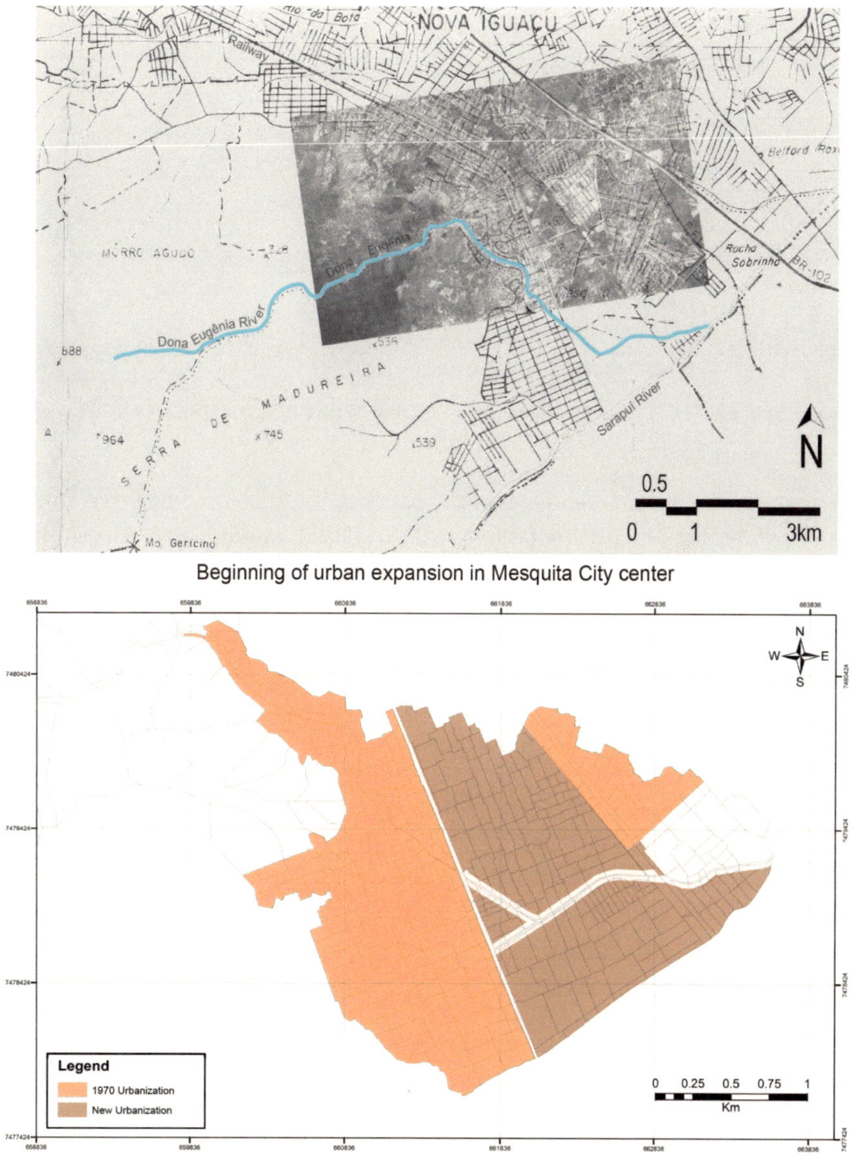

Figure 3. Past situation (1970) (Source: see Ref. [42]).

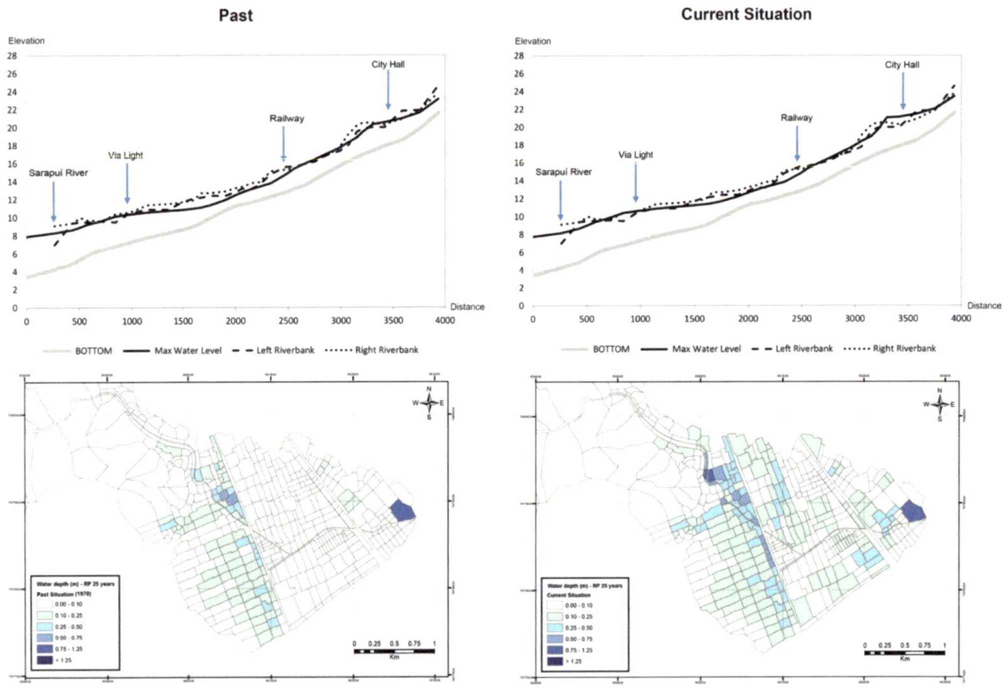

Figure 4. Current and past (1970) situation—flood maps—recurrence period 25 years (Source: see Ref. [42]).

using the open space system to reorganise flow patterns, but also providing river improvement and urban revitalisation, increasing both natural and built environment values. This alternative combined River Restoration (RR) solutions and Sustainable Drainage (SD) techniques, acting through the spaces provided by the open spaces system. In this alternative (RR + SD), the river was analysed and divided in three stretches in the urban area, identified as different landscape units: an upper reach near the natural protected area (area 1—upstream); a central area in the core of the city (area 2—middle reach); and a wetland in the river outfall (area 3—downstream), as schematically shown in **Figure 5**.

All the three areas suffer from informal settlements, substandard habitations and flood risks. Taking into account that it is a consolidated urban watershed, the open spaces shall be considered with multifunctional characteristics (landscaping, recreation, biodiversity and flood control, for example).

The design project proposals are listed in **Table 1** and **Figures 6–9** show the concepts supporting the proposals. The total storage volume offered by the three areas, in the combined RR + SD alternative, sums approximately 270,000 m^3, with additional contribution of infiltration measures and on-site detentions (adding 106,000 m^3).

Figure 10 shows the flood map for the alternative proposed. It is noticeable that flooding levels reduce all over the basin. The modelling system did not represent minor drainage, what

http://dx.doi.org/10.5772/intechopen.71658

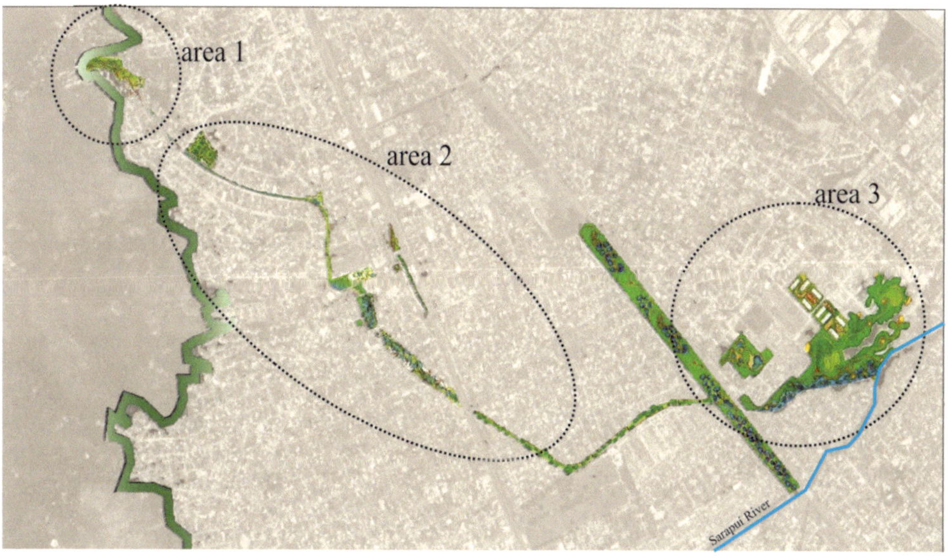

Figure 5. RR + SD alternative (Source: see Ref. [66]).

is probably the explanation to the remaining flooding areas. These water depths are low and the river does not overflow anymore. Additionally, the city of Mesquita currently marked by a grey landscape has the opportunity to revitalise degraded urban areas, also improving environmental connections.

Watershed Reaches	Characteristics	Vocation	Proposal
Area 1 (upstream)	This area is the most upstream urban reach. Urban growth is spreading and forcing the occupation of the valley that leads to environmental protection area. Substandard houses, poverty and an informal growth mark this area. Natural and built environment degradation is spreading. Floods are not observed in this area	This area may act as a transition from the city to the upstream protected areas, working as the entrance door linking the city with the upper green park. It is possible to use the hills and the steeper slopes to propose a dam for temporary flood storage purposes.	• Control urban occupation, diminishing urban density; • Avoid increasing imperviousness; • Escalate a set of different park typologies to allow a smooth transition between the city and the upstream green areas. This way, the park set in the sequence of the controlled urban growth area should begin with urban equipment, then becoming green, possibly exploring fruticulture (as an economic possibility for the local population), until reaching the protected environmental area. • A fluvial park, along the river course is also proposed. This park will begin here, connecting the upper forest to the city, and it will continue throughout the watercourse, crossing the city from one end to another.

Watershed Reaches	Characteristics	Vocation	Proposal
Area 2 (middle reach)	This area is the most floodable one. It is also the densest occupied area. The Major hall is here, in the city centre, near the river. There are also a line of houses, stacked on both riverbanks, strangling the river, discharging their sewage and hiding the river from urban landscape.	This area should provide space for flood storage. However, multifunctionality here is a key concept, once free space is rare.	• Continue with fluvial park, opening space where possible or necessary • Relocation of the houses that are directly settled on the riverbanks, giving space to the river. New housing typologies must be explored, using multi-floor buildings on stilts or pilotis. • Fluvial parks store temporary floods. • A number of soccer fields, public squares and free open areas near the river shall be incorporated in the blue-green way that is being created. These areas will also assume multifunctional characteristics, helping in the flood control.
Area 3 (downstream)	This is the lowest area, near the outfall on the Sarapuí River. It is a floodable area, and there are slums spreading fast. It is also a swampy area.	This area can be used as a humid park, also contributing in improving water quality. Urban occupation should be avoided	• The fluvial park finishes in a constructed wetland. • The houses settled in a very low elevation should be relocated.

Table 1. Characteristics of the design areas and proposal.

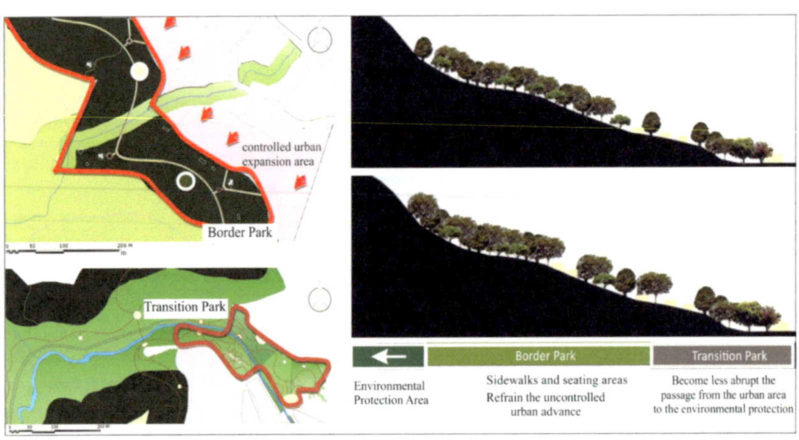

Figure 6. RR + SD alternative – area 1 (upstream) proposals (Source: see Ref. [66]).

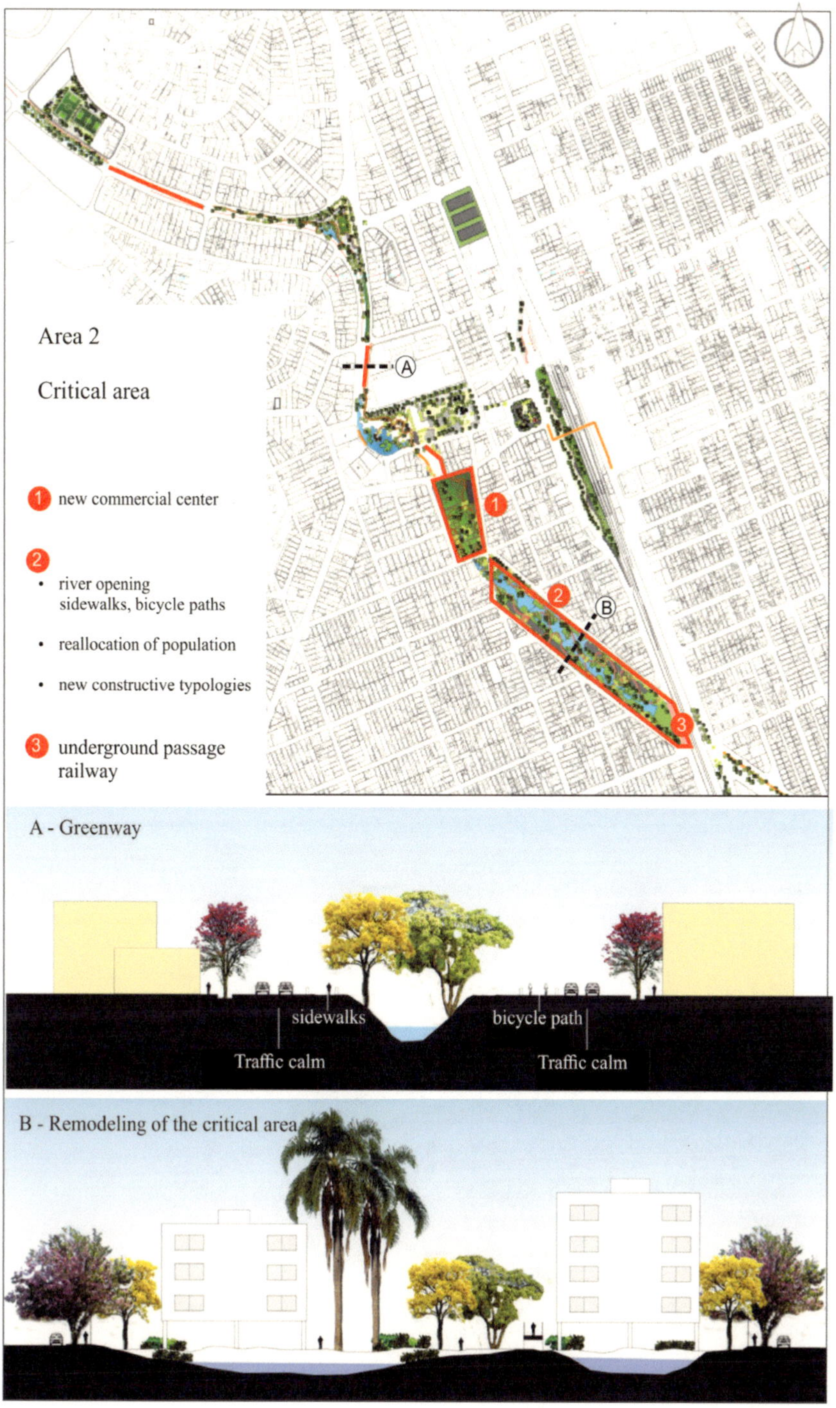

Figure 7. RR + SD alternative – area 2 (middle reach) proposals (Source: see Ref. [66]).

Figure 8. RR + SD alternative – area 2 (middle reach) – details (Source: see Ref. [66]).

Figure 9. RR + SD alternative – area 3 (downstream) proposals (Source: see Ref. [66]).

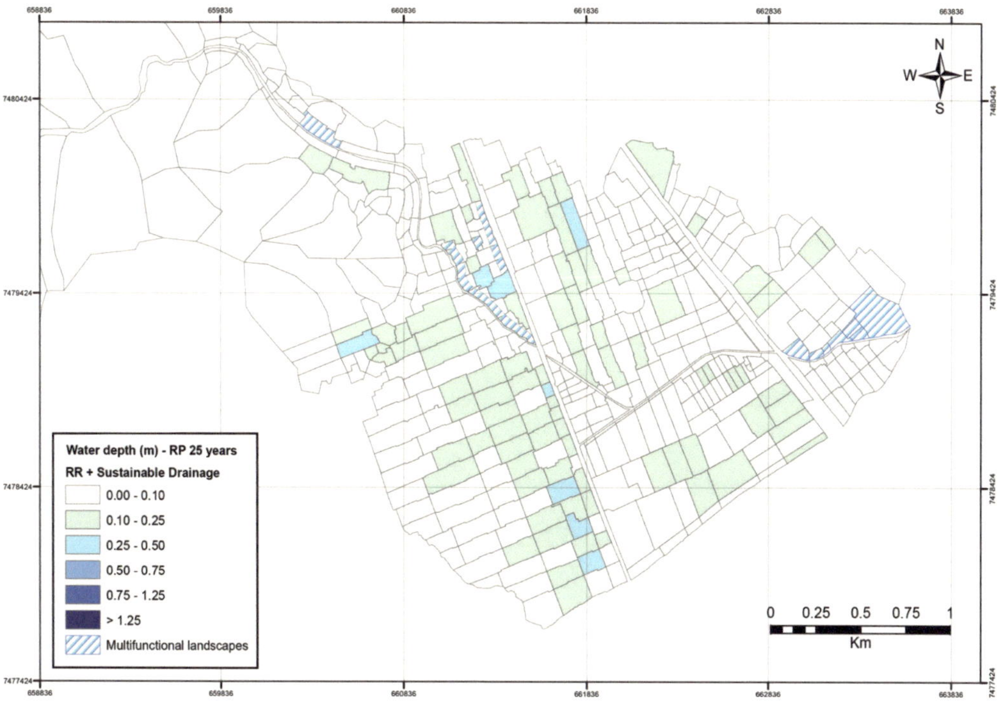

Figure 10. RR + SD alternative – flood map – recurrence period 25 years (Source: see Ref. [42]).

7. Concluding remarks

The urban flooding problem can be seen as problem of space allocation. In using areas that should be preserved for inundation purposes and encouraging a land use that generates a major quantity of runoff, the process of urbanisation must foresee new and alternative spaces to temporarily store floods. In this way, the drainage system may continue working, and, consequently, the whole city may preserve its functions.

Therefore, one is aware of the necessity of utilising open spaces as fundamental pieces for mitigating floods in the cities. By doing this, the limited discharge capacity of the traditional drainage system, whose result is converted into a flooding volume spread throughout the city, can be resolved when the open spaces are considered as a possible reservoir to store exceeding volumes of the drainage system. In this approach, open spaces act as multifunctional landscapes. In this sense, the creation of multifunctional spaces can permit that the remaining open spaces in the already consolidated urban centres incorporate hydrologic functions to their original purposes.

It needs to be pointed out, nevertheless, that the city growth, in an uncontrolled form, leads to urban sprawl, on the one side, and the discussion of sustainability considering the concept of compact cities, on the other side, can both lead to a suppression of open spaces. Therefore, the urban and environmental planning, taking into consideration the supporting capacity of the natural environment in which the city is inserted, make it possible to improve the quality of life in cities.

The set of actions in the urban tissue, utilising open spaces to support drainage solutions, acting in the fluvial corridor and combining efforts to revitalise the urban environment, poses a sustainable approach for facing urban floods within an agglomeration trend.

The case study chosen to illustrate this discussion, and shown in the previous section, confirms the proposed concepts, collaborating for an urban sustainable development, through a model of interventions capable to embrace environmental, economic and social aspects.

Author details

Marcelo Gomes Miguez*, Aline Pires Veról, Andréa Queiroz da Silva Fonseca Rêgo and Ianic Bigate Lourenço

*Address all correspondence to: marcelomiguez@poli.ufrj.br

Federal University of Rio de Janeiro (UFRJ), Rio de Janeiro, Brazil

References

[1] The Club of Rome. About the Club of Rome. 1968. Available from: http://www.clubofrome.org [Accessed: Aug 25, 2017]

[2] UNEP – United Nations Environment Programme. Stockholm, 1972. Report of the United Nations Conference on the Human Environment. 1972. Available from: http://www.unep.org [Accessed: Jun 25, 2012]

[3] United Nations. Our Common Future-Brundtland Report. Oxford, UK: Oxford University Press; 1987. p. 204

[4] UNCDE – United Nations Conference on Environment and Development. Earth Summit. 1992. Available from: http://www.un.org/geninfo/bp/enviro.html [Accessed: Jul 25, 2017]

[5] UN – United Nations. Agenda 21: Earth Summit – The United Nations Programme of Action from Rio. Rio de Janeiro: United Nations, Department of Public Information; 1993

[6] Lindholm G. 'Visible gestures': On urban landscape perspectives in planning. Planning Theory. Swedish University of Agricultural Sciences, Sweden. 2011;**11**(1):5-19

[7] Johannessen Å, Wamsler C. What does resilience mean for urban water services? Ecology and Society. 2017;**22**(1):1. DOI: https://doi.org/10.5751/ES-08870-220101

[8] Andoh RY, Iwugo KO. Sustainable urban drainage systems: A UK perspective. In: Global Solutions for Urban Drainage. Portland, Oregon, United States: American Society of Civil Engineers; 2002. pp. 1-16

[9] Sayers P et al. Flood Risk Management – A Strategic Approach, Part of a Series on Strategic Water Management. Paris: Asian Development Bank, GIWP, UNESCO, and WWF-UK; 2013

[10] UNDP-United Nations Development Programme. Human Development Report 2014–Sustaining Human Progress: Reducing Vulnerabilities and Building Resilience. New York, USA. 2014

[11] European Commission. The EU Approach to Resilience – Learning from Food Crises, COM(2012) 586 final, Communication from the Commission to the European Parliament and the Council. Brussels, Belgium. 2012

[12] IPCC. Managing the Risks of Extreme Events and Disasters to Advance Climate Change Adaptation. A Special Report of Working Groups I and II of the Intergovernmental Panel on Climate Change In: Field CB, Barros V, Stocker TF, Qin D, Dokken DJ, Ebi KL, Mastrandrea MD, Mach KJ, Plattner GK, Allen SK, Tignor M, Midgley PM, editors. Cambridge, UK, and New York, NY, USA: Cambridge University Press; 2012. 582 pp

[13] McBain W, Wilkes D, Retter M. Flood Resilience and Resistance for Critical Infrastructure. Publication C688. CIRIA, London, UK; 2010

[14] UNISDR. UNISDR Terminology on Disaster Risk Reduction. Geneva, Switzerland: United Nations International Strategy for Disaster Reduction; 2009

[15] Brown R, Keath N, Wong T. Transitioning to water sensitive cities: Historical, current and future transition states. 11th International Conference on Urban Drainage. IAHR/IWA, Edinburgh, Scotland, UK; 2008

[16] Abhas KJ, Miner TW, Stanton-Ge Z. Building Urban Resilience – Principles, Tools and Practice. Washington DC: International Bank for Reconstruction and Development/The World Bank: 2013

[17] UNISDR. Making Cities Resilent: My City is Getting Ready. 2012. [Online] Available from: http://www.unisdr.org/campaign/resilientcities [Accessed: Aug 19, 2017]

[18] Miguez MG, Veról AP. A catchment scale integrated flood resilience index to support decision making in urban flood control design. Environment and Planning B: Urban Analytics and City Science. 2016;**44**(5):925-946. DOI: 10.1177/0265813516655799

[19] Meerow S, Newell JP, Stults M. Defining urban resilience: A review. Landscape and Urban Planning. Mar 2016;**147**:38-49

[20] Xu L, Marinova D, Guo X. Resilience thinking: a renewed system approach for sustainability science. Sustainability Science. 2015;**10**(1):123-138. DOI: https://doi.org/10.1007/s11625-014-0274-4

[21] Folke C, Carpenter SR, Walker B, Scheffer M, Chapin T, Rockström J. Resilience thinking: Integrating resilience, adaptability and transformability. Ecology and Society. 2010;**15**(4):20. DOI: https://doi.org/10.5751/ES-03610-150420

[22] Turner BL. Vulnerability and resilience: Coalescing or paralleling approaches for sustainability science? Global Environmental Change. 2010;**20**(4):570-576. DOI: https://doi.org/10.1016/j.gloenvcha.2010.07.003

[23] Lélé S. Resilience, sustainability environmentalism. Environment and Development Economics. 1998;**3**(2):221-262. DOI: https://doi.org/10.1017/S1355770X98260128

[24] Mcharg IL. Design with Nature. New York: Natural History Press; 1969

[25] Rogers R. Cidades Para um Pequeno Planeta, Barcelona: GGili; 2001

[26] Harvey D. The Right to the City. New Left Review 53, London, UK. September-October 2008

[27] Forest Stewardship Council (FSC). Sustainable Cities. Sustainable Development for Urban Infrastructures. 2010. pp. 1-38. Available from: www.siemens.com/cities

[28] Allen A. Sustainable cities or sustainable urbanisation? Journal of Sustainable Cities. 2009;**34**(10):1667-1691. DOI: https://doi.org/10.1080/0042098975394

[29] Zetter R, Butina Watson G. Designing sustainable cities. In: Zetter R, Butina Watson G, editors. Designing Sustainable Cities in the Developing World. Hamphire: Routledge Taylor & Francis Group; 2006. pp. 3-18

[30] Shi L, Gao L, Yang S. Effects of a compact city on urban resources and environment. Journal of Urban Planning and Development. 2016;**142**(2014):1-9. DOI: https://doi.org/10.1061/(ASCE)UP.1943-5444.0000324

[31] Seya H, Yamagata Y, Nakamichi K. Creation of future compact urban form scenarios of Tokyo in combination with adaptation policies. In: Energy Procedia. Vol. 61. Amsterdam, Netherlands: Elsevier Ltd; 2014. pp. 357-360. DOI: https://doi.org/10.1016/j.egypro.2014.11.1124

[32] Woodcock K, Dovey S, Wollan I, Robertson I. Urban policy and research. Speculation and Resistance: Constraintson Compact City Policy Implementationin Melbourne. 2011; **29**(4):343-362

[33] Catalán B, Saurí D, Serra P. Urban sprawl in the Mediterranean?. Patterns of growth and change in the Barcelona Metropolitan Region 1993-2000. Landscape and Urban Planning. 2008;**85**(3-4):174-184. DOI: https://doi.org/10.1016/j.landurbplan.2007.11.004

[34] Dieleman F, Wegener M. Compact city and urban sprawl. Built Environment (1978–). 2004;**30**(4):308-323. DOI: https://doi.org/10.2307/24026084

[35] de Roo G. Environmental planning and the compact city a dutch perspective. Studies in Environmental Science. 1998;**72**(C):1027-1042. DOI: https://doi.org/10.1016/S0166-1116(98)80061-7

[36] Burgess R, editor. Compact Cities: Sustainable Urban Forms for Developing Countries. London and New York: Spon Press; 2001

[37] Rogers R. Cidades para um pequeno planeta. Barcelona: Gustavo Gilli; 2005

[38] Andersen TK, Shepherd JM. Floods in a changing climate. Geography Compass. 2013; **7**(2):95-115

[39] Barbedo J, Miguez M, Van Der Horst D, Marins M. Enhancing ecosystem services for flood mitigation: A conservation strategy for peri-urban landscapes. Ecology and Society. 2014;**19**(2):54. DOI: http://dx.doi.org/10.5751/ES-06482-190254

[40] Grizzetti B, Lanzanova D, Liquete CA. Assessing water ecosystem services for water resource management Reynaud, A.C. Cardoso European Commission Joint Research Centre (JRC), I-21027 (VA). Italy Environmental Science & Policy. 2016;**61**:194-203

[41] Miguez MG, Veról AP, Rezende OM. Drenagem Urbana: do Projeto Tradicional à Sustentabilidade. Rio de Janeiro: Elsevier; 2015

[42] Veról AP. Requalificação Fluvial Integrada ao Manejo de Águas Urbanas para Cidades mais Resilientes. Tese de Doutorado: COPPE/UFRJ, RJ; 2013

[43] Miguez MG, Rezende OM, Veról AP. City growth and urban drainage alternatives: Sustainability challenge. Journal of Urban Planning and Development. 2015;**141**(3): 04014026-1-04014026-10

[44] Miguez MG, Veról AP, de Sousa MM, Rezende OM. Urban floods in lowlands—Levee systems, unplanned urban growth and river restoration alternative: A case study in Brazil. Sustainability. 2015;**7**:11068-11097

[45] Comissão das Nações Unidas para o Desenvolvimento Sustentável. COMISSÃO EUROPEIA. Cidades europeias sustentáveis. Bruxelas, 1996. Available from: ec.europa.eu/environment/urban/pdf/rport-pt.pdf. [Accessed: Sep 10, 2017]

[46] Kyushik O, Yeunwoo J, Dongkun L, Wangkey L, Jaeyong C. Determining development density using the Urban Carrying Capacity Assessment System. Landscape and Urban Planning. 15 August 2005;**73**(1):1-15. DOI: https://doi.org/10.1016/j.landurbplan.2004.06.002

[47] Chung S. A conceptual model for regional environmental planning centered on carrying capacity measures. Korean J. Region. Sci. 1988;**4**(2):117-128

[48] Schneider DM, Godschalk DR, Axler N. The Carrying Capacity Concept as a Planning Tool. Planning Advisory Service Report 338. Chicago: American Planning Association; 1978

[49] Oh K, Jeong Y, Lee D, Lee W. An integrated framework for the assessment of urban carrying capacity. Journal of Korea Planners Association 2002;**37**(5);7-26

[50] Kozlowski JM. Sustainable development in professional planning: A potential contribution of the EIA and UET concepts. Landscape Urban Plan. 1990;**19**(4):307-332

[51] Tuan YF. Espaço e lugar. São Paulo: DIFGL; 1979

[52] Schlee MB, Nunes MJ, Rego AQ, et al. Sistemas de espaços livres nas cidades brasileiras – um debate conceitual. In: Tângari VR, Andrade R, Schlee MB. (org.). Sistemas de Espaços Livres – o cotidiano, apropriações e ausências. Rio de Janeiro: PROARQ; 2009

[53] Meining DW, (org.). The Interpretation of Ordinary Landscapes – Geographical Essays. Oxford: Oxford University Press; 1979

[54] Macedo S, Queiroga EF, et al., (org). Sistemas de espaços livres – conceitos, conflitos e paisagens. São Paulo: Laboratório QUAPÁ-SEL; 2011

[55] Rego AQ, Tângari VR, Gomes RB. Paisagem e Planejamento do Sistema de Espaços Livres sob a Influência do Arco Metropolitano do Rio de Janeiro: Magé. Anais do VI Colóquio QUAPÁ-SEL. São Paulo: FAUUSP; 2011

[56] Tardin RH. Espaços Livres: Sistema e Projeto Territorial. Ed. 7 Letras, Rio de Janeiro, Brazil. 2008

[57] Chiesura A. The role of urban parks for the sustainable city. Department of Leisure, Tourism and Environment, Wageningen University General Foulkseweg. Landscape and Urban Planning. 2004;**68**(2004):129-138

[58] Spirn AW. O Jardim de Granito. São Paulo: EDUSP; 1995

[59] Naveh Z. Ten major premises for a holistic conception of multifunctional landscapes landscape and urban planning 57. In: Naveh Z, editor. Transdisciplinary Challenges in Landscape Ecology and Restoration Ecology.Org.por, Holanda, Netherlands: Springer; 2001. pp. 269-284

[60] CIRF – Centro Italiano per la Riqualificazione Fluviale. In: Nardini A, Sansoni G, editors. La riqualificazione fluviale in Italia: linee guida, strumenti ed esperienze per gestire i corsi d'a cqua e il territorio. 1st ed. Venezia: Mazzanti; 2006

[61] Gusmaroli G, Bizzi S, Lafratta R. 2011. L'approccio della Riqualificazione Fluviale in Ambito Urbano: Esperienze e Opportunittà. In: Acqua e Città – 4° Convegno Nazionale di Idraulica Urbana, Venezia, Italia. June 2011

[62] Batista M, Nascimento N, Barraud S. Técnicas Compensatórias em Drenagem Urbana. Porto Alegre, Brazil: ABRH; 2005

[63] COPPETEC. Plano Diretor de Recursos Hídricos, Recuperação Ambiental e Controle de Inundações da Bacia do Rio Iguaçu-Sarapuí, Laboratório de Hidrologia e Estudos do Meio Ambiente. Rio de Janeiro: COPPE/UFRJ; 2009

[64] IBGE–Instituto Brasileiro de Geografia e Estatística. Censo Demográfico, Brazil. 2010. Available from: https://censo2010.ibge.gov.br/resultados.html

[65] IBGE–Instituto Brasileiro de Geografia e Estatística. Censo Demográfico, Brazil. 1970. Available from: https://biblioteca.ibge.gov.br/visualizacao/periodicos/69/cd_1970_v1_t16_rj.pdf

[66] Bigate I. Rios Urbanos e Paisagens Multifuncionais: O Projeto Paisagístico Como Instrumento de Requalificação Urbana e Ambiental. Dissertação de Mestrado. Rio de Janeiro: PROURB/FAU/UFRJ, RJ; 2013

Chapter 2

Characteristics of Urban Agglomerations in Different Continents: History, Patterns, Dynamics, Drivers and Trends

Wolfgang Loibl, Ghazal Etminan,
Ernst Gebetsroither-Geringer,
Hans-Martin Neumann and
Santiago Sanchez-Guzman

Additional information is available at the end of the chapter

http://dx.doi.org/10.5772/intechopen.73524

Abstract

Urban agglomerations show different development patterns and stages. Here, we describe, discuss and compare urban agglomerations in different continents. The introduction section gives a general overview of specific issues of urban agglomerations. Different characteristics in Europe, Asia and America are discussed as experienced by the article's co-authors, living in or working for urban agglomerations in these continents. First, the history of urbanization and agglomeration evolvement is described, then patterns, functional structures and relations, drivers as well as social and demographic characteristics are discussed (e.g. migration, aging, household structure, housing patterns, workplaces, etc.). Transportation infrastructure (roads, public transport systems) is also addressed as trigger for spatial dynamics causing certain effects (floor space, office and apartment rents releasing urban sprawl or hyper-densification), as well as gentrification. Further topics are urban governance and its impact on agglomeration development. Recent state and future trends will be debated, if important. A conclusion section summarizes the comparison of state, dynamics, drivers and trends.

Keywords: urban agglomeration, trends, drivers, patterns, functions and relations

1. Introduction

Urban agglomerations can be defined as contiguously built-up area, shaped by one core city or by several adjacent cities, sharing industry-, infrastructure- and housing-land use with high-density levels as well as embedded open spaces. Urban agglomerations can be addressed in different ways ranging from "large cities," "urban economic zones," to "integrated groups of cities that share a common interest and fate," Fang and Yu [1] gave a comprehensive overview on the different contexts. Urban agglomeration is a somehow technical term emphasizing extent and morphology. We prefer the term "metropolitan region," which better labels socio-spatial systems addressing activity and interest sharing of urban actors as well as the drivers, dynamics and functions which result in certain pattern and shape.

1.1. General triggers extending urban agglomerations

The utmost driving force of urban agglomerations is growth! The major reason for growth is population migration towards cities and metropolitan regions, driven by the population expectation, to find attractive living conditions (c. f. [2, 3]). Urban population growth and related housing demand are mostly caused by migration of rural population from rural areas or from remote (emerging) countries. There exist various migration theories. Bogue [4] addresses push

Topics	Push factors	Pull factors
Economic sphere	• Lack of resources (water, food, goods) • Economic disadvantages (salary, lack of services, unemployment)	• Resources availability • Economic advantages (better employment opportunities, salaries) • Good accessibility (enabling faster commuting)
Housing/living condition sphere	Dissatisfaction with present living conditions: • Lack of housing • Low neighbourhood quality • Few social, educational, leisure opportunities	Pleasant neighbourhood and surroundings • Sufficient housing, building land provision • Better neighbourhood quality • Good social, education, leisure facilities
Social sphere	• End of professional education • Change of job, increase of personal income • Family status change (marriage, children…)	• Appropriate social neighbourhood structure • Former neighbours and friends moved earlier to the target area, which evolves as homogenous ethnic district • Building homogenous ethnic districts "Ghettos"

Source: Adapted from Refs. [3, 4].

Table 1. General driving forces of urban migration.

and pull factors triggering migration decisions. In **Table 1**, these factors have been distinguished into three spheres.

Important push and pull factors for urban migration refer to the economic and the housing/living condition sphere. The social sphere is only considered if the other two spheres are fulfilled. Social sphere-related desires vary much with respect to personal needs.

1.2. History and dynamics: patterns and triggers shaping growth

1.2.1. Economy and employment

Economically driven expectations of newcomers are triggered by the economic attractiveness of the agglomeration (c. f. [5]). Larger agglomerations—in developed as well as in emerging countries—show higher chances appealing more customers to gain income, higher salary or profit.

The economic sector depends on location factors. High quality services and administrations allow occupying the central core of a metropolitan region resulting in very high rents. The demand for office space in central areas triggers density increase in the central business districts (CBDs) within urban agglomerations. Agglomeration advantages attract businesses of different size, especially globally acting, powerful companies, establishing new offices to benefit from contact opportunities, good traffic connections, the large number of potential employees and clients. These agglomeration economies of scale are result of spatial proximity and frequency of opportunities, which are major contributing factors to the growth of cities [6]. Benefits arise from:

- low transportation costs
- big (local) markets
- large supply with labour and reducing search costs
- accumulation of knowledge and knowledge spillovers between companies [7].

Hiring specialized staff is easier in large agglomerations. More people undertake professional training and education, and the number of highly skilled employees is larger, since new academic personnel leave the city's universities and more already skilled personnel work in competing companies nearby, which may change the workplace for higher salary [8].

While the highly specialized service sector (banking, insurance, ICT, R&D, governmental and urban administration) has its allocation focus in the centres, increasing densification, space intensive enterprises (logistics, retail- and large production sites), relocate to cheaper areas outside the urban cores, thus contributing to urban sprawl.

1.2.2. Housing

Despite high pressure on building land and rents, in-migration of new population to the cities is a steadily observed trend. Migration patterns in developed countries depend on the wealth of the potential movers able to rent or buy an apartment, where wealthy people are gradually displacing less wealthy ones, forcing them to move to cheaper outskirt areas, living often in low quality social housing blocks. Furthermore, central areas evolve as attractive neighbourhoods for the wealthy with short distance food supply, leisure opportunities, education facilities and social care.

Housing development in emerging countries shows extreme patterns. Displacement of poor people triggers informal housing in many ways.

1.2.3. *Urban sprawl versus "leapfrogging" urban development*

Urban sprawl "flooding" opens spaces around single core cities or between several cities and is the major dynamic to build urban agglomerations. Urban sprawl and its reasons have been discussed by many authors (e.g. by Bhatta [9] who provided a table of 30 reasons). In the nineteenth century during industrialization, the cities' density increased and living conditions declined because of coal fire use, leading to heavy air pollution in the cities and their industrial belts. Expansion of urban regions was driven in Europe during this century by new railway lines resulting in a "leapfrogging" urban development: With growing economic prosperity and expanding railway networks, the middle class move towards the suburban towns with better quality of life, while working in the cities, evolving commuting patterns along the railway lines. In London, this trend started around 1850, in Paris in 1890 and in New York City in early 1900 [10].

This urbanization trend was initiated during the 19th century by railway lines and train stations. In the 20th century, it was transformed to uncontrolled urban sprawl, which was triggered by private cars, allowing to settle in open land near already built-up areas, which provides easy access to the core city as well as easy connection to infrastructure supply networks. The trend accelerated in the United States in the 1930s, through the National Housing Act of 1934, promoting the "free-standing, owner-occupied, single-family home in the suburbs" [11]. Since the 1950s in the USA, the 1960s in Europe and the 1980s in parts of Asia and Latin America, urban sprawl increased even more with car growing ownership allowing to settle wherever one wants, fostering additional extension of street networks, again pushing urban sprawl. Industrial deconcentration contributes also to sprawl in the large urban agglomerations [12].

In emerging countries, urban sprawl started as a mass phenomenon in the late twentieth century, building huge new metropolitan regions, often within one decade—triggered through massive rural population migration to the cities (c. f. [13]). This leads to the growth of informal settlements resulting in large, sometimes rather dense, squatter dwellings close to the city centres (especially in Latin America) or at the outskirts of the cities [14]. Squatter dwellings were first built with sheds, which were particularly transformed later into basic buildings, partly accepted, legalized and supplied with regular roads, water pipes and power lines (c. f. [15]).

A major pull factor of urban sprawl is, no doubt, the lack or insufficiency of strict planning guidelines and urban development policies, which do not control land consumption in an appropriate way.

1.2.4. *Densification*

Densification in centres is triggered by the land price gradient. Increasing densification is constantly accelerating the property prices (c. f. [16]). Expansion of the service sector—banking, insurance, ICT and governmental and urban administration—is enhancing densification. The most densified blocks often evolve in areas of good accessibility, close to higher capacity transportation hubs, with high visiting frequency and increasing opportunity to attract new employees, business partners and clients or to allow fast traveling to business partners.

While density is increasing in the centres, in the entire urban agglomerations, the density is decreasing due to accelerated dispersion. Bertaud [17] showed this by an impressive image (see **Figure 1**), comparing population distribution in seven agglomeration areas. The distribution of the working population would show a similar pattern although the density gradient will be higher as office space is even more concentrated in urban cores.

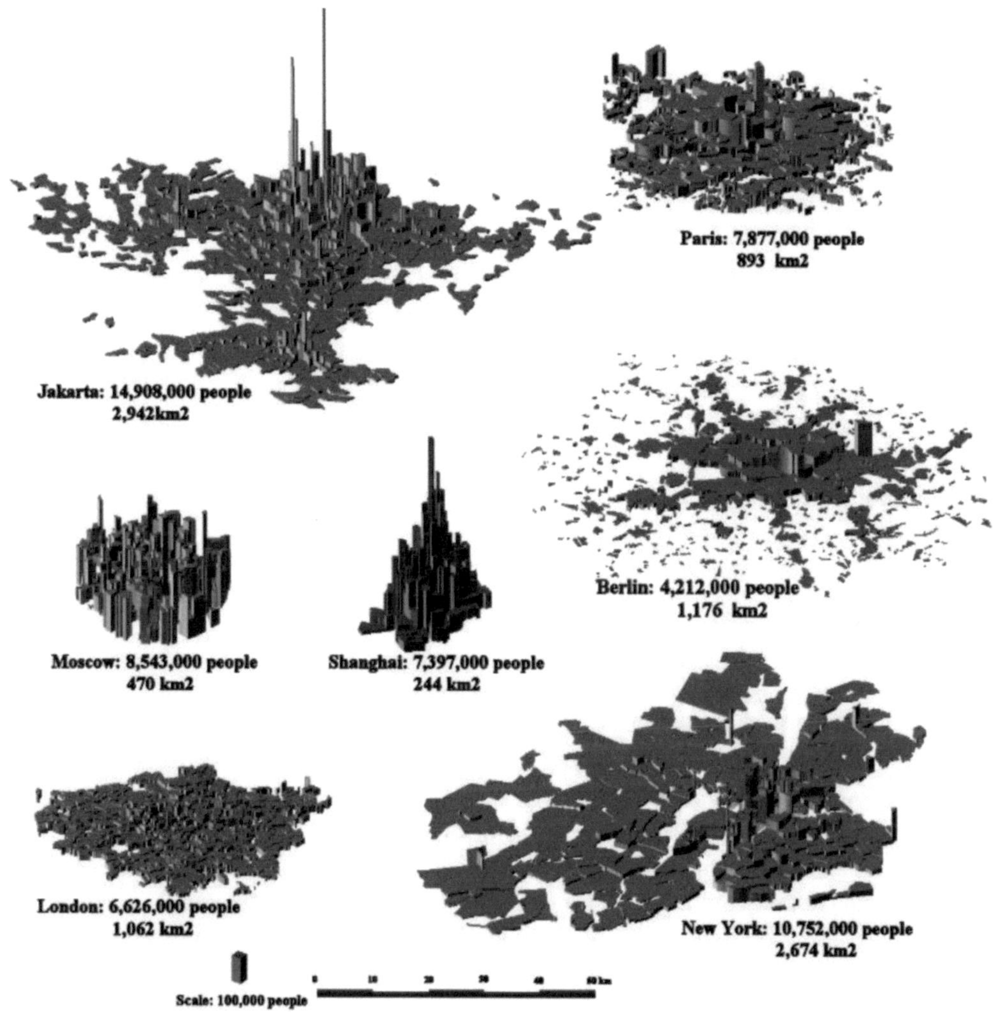

Figure 1. Population distribution in seven urban agglomerations. Source: Bertaud [17].

1.3. Socio-spatial structures and dynamics: segregation, gentrification and aging

1.3.1. Segregation

Segregation—unequal distribution of social groups—is common in urban agglomerations. In small cities, only little segregation of national or ethnic groups is observed, as the number of newcomers is too small to establish a separated community. Metropolitan regions are migration targets for many foreign immigrants as they expect there higher chances to meet fellow citizens and find familiar infrastructure that makes it easy to settle, to get support and advice, to find jobs, to adapt and to integrate (c. f. [18]). There ethnic and national differences play a more important role: as the size of national and ethnic groups is big enough to build clusters with own identity (c. f. [19]). Thus, national or ethnic quarters occur with own shops, restaurants, leisure opportunities, social infrastructure, churches or mosques and finally services like native speaking lawyers, doctors and even newspaper publishers.

1.3.2. Gentrification

Gentrification is the process by which deprived urban neighbourhoods experience a reversal, reinvestment and in-migration of middle class to upper middle class population (c. f. [20, 21]). It is the effect of more wealthy people who show new demands for high quality apartments and living areas. This process is more distinct in urban agglomerations as migration pressure of wealthy people is higher and social mobility (through raise of income and change of status) is higher too, because of more earning opportunities in these larger economic zones.

Gentrification can be seen as a negative or positive process: as displacement of the urban poor and destruction of indigenous social communities (c. f. [20]) or as economic revitalization of urban spaces (c. f. [21]), driven by wealthy classes and property developers. An issue, which is in urban agglomerations, with a vibrant housing market is more important. In smaller towns, gentrification plays a minor role, as there is less demand of wealthy classes for high quality housing.

1.3.3. Age distribution

The stages of life course (childhood, youth, working age and aging) play a certain role in todays' urban agglomerations. Both—increase of the number of the elderly and of young ones—can be observed in cities, depending on the age structure of the nation and the immigration pressure. In mature industrialized countries with small birth rates, population structure is frequently aging if only little immigration takes place. In emerging countries with high birth rates and in urban areas of developed countries with high immigration rates, the share of young population is growing, attracted by urban areas when searching for higher education, better housing possibilities and workplaces (c. f. [22]). Age group distribution patterns are fluid and differ over time. New housing areas lead during growth periods to homogenous age structures with young singles or families in their early 1930s and small children. In cities where less relocation within a life course takes place, like in central Europe, the age structure pattern remains may cluster into homogenous groups over several decades (c. f. [23]).

1.4. Functions, interactions and traffic

The functional urban region—and urban agglomerations are highly interlinked functional areas—provides complementary functions of different levels at different places to supply the population with all necessities—ranging from residential function to workplaces, to education, shopping and using various services. Making use of these functions requires communication or traveling between the places where those functions are provided.

1.4.1. Functions

Referring to Christaller's Central Place Theory [24, 25], explaining spatial arrangement, size and function provision of settlements, there exists a hierarchy and pattern of towns and cities, which provide, depending on size, a range of services for the city and the adjacent service, administration and market area. Simple everyday services like food stores, kindergartens and elementary schools can be found in low-ranked Central Places: settlements, serving the local population and the surrounding area within small distance. Occasionally, demanded specialized services—hospitals, colleges, province administration—are located in higher ranked Central Places: mid-sized and larger towns, serving the population of a larger area—e.g. a province—requiring some traveling. In highest ranked Central Places: large cities with high level services—specialized clinics, universities, national museums, ministries and central governmental bodies—which supply the population of a large area, even a country—are found.

Urban agglomerations allow providing services of all hierarchies up to the highest one in an efficient way, as the different hierarchical services are located within small distance, and supply a large number of people, living or working closely together.

1.4.2. Interactions, infrastructure and traffic

Distributed functions in metropolitan regions require more interactions—people flows (shopping, learning, commuting, relocation, etc.), business linkages (supply chains, goods flow, retail, services)—and infrastructure to enable these interactions: Traffic and communication infrastructure, energy and technical services infrastructure and shared social and administrative services. As urban agglomerations are faster growing than standard cities, these services must grow fast too and must be well maintained for sufficient supply, which is again increasing the attractiveness of the agglomerations, at least today.

Urban agglomerations observe high traffic loads frequently causing heavy congestion problems [26]. Compared to the number of inhabitants, the traffic volume is less as the density is higher, the functions are located closer, travel distances are shorter and the public transport share is much higher, at least in the central areas of those metropolitan regions. So, larger and denser metropolitan areas do not mean more car traffic [27]! Building density and thus functional density can significantly reduce traffic, as closely allocated functions provide sufficient supply within short distance, which allows walking for shopping, leisure facilities, schools, etc. instead of driving. Central areas are good examples to enable car-free living. Especially in Europe, remaining medieval structures in many town centres have often hindered to evolve "car-friendly" cities.

Local traffic varies depending on the layout of neighbourhoods and function allocation. Transregional traffic is the effect of large distance mobility demand. Here, it is important to distinguish between monocentric and polycentric agglomerations—the monocentric ones target more the centre, but show less traffic loads because of less transregional transportation requirement, while the polycentric ones show distributed traffic loads because of wider distance and higher transportation requirement (c. f. [17]).

1.5. Policy: spatial impact on urban expansion patterns

Urban agglomerations are driven by administrative structures distinguishing the areas of responsibility and by planning policies establishing the administrative structures.

The government systems and planning policies differ substantially between countries depending, e.g. on the number of intermediary planning authority levels, which has some effect on the development of urban agglomerations. Political and administrative borders may thus lead to lacking planning guidelines, which may result in uncoordinated urban region development—urban sprawl is one severe problem directly caused by uncoordinated spatial planning or lack of interplay between government levels.

Based on the planning power of local and regional government levels, one can distinguish between centralized and decentralized planning systems. The interplay of government systems and overarching planning policies show the potential of the public role to manage and control urban agglomeration development, which can range from no control to full planning control covering the entire region (c. f. [28]). Agglomerations cover usually not just one administrative entity as they are growing beyond the city borders. They cover clusters of administrative entities: in case of monocentric agglomerations, the core city and surrounding peri-urban municipalities and in case of polycentric agglomerations, several core cities and secondary cities.

In countries with weak national to regional planning power and strong local power, local policy makers focus on their own territory resulting in less coordination to develop the parent region. The focus on municipal policies leads also to less transboundary or to divided public transportation networks. This triggers mobility behaviour towards private car use, which again accelerates urban sprawl.

The authorities of several cities embedded in urban agglomerations have learned from the past and established common governance bodies for the agglomeration region as well as regional transportation organizations providing a better framework for regional development (c. f. [29]).

2. Urban agglomerations in Europe

Urban agglomerations in Europe show frequently transnational structures, which grew beyond national borders due to dense distribution of cities in small nations. Despite this transnational growth, where cities profit from each other, being part of a large conurbation, they are also urban economies competing with each other while they collaborate.

The "Blue Banana," the name for Europe's largest transnational conurbation, created by Brunet [30], comprises Greater London; the Dutch Randstad region in Amsterdam; The

http://dx.doi.org/10.5772/intechopen.73524

Hague, Rotterdam; the Brussels agglomeration; the German Rhine-Ruhr conurbation from Dortmund via Frankfurt to Stuttgart, the Swiss Basel-Zurich agglomeration and, finally, the Milan urban region in Italy. Recently, the "Sunbelt" from Milan to Valencia and the "Yellow Banana" from Paris to Warsaw were identified as new transnational urban agglomeration axes [31].

Besides these "bananas," many more urban agglomerations and functional urban areas (FUAs) can be found in Europe although they are smaller than the transnational ones. Nearly each European larger capital city has evolved as urban agglomeration (see **Figure 2**).

2.1. History of urbanization: the nuclei of urban agglomerations in Europe

About 1000 years ago, Europe starts becoming urban. Between 1000 and 1300, Europe acquired an urban system dominated by typical producer cities, which prospered in spite of Europe's political fragmentation. In fact, this fragmentation was enhanced during the rise of independent city-states or cities with a large degree of local authority. The industrial revolution developed rapidly in England during 1750–1850 and stimulated growth of cities and urban agglomerations. In 1801, 17% of the European population lived in cities; by 1851, it was 35% and by 1891, it was 54% [32]. All cities and agglomerations show similar dynamics: moderate growth inside the core areas and extended growth of the outer urban region.

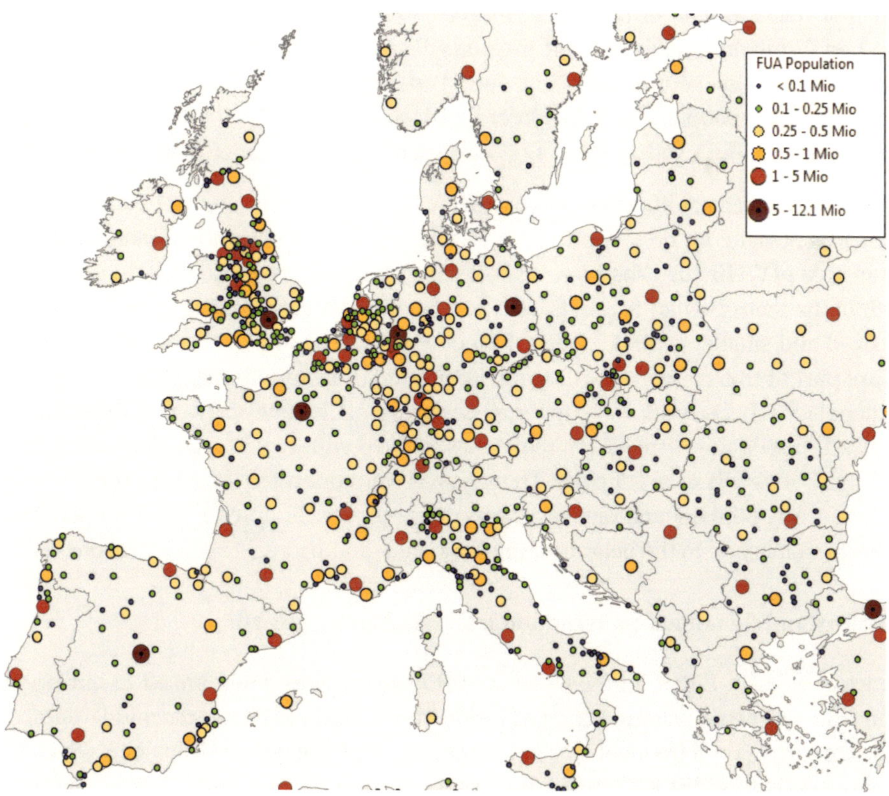

Figure 2. Functional urban areas (FUAs) in Europe by population numbers. Sources: Various European Statistical Data; mapping by Mario Köstl, AIT.

London and Paris are the two European agglomerations of outstanding size, forming a monocentric core area incorporating during their growth a range of small towns within the outskirts building the agglomeration.

London was rapidly rising after 1500 in importance among Europe's commercial centres. Trade expanded beyond Western Europe, and monopoly trading companies were established in London. Immigrants arrived from Britain and from abroad. The growth of population and wealth of London was fuelled by coastal shipping. The population rose from 50,000 in 1530 to about 225,000 in 1605 [33]. During the 19th century, a network of railways accelerated the growth of the suburbs of Greater London, from where middle class people commute to London centre, while the low-income remained living in the inner city areas. As capital of a large empire, London became also a target for immigrants from the British colonies and Europe [34]. In 1801, Greater London reached around 1.2 million inhabitants, where the outer London's share (the population living outside the London city borders) was 15%. In 1901, the city grew to a population of 6.2 million, where the outer London's share reached 22%. In 2011, Greater London hosted 8.2 million people, where the outer London's share reached around 60% [35].

Paris show similar growth dynamics. During the middle ages, Paris was an important commercial centre and the largest city in Europe. Under Georges-Eugène Haussmann, Prefect of the Seine, the Paris centre was rebuilt until 1870 with wide avenues over forming the earlier street layout. In 1860, the city expanded to its present administrative area. Around 1800, Paris had 550,000 inhabitants; after the extension of the city boarders, Paris inhabited 1.7 million people, Paris urban region had 2 million inhabitants and the outer Paris share reached 15%. In 1911, Paris urban region had 4.5 million people, where the outer Paris' share was 35%. In 2010, Paris urban region reached 10 million people, where the outer Paris' share was ca. 70% (numbers reconstructed by Ph. Julien from INSEE). Today, Paris' agglomeration is 27 times larger as the city itself [36].

The Rhine-Ruhr metropolitan region is one of the large polycentric metropolitan areas with several cities as nuclei with no real prime city, hosting more than 11 million inhabitants and covering an area of 7.110 km^2. The population dynamics history is not so well documented as the growth of the metropolitan region is based on the moderate but steady growth of many moderate sized and smaller centres. Today, 10 moderate-sized cities (with several 100.000 inhabitants) are part of this metropolitan region: Dortmund, Bochum, Essen, Duisburg (Ruhr Area), Mönchengladbach, Düsseldorf, Wuppertal, Leverkusen, Cologne and Bonn (Rhine Area). In the south, the Rhine-Main metropolitan region has evolved with Frankfurt, Offenbach, Wiesbaden, Mainz and Darmstadt as larger cities. The region is well connected through a very dense motorway and railway network and interrelated by dense commuter flows. Further close connections have been established to the neighbouring Randstad- and Brussels agglomerations [37].

2.2. History and dynamics: patterns and triggers shaping growth

The ancient cities in Egypt, in the Greek and Roman empires, often linked to military camps, show a strict geometric pattern—gridded inside their boundaries. Since the middle ages, circular boundaries turn up as a frequent shape representing a natural bound of any city, which is based on some central focus, distorted by terrain, rivers and large distance routes, connecting cities. [38]. Towns started evolving as fortresses or were built around fortresses with walls to protect the citizens. Inside the walls, the cities grow and densify, building irregular shaped blocks and narrow streets around the centre, which is the market place or the main church place. The old city was a walking city, allowing to buy the necessary things and to work in close distance to the living place.

Cities have been connected though a star-like road network with neighbouring villages and further cities in wider distance. The radial major road network have been superimposed by secondary places, which serve as sub-centres for neighbourhoods or outskirt areas, being the origin of secondary again star-like street networks inside the built-up area. This structure strengthens the centre function of urban cores as well as functions of the sub-centres and triggers later the layout of public transport systems. The secondary street network provided new connections between the sub-centres, connections which are the origins of the ring road systems around the centre.

In the late nineteenth century, the local street network shaping the villages outside the ancient boundaries of the large cities was replaced frequently by a gridded block system to develop efficient housing for the growing population. Between the 1950s and the 1980s, these structures have been further superimposed by large ring roads or tangential roads outside the core cities (e.g. in London, Paris, Munich, Vienna).

The railway systems, set up in the nineteenth century by private companies, who developed usually one certain network branch, supply a particular suburban area, each railway line with its own terminus, not inside but around the city centre. These systems triggered the development of inner city tram and metro lines, which connected the terminus stations in all larger cities—e.g. in London, in Paris, in Vienna or Berlin (c. f. [39]). At the "rural end" of those network branches, the railway terminals in the peri-urban area provide connectivity with the core city triggering the transformation of former rural villages to small towns, securing access to workplaces and functions of the core city before the era of private car ownership.

Typically, Europe's quite public transport-oriented metropolitan regions are now public transportation organizations with a common tariff system, which allows using one ticket for traveling in the region, while using trains, metro lines, busses, etc. from different companies.

2.3. Socio-spatial structures and dynamics

In the European metropolitan regions, the structures are similar as described in the introduction. Wealthy classes move towards refurbished prestigious blocks or high quality, green areas close to the city centre. Inner city gentrification can be observed triggered by upper middle-class population to benefit from close central core services and various location factors (e.g. restaurants, shops, leisure areas, transportation hubs). The lower middle class moved to the attractive outskirt areas—either in sub-centres with sufficient infrastructure and public transport connection or in case of less income "just" to any green area in quite far distance to the city and with weak access to public transport infrastructure (c. f. [3]).

Segregation is a particular topic: the more people from abroad are moving to the city, the more mono-ethnic and mono-national clustering takes place, chasing away the origin population.

Informal housing was long time not an observed issue in Europe's urban agglomerations. Only in some Balkan countries (most affected are Albania and Serbia, but also Romania and Greece) and in the Caucasus countries (e.g. Moldova, Armenia, Kazakhstan, Kyrgyz Republic), this phenomenon occurs (c. f. [14]).

2.4. Functions and relations, communication and interaction

While monocentric agglomerations include all functions of a traditional city within its borders, polycentric agglomerations show some distribution of functions between the centres. For instance, the adjacent Rhine- Ruhr and Rhine-Main metropolitan regions show a certain distinction into heavy industry production areas in the northern Ruhr valley and in service centres in the central area—among them Frankfurt as a European banking centre or Bonn as former administrative centre of (West-) Germany and high value industry (cars, machinery, ICT) in the southern part (c. f. [37]).

The distribution of functions over a wider area requires more interactions, which are raising the traffic load in the metropolitan region. Therefore, the road traffic loads are high and heavy congestion problems are frequently observed—on the highways as well as on the inner city major street networks, although the urban public transport share is relatively high.

2.5. Policy: spatial impact on urban expansion patterns

European urban agglomerations are particularly driven by administrative structures and planning policies related to these administrations.

In the introduction, the relative power of the local and the sub-national (regional) levels of government has been addressed as effect of different centralized or federal planning systems. This is of particular interest in the Europe with its small-scale system of many small countries—all with different governmental systems, which show a range of centrality to federalism. Countries with central administration like France allow easy development of common regional planning strategies for the entire urban agglomerations (c. f. [28]). However, even there, city planning is quite independent and neighbouring villages follow their own strategies, although they are forced to respect the national or provincial planning framework. In countries with a strong power for local political entities, like Germany or Austria, the local policy makers feel responsible for their own territory, which leads to more competitions and less collaboration, where it is hardly possible to develop strategies for the entire urban agglomeration. Details can be found in Section 1.

In countries of the former Soviet Union influence sphere like Poland or Hungary, liberal governance structures turn up as the opposite of the former totalitarian framework conditions. This leads to some extent to a reduction of public planning power allowing private developments, not all time in line with the national legal systems, by ignoring zoning regulations and creating gated communities (c. f. [40]) and uncontrolled sprawl.

3. Urban agglomerations in Asia: focus Iran

3.1. History of urbanization

Iran is one of the oldest and culturally richest civilizations in the world. Shoush, Hegmataneh, Pasargad and Takht-e Jamshid were well-developed ancient Iranian cities, which can be perceived as the origin of Persian urban planning resulting as a major influence in shaping an urban tradition addressed today as the Islamic city (c. f. [41, 42]).

Iran experienced of expeditious urban population growth after the 1940s, which influenced the process of planning, development and new cities (c. f. [43]). Today, Iran is a modern developing country with the second largest population and the second largest area within the Middle East. Iran hosts the greatest number of cities above 1 million inhabitants in the region: 7 out of 28 [42].

During the past 5 decades, Iran is experiencing the population shift from rural areas along environmental, social and economic development by 68.5%. While there are many benefits of efficient cities, we need to recognize this rapid, often unplanned urbanization that brings different risks such as critical infrastructure, cultural heritage and the potential for devastating spread of disease [42]

When discussing urban agglomerations in Iran, Tehran with 8.4 million inhabitants within the city and 14 million within the agglomeration [43] is the outstanding example.

The city of Tehran was first mentioned around 1100 AD in the book Farsname. Its origin go back to an oasis at the foot of the Alborz mountains. Tehran is located 20 km north of the city of Ray, which was destroyed in 1220 by the Mongols, causing a shift in the settlement focus in favour of Tehran. Tehran, embedded in a favourable climate and supplied with adequate water resources and, moreover, located along the formerly most important West-East trading route of the Middle East, the old silk road, emerged after its appointment to the capital of Iran in 1798 (instead of Shiraz) during short time as centre of administration and finance of the country [44].

The historic Iranian cities are, like many Arabic cities, walled cities—because of defense and environmental reasons (shading through, protection against desert sand). In 1874, Tehran was provided with a new city wall covering an area of around 20 km^2. With this extension started a functional differentiation both within the city and in its surrounding area. While the south manufactories and smaller industries have been established, the northern districts show residential areas with luxury shops for the relatively wealthy upper class. Urban growth accelerated in Tehran after the destruction of the old city wall in 1934 during the first Pahlavi era.

After the 1960s, strong increase in population numbers lead to a big construction boom in Tehran and thus a change of the old structure. This contributed to the intensification and production expansion of the economic sectors and to intensify the building activities and industry's desire for concentration. Originally pure residential and commercial settlements were by the immigration to industrial zones, whereby the growth direction through the arterial roads (primarily in west direction, towards Karaj). Old villages embedded in the area were overrun by this development and are today limited "historic islands" in a purely industrial-commercial environment [44].

After the 1978, Khomeini revolution peripheral developments took place by copying western development models—in Tehran, but more or less in all Iranian cities (c. f. [45]). Standardization of urban systems, widening of streets to facilitate vehicular movement, density increase and construction of elevated highways and bridges which penetrated the urban fabric and public and private realms are some examples.

Tehran is now among the world's fastest growing cities. In the early 1940s, Tehran's population was about 700,000. By 1966, it reached 3 million, and by 1986—during the Iran-Iraq war—intensive migration lead to a population increase of up to 6 million. Today, Tehran's extent is around 1500 km^2 and its population is 8.4 million, but the metropolitan region has more than 14 million residents [43], as addressed above.

3.2. Development drivers for urban agglomerations

Under the current economic liberalization process, Iranian cities and especially Tehran have been subjected to dramatic development challenges, which have impact on the urban fabric, and the identity of communities. Urban agglomerations evolve leaving the early walled city concepts.

As effect of insufficient urban planning focusing on road infrastructure and housing, Tehran's downtown area show a weakly organized structure of neighbourhoods providing various functions. A proper centre with a meaningful distribution of functions is missing.

The high-rise building area of Tehran is not in the downtown area, although there are some larger buildings—particularly hotels of international companies built in the 1960s. The most new high-rise buildings are erected as apartment buildings for upper middle-class residents in the northern, elevated area of Tehran, along the Alborz slopes with great views, lower temperature, more humid and green environment.

The pressure on new housing leads to various housing development projects in the western, eastern and southern flat areas around Tehran. The new planned developments are designed for basic residential purposes and do not provide proper further infrastructure and have little access to public transportation, as the private developers consider only return of their investments and do not take into account the entire urban system with functions, relations and thus traffic.

The city is little involved in the development of these areas—public investments refer to roads and power infrastructure. The only social infrastructure erected intensively are shopping malls, which serve as communication and entertainment areas with shops, cafes, cinemas, etc.—again financed with private capital.

As urban development is left to private entrepreneurs, the urban agglomeration shows no proper development direction and is growing more or less randomly. This is somehow supported by the physical framework condition, which is just a flat dry area with no barriers and no attractive features like rivers, lakes, forests, parks, etc. in the south.

Urban development is little regulated, as there exist no zoning maps for the city. High-rise buildings can be erected close to single-family homes, which can be observed frequently in selected districts.

The attractive parts of Tehran are in the north, which are constantly densifying through new high-rise buildings, while the west, south and east of Tehran show urban sprawl, not so much triggered by single-family homes but by medium-sized housing developments, production sites, storage and logistics functions as well as transportation infrastructure.

3.3. Socio-spatial structures and dynamics

One of the main objectives of the Iranian revolution was to improve social equity. After the revolution, the composition of society changed to some extent, which was in line with the targets of the revolution: the share of the middle-class composition doubled from 15% (1979) to 32% (2000) [46].

Urbanization is rapidly progressing: in 2005, already 68% of the population was living in cities, and in 2015, the share reaches 73%. Over the past three decades, the country

http://dx.doi.org/10.5772/intechopen.73524

experienced rapid socio-demographic and economic changes. New patterns of internal migration, declining fertility, the current youth peak, an upcoming aged population, as well as the growing number of female-headed households are some of the broad socio-economic changes (c. f. [47]).

3.3.1. *Social structure and segregation*

In Tehran, one do not find ethnic but social segregation, not forced by some unspoken rules but just by affordability. Tehran shows, similar to the Iran average, a small upper class whose share is stable since the Pre-Revolution era. As mentioned above, this class is living since ever in the north in the hilly areas along the mountain slopes because of attractive landscape, climate, environment and view. Except some bus lines, one can hardly find any public transportation because of the complex terrain requiring to use private cars, which is anyway the favourite mode of transport.

A quite big share of middle-class residents is observed living in the central district of Tehran, where also the governmental and company offices and banks are located.

Quite large fraction of lower middle to low-class inhabitants are living in the outskirt areas particularly in the west and the east were the newcomer residents find affordable housing opportunities.

3.3.2. *Gentrification*

Gentrification is observed in a moderate way, triggered by middle-class population, to benefit from close central core services and various location factors (e.g. restaurants, shops, leisure areas, transportation hubs).

The rents and prices rise and force the lower middle class to leave the area, moving to the west and east outskirt areas.

3.4. Functions and relations, communication, interaction and traffic

Iranian cities are very much organized in neighbourhoods, which provide all functions within small distances—this has not changed in general, but the habits of the population have changed—not just using the local functions in close distance but different ones because of various reasons—e.g. distance to working place, different supply structure, different quality, etc. (c. f. [48]).

As mentioned above, markets, bazaars and the shopping malls as descendants are part of the social infrastructure, which serve as communication and meeting areas. While bazaars are traditional, narrow places with a focus on trade, which provide cooling only by shading, air-conditioned, spacious shopping malls, usually with own parking facilities, provide convenient and cooled environments for shopping and dining and thus invite customers to stay (and consume).

Thus, the walkable neighbourhood and city are of less importance. The residents make use of their cars or taxis to reach the different places they want to go. Public transport is less relevant—maybe also because of the supply quality. Public transportation infrastructure supply in the Tehran agglomeration is rather weak: eight metro lines, some BRT routes and a considerable number of bus lines, which are supplemented by a large taxi fleet, again privately organized.

In contrast to the weak public transport system, the road network is very dense. The local road structure providing access to the neighbourhoods is superimposed by a wide network of elevated multi-lane motorways in the city with rather few exists and some U-turns (!), which require to drive long distances on the motorway network to reach quite close areas one can nearly reach by walking.

Most people who can afford their own car use it, others use the rather cheap taxis than public transport. The intense use of private cars and taxis causes heavy congestion not only during peak hours but also throughout the day. So, the transportation system is extremely inefficient and it takes much longer to go by car than to take a metro or BRT which their separate bus lanes.

However, public transport is experienced to be less attractive than private transportation, which has also cultural/religious reasons, leading partly to a social divide of men and women in public life where women avoid using public transport.

The effect of the intensive private car use is that Tehran is heavily suffering from smog and dust.

3.5. Policy: spatial impact on urban expansion patterns

Urban management is since 20 years organized capacity of communities and grassroots in the management of cities in Iran. After the revolution, Islamic councils have been established both in cities and in rural areas. The councils are directly elected by the people and have selected municipal functions. The countries' policy is framed in 5-year development plans (c. f. [49]).

The history of development planning in Iran started in 1949 with the first development plan. Within this period, eight development plans have been implemented, five of them before and three of them after the 1979 Revolution [50].

With the beginning of the Islamic Revolution, urban development planning was disrupted. The first development plan after the revolution, set up in 1982 and approved in 1988 after a number of changes, stipulated construction of 10 million housing units and overcoming the housing shortage within 20 years [50].

The measures defined by the "Development Plan 1988–93" include granting of easy accessible loans, provision of construction materials at government prices, sale of land at low prices and approval of new housing cooperatives. These measures reduced the housing development costs, increased the number of housing units and led to a horizontal expansion of the cities, reducing housing density [51].

The recent development plan addresses again the rapid urbanization announcing access of all groups of the society to "adequate and decent housing and provision of basic services for all households" as focal points [52].

The government's current planning policy, called PAK (from pasandāz [saving], anbuhsāzi [collective housing], and Kuchaksazi [building small]) which aimed to optimized housing (c. f. [50]).

It consists of three main doctrines: (1) developing housing provision plans within the framework of a comprehensive social welfare policy, (2) optimum use of market mechanism with little government intervention only in case of market failure, and (3) promoting savings and loans system to enhance the access of middle-income groups to adequate and decent housing (c. f. [49]).

As urban development focuses only on housing, which shall be triggered "by the market," there is little opportunity for strategy-based urban development (c. f. [50]).

4. Urban agglomerations in Asia: focus China

4.1. History and dynamics of urbanization

China's cities are among the oldest in the world. Banpo, the first Chinese city appeared on the Central plain around the Yellow River more than 6000 years ago. During the Iron Age, or Eastern Zhou Dynasty, a differentiated urban system was established for the first time. The role of the cities, at this time, was defined by the local economy and by their military and administrative functions. In addition, principles for governmental administration and for urban planning were written down for the first time in this period. During, the Qin Dynasty, an administrative system was established, which divided the country into provinces, prefectures and counties. On the top of the hierarchy was the imperial capital, which had not only economic and military functions but was also of symbolic and religious importance (c. f. [53]).

Although, in ancient China, some capital cities had more than one million inhabitants and were bigger than European cities of that time, ancient China was a predominantly rural country. When the People's Republic of China was founded in 1949, the urbanization rate was at 10.6%. Urbanization increased slowly, but steadily during the 1950s, and stagnated during the 1960s and 1970s, when Chinese society was shattered by social experiments like the "Great Leap Forward" (1958-1962) and the "Cultural Revolution" (1966–1976). When the political and social situation developed after Mao's death, the urbanization started to rise again. The incremental transition of the Chinese economy from a planned system to a "socialist market economy" accelerated the growth of the urban agglomerations in and unprecedented way. In 1990, around one quarter of the Chinese population lived in urban area. In only 20 years, this share doubled: in 2011, already more than 50% of the population was living in urban areas, and the urban agglomerations continued growing (c. f. [54]).

The Chinese government defined the Pearl River Delta, Shanghai and Beijing as future megacities. These megacities were supposed to attract the lion's share of Foreign Direct Investment and to become thus the powerhouses of the Chinese economy. The concept turned out to be very successful. A couple of years later, the Szechuanese cities, Chongqing and Chengdu, were selected to become two inland mega cities, following the example of Beijing, Shanghai and the Pearl River Delta.

Despite negative social and ecological consequences, such as very high levels of air pollution, huge land consumption and loss of cultural and natural heritage, the pace of Chinese will stay rapid in the coming years.

4.2. History and dynamics: patterns and triggers shaping growth

Beijing was founded around 3000 years ago and became capital of China in 1152. Its current urban structure traces back to interventions carried out in the early fifteenth century, when emperor Yongle built the Forbidden City and the Temple of Heaven as the governmental and the spiritual centres of old Beijing (c. f. [55]).

Old Beijing was planned as a system of circular rings: the Forbidden City was in the centre and served as imperial palace until 1911. The Forbidden City was surrounded by the Imperial city as administrative centre, which itself was surrounded by the inner city and the outer city, both of them areas where artisans, craftsmen and traders lived. All circular rings were walled. This urban morphology was comparatively stable over a long time (c. f. [55]).

In the 1960s, the city walls were demolished and later replaced by an urban expressway, the so-called 2nd Ring Road. With this ring road, the concentric development pattern of the historical city was replicated with modern means. Following the increasing urbanization, a 3rd, 4th, 5th and 6th ring road were added, which structure the urban metropolis. The space between the ring roads was quickly filled up with new developments.

The development of an efficient public transport system started rather late. Before the Beijing Olympics of 2008, there were only few subway lines in operation, thus the urban development could not be concentrated on public transport lines. It rather filled up the space between the ring roads, continuing the concentric development typical for Beijing.

In the past 10 years, the metro system expanded rapidly. It is now, after Shanghai, the second longest metro system in the world, and is expected to further grow in the future. However, the system does not have the typical star structure like European cities, but a grid layout, which is superimposed by two circle lines, reflecting the agglomeration's concentric development.

Shanghai, China's second city, is much younger than Beijing. It was a city of less importance until it became the gateway of several colonial powers to China in the late nineteenth century. Several foreign settlements, so-called concessions were built by the colonial powers adjacent to the old Chinese town. Together, they now form the downtown of Shanghai. A good part of the urban expansion of Shanghai followed the concentric model of Beijing. Also in Shanghai, an efficient public transport network evolved quite late, but it has been widely expanded in a comparatively short period.

A significant difference is however the spatial direction of the urbanization process: As Shanghai is confined by the sea in the south and the east and by the Yangtze River in the north, urban expansion can only move towards west. In this direction, there are several medium-sized cities (like Suzhou, Hangzhou, Changzhou, Wuxi, Nantong) with a strong economy and their own culture and history. Many of these cities are already connected by high-speed railways. They build together with Shanghai, the Yangtze River delta region.

The third urban agglomeration in China presented in this article, the Pearl River Delta, differs significantly from Beijing and Shanghai, as it is a clearly polycentric region called a city network. It comprises several cities in the Guangdong province and the Special Administration Zones of Hong Kong and Macao with in total 63 million inhabitants. The cities in the region have very different characteristics regarding history, culture and economic structure. The strong, mainly manufacturing-based economy in the region and the urban growth lead to increasing regional integration and the formation of an agglomeration. Creating connectivity and access is a particular challenge due to the polycentricism. As expressways are frequently congested, there are now plans to connect the metro networks of different cities in the region and to establish a regional network of high speed and commuter trains called Pearl River Delta Rapid Transit.

4.3. Socio-spatial structures and related dynamics

In 2016, the Chinese per capita income was 7.5 times higher than in 1996 [56]. This is a spectacular increase in wealth in just 20 years. Nevertheless, there is still a very big difference in the regional distribution of wealth: The household incomes in the urban agglomerations at the coast are much higher than in the hinterland. This leads to strong internal migration flows from the hinterlands to the urban agglomerations.

To control the influx of people into urban agglomerations, the Chinese government uses the so-called Hukou (household registration) system [57]. This system divides the population into urban and rural dwellers according to their place of birth. Citizens with a rural Hukou can under certain circumstances live and work in urban areas, but will not have access to social services in the cities and usually receive less salary than urban dwellers. This creates a big polarization in Chinese cities, as the share of persons holding rural Hukous reaches 30% and more in some Chinese cities. There is a very lively debate in China on how the Hukou system can be reformed and eventually abandoned. Many Chinese regard it as highly unfair, but it is an open question how urban agglomerations could deal with the influx of more internal migrants once the system is removed.

The Hukou-holding migrant workers are at the bottom of the urban societies in China. Many migrant workers still live in containers on construction or factory sites, physically separated from the city dwellers.

4.4. Functions and relations, communication and interaction

In case of the monocentric urban agglomeration of Beijing, nearly all urban functions are concentrated in the core city. The Jing Jin Ji plan aims at reducing the pressure on Beijing by establishing a 130-million capital region, covering three provinces with the city triangle Beijing-Tianjin-Baoding as its core. The new Beijing Daxing airport and the Xiong'an New Area south of Beijing, which shall absorb a significant share of the future urban growth of the agglomeration, are key measures in the development of Jing Jin Ji (c. f. [58]).

As the distances between Beijing, Tianjin and Baoding are between 130 and 200 km, high-speed railways will play an important role in connecting the three cities. The use of high-speed transport to create connectivity and access not just at national level, but also at regional scale is a new strategy for shaping urban agglomerations in China. To a certain extent, it has already been implemented in the Yangtze River Delta around Shanghai, and it will be applied in Jing Jin Ji and the Pearl River Delta. Eventually, the network of high-speed trains will lead to an integration of concentric or star-shaped urban agglomerations into a network of cities that will cover most of China. Chinese researchers expect this new development model of a network city to be more efficient than star-shaped or concentric urban agglomerations (c. f. [58]).

4.5. Policy: spatial impact on urban expansion patterns

Chinese centralized policy had great influence on shape and development of cities and later agglomerations. The ring structure of Beijing, the economic liberalization with the set-up of special economic zones and the Jing Jing Ji plan are impressive examples of the effects of the national policy and planning strategy.

After almost 30 years of high-speed urbanization, the Chinese government has announced a new era for the development of the urban agglomerations in China called the New Normal. Lower but more qualitative, efficient and innovative growth shall characterize it. In addition, the Chinese National New-Type Urbanization Plan 2014–2020 and the 13th 5-year Plan 2015–2020 suggest a new pathway for urban development in China, with a stronger focus on quality of life and on environmental quality than in the past (c. f. [59]). A better planning and management of urban agglomerations will be of utmost importance for reaching these goals.

5. Urban agglomerations in Latin America

5.1. Development drivers for urban agglomeration in Latin America

During the past 60 years, Latin-American has experienced very strong structural changes in its social, economic and spatial dimensions, shifting its settlement patterns from dispersed rural arrangements into urbanized or agglomerated concentrations (see **Figure 3**). Today, 80% of the Latin American population concentrate in urban agglomerations and this share seems to continue growing (c. f. [60]).

Despite the difficulty to generalize such a diverse (sub-)continent, independent from the Spanish colonial administration since more than 200 years, similar urban patterns of agglomeration and development trajectories and processes can be identified from the second half of the last century until our days. On the one hand, socio-economical drivers, as the very rigid and traditional system based on the socio-economic status differentiation of the population, have been spatially settled, leading to urban fragmentation forms in both sides, the rich protected by fences in gated communities and the poorest hiding in segregated informal settlements. This reality has shaped not only the spatial organization of almost all Latin American cities but also a continental feature, positioning the whole region as the one with the highest level of income inequality of the world.

On the other hand, although exponential urban growth had several positive economic effects, this also caused various negative environmental impacts. It has not been successful per se in influencing political concerns and policy measures oriented to reduce vulnerabilities or improving living conditions for urban citizens in an effective manner.

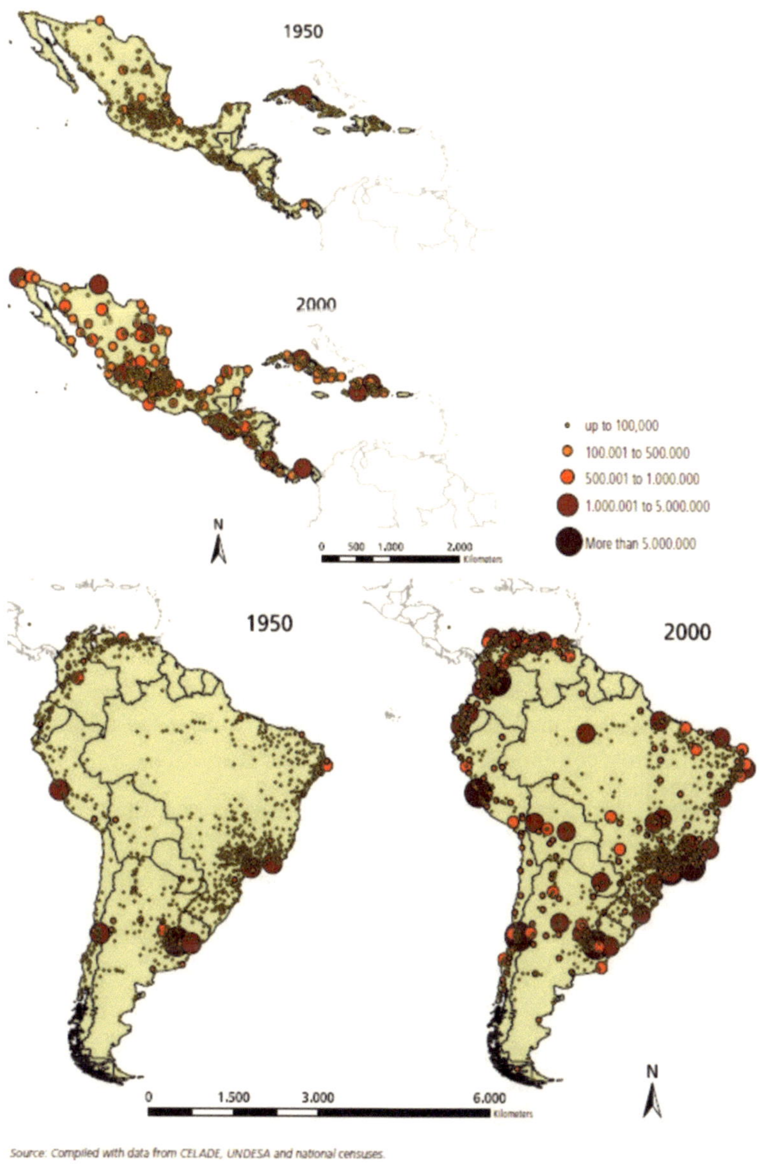

Figure 3. Latin America and the Caribbean. Cities with over 20,000 inhabitants, 1950 and 2000. Source: UN Habitat [72].

This, in fact, questions the capacity of governmental institutions, particularly in similar spatial distribution and accessibility to public transport, education, health, justice, security and/or even to trade for urban dwellers. "One thing is clear: Latin American nations continue to tackle with many of the same problems they have tackled with for decades - uneven growth with high levels of poverty and the world's highest levels of inequality; ineffective state institutions that deliver limited benefits to often a highly restricted minority of the population" [61].

However, evidences from specific institutional and private efforts to update and implement projects and measures oriented to reduce inequality and improve urban environments have been internationally recognized and awarded from Brazil, Ecuador, Chile and Colombia during the last 15 years and should be mentioned due to its relevance and deep positive impacts.

5.2. Patterns of spatial development in Latin American urban agglomerations

AAlthough the urban agglomeration process in the region started 65 years ago, it continues expanding at a rate of 20 m^2 per minute [62]. During the last years, the development speed started to decline.

Thus, despite there seems to be a deceleration in in-migration rates, while "...in the 1980s, migration from the countryside contributed 36.6 per cent to the growth of urban population in the whole region, in the 1990s this share had fallen to 33.7 per cent" [63], net internal population growth and physical expansion continue rising.

Empirical evidences from different Latin-American cities show similarities in their spatial development patterns and processes, evolving diffuse peripheral fragmentation due to informality development (c. f. [64, 65]) and increasing tendencies of urban sprawl (c. f. [66]), as well as social and spatial fragmentation due to gated communities (c. f. [67–71]).

Although urban sprawl and/or suburbanization phenomena have constantly increase in Latin America ([62]), compact horizontal expansion and densification of the existing built environment keep being the most common growth patterns [73].

In fact, Latin America is still more or less compact, built areas have densified and intensify it uses at the periphery in higher proportion as in the urban centres (c. f. [74, 75]) in "... open-fragmented peri-urban fabric where autonomous elements do not relate to each other anymore..." [76].

This uncontrolled and fragmented expansion process has been observed in almost all countries of the continent, passing the boundaries of the administrative municipalities absorbing and agglomerating other close located settlements and towns in conurbation processes.

Building urban agglomerations through sprawl is in fact the most common urban development pattern in Latin America, and it is one of the reasons of the polycentric structure of the Latin American urban landscape. Metropolitan regions in Latin America bring new challenges at the spatial, the sustainability, the planning and administrative levels, as long as there are almost no mechanisms for integrated governance between different municipalities and territorial institutions (**Figure 4**).

5.3. Socio-economical spatial dimensions and vulnerability

Some roots for higher spatial fragmentation levels, segregation, as well as for the increasing urban inequality can be understood as a spatial effect of a fragmented, stratified society. Stratification as a spatially structured result of socio-economic class differentiation relates to colonial times, where social strata are determined by the person's position at the moment of birth, with little or no social mobility possible (c. f. [77, 78]).

Figure 4. Urban sprawl in Bogota's southeast periphery. Source: Santiago Sanchez Guzman.

With almost one fourth of the entire urban population still living in informal settlements in the region in 2014 [79], levels of urban environmental vulnerabilities thread exactly those poorer and denser populated areas in the outskirts.

Informality, as a result of postcolonial socio-economical dynamics (c. f. [80, 81]), has only consolidated and spread longer since the introduction of neoliberal policies and measures tended towards privatization and economic efficiency (c. f. [69]).

The implementation of neoliberal policies, understood as "the set of policies that were prescribed together as part of a strategy to enhance the role of markets in economic development" [61] in countries like Chile and Bolivia, determined some of the patterns of privatization of infrastructure, social services and even natural resources as tap water provision.

"The process of water privatisation in Chile which began in 1981 under General Pinochet established a model for water management that strengthened private water rights, adopted a market-based allocation system and reduced state oversight. That model became emblematic of neoliberal reforms heavily promoted by the World Bank and International Monetary Fund" [82].

Latin America, as a playground for experimentation of neoliberal reforms since 1976, did not achieve reducing socio-economical gaps between class groups, but on the contrary, contributed to consolidate monopolies and disparities in cities as well as asymmetrical distribution of its services and wealth.

5.4. Political economic perspective

"Economic growth is not a sufficient condition for improving the quality of life. Millions can be and have been left out systematically from the benefits of economic growth" [61]. Latin

American urban agglomerations are a good example of how financial and physical urban growth per see has not been strong enough to reduce poverty and inequality, but on the contrary, has raised uneven development. Development, understood "as a process of expanding the real freedoms that people enjoy" [82], has unfortunately not occurred at the same velocity of the growth and urban expansion rates.

Although the Human Development Index of the region shows still better performance than African Regions, they share the same levels as most Eastern-European and some Asian countries, but are far lower than OECD levels (c. f. [79, 83]).

Although Latin American political concern has centered during different periods on the discussion of import substitution versus market liberalization, the central problem has not been addressed in improving effectiveness of institutions. "The problem is not 'too much market' or 'too much government'. Rather, it is poorly functioning markets and poorly functioning governments. Both markets and governments need effective institutions to perform well" [61].

Institutional weakness is, by far, one of the most important reasons of the difficulty in monitoring, planning and designing urban growth and has contributed to a lack of participation from civil society in governance processes as well as created the perfect context for illegality.

5.5. Public transport, housing and social infrastructure

Latin American agglomerations face big challenges in shortening the distance between socio-economic classes improving living conditions of its most vulnerable communities and turning actual institutions into more effective active actors for development. Political continuity, or better, the continuity of development policies and projects, is also a key element to accomplish heavy development endeavors and plans as seen in cities like Curitiba, Quito, Medellin and Bogota during different periods. Innovative measures and projects related to urban systems, such as bus rapid transit (BRT), cable cars, improving the accessibility of informal housing areas, public libraries, schools and kindergartens and improving education in those cities, have been only possible due to the political compromise of different administrations to achieve and implement the projects rather than use their own positions to brand themselves.

However, despite those very relevant achievements, the region is still facing big challenges, aiming at sustainable solutions for housing provision in an integrated way, reducing fragmentation and transforming agglomerations into more inclusive, resilient and healthy urban environments.

6. Urban agglomerations in North America: focus USA

6.1. Background: the start of urbanization

In North America, large urban agglomerations can be observed along the east coast where the first colonial settlements, evolved as trading centres and harbours as transportation hubs to the old world. They have been later connected with the hinterland and the west coast through

http://dx.doi.org/10.5772/intechopen.73524

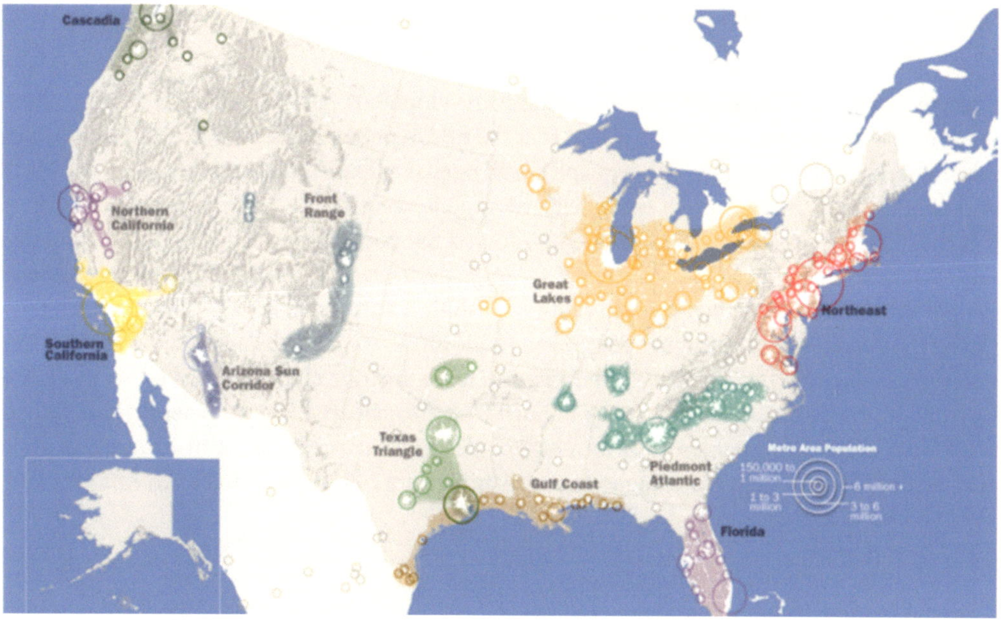

Figure 5. Emerging American-Canadian megaregions as defined by America 2050. Source: Ref. [84].

railways. These settlements with attractive location factors build the nuclei of cities, which steadily grow and merge into large polycentric conurbations. In the mid-west, these dynamics and the drivers behind—harbours at the Great Lakes, trading centres and transportation hubs can be also identified—created e.g. the Chicago-Pittsburgh conurbation ("the Great Lake Megalopolis") as an agglomeration covering several cities like Chicago, Milwaukee, Detroit, Cleveland and Pittsburgh. Along the West Coast, the dynamics turned out in the beginning similar with harbour cities like Seattle, San Francisco, Los Angeles and San Diego, later connected with the hinterland and the entire country through railway lines, which grew first through housing areas and later through industry centres—where the Silicon Valley is the most famous one. As these cities are located in quite distance, they do not evolve polycentric structures but stay as monocentric agglomeration on its own. As Fang and Yu [1] addressed, agglomerations turn out as urban systems with self-sustaining hierarchical structures and functions. This can be observed in the huge Washington, Baltimore, Philadelphia, New York and Boston agglomeration, where each city itself has different core functions: Washington D.C. is the political and administrative centre with historical and cultural heritage, Philadelphia the manufacturing centre, New York is the commercial, service and banking centre, while Boston can be seen as academic centre. **Figure 5** gives an overview of the most important urban agglomerations in North America.

6.2. History and dynamics: patterns and triggers shaping growth

In the 19th century, growth of cities in Europe and North America accelerated, but reached their peak by the mid-20th century. The West Coast of the USA grew substantially during the years until to 1990. North America is now one of the most urbanized regions of the world with about 82% of people living in urban areas (c. f. [85]).

Growth was always closely related to economic development. Analyses show that larger cities tend to perform economically better and that the future patterns of economic output, the current regional differences in GDP per capita, will persist at least until 2030 for the global north and south. Until the mid-nineteenth century, the city centre was a most fashionable living location. Changes in the transport systems allowed the population not to live in walking distance to their working places, which had a significant impact on the dwelling patterns where people started to seek for neighbourhoods with similar social structure to show and pretend their own status. The wealthy people moved to the attractive suburbs around the urban cores and the low-income population concentrated in less attractive cheaper areas either in downtown or in far distance less attractive suburbs. Due to cheaper but sufficiently still skilled workforce abroad companies start moving to develop areas e.g. in Latin America or Asia and, as jobs disappeared, cities began to shrink in the late 20th century in North America [86].

6.2.1. *Densification*

A reason for densification is often the revitalisation of old historic centres of those cities, which constantly have deteriorated, in the second half of the twentieth century. Major public and private investment was needed to restore the buildings and places to attract new residents, but these investments lead often to a significant increase of lot-prices and reduce the affordability, leading to gentrification and segregation. This dynamic also brings up new pressure to the transport sector and increases the energy demand. The support for housing was mainly given to expand and improve already existing units, which led to a lack of appropriate design principles causing problems such as overcrowding or low privacy. These housing units are also often very vulnerable to environmental disasters. This causes the important challenge that how to raise the social and spatial living conditions in the already consolidated settlements with improved design.

6.2.2. *Segregation*

In the USA, segregation is traditionally associated with racial segregation. Officially, it is forbidden but can be still observed. Racial segregation starts centuries ago and continues during the rapid urbanization in the early twentieth century, which was associated with a significant increase of European immigrants and African-Americans, both clustered because of their limited income in neighbourhoods with poor housing. A further driver was the school segregation, which was legal until 1954 in the USA, but still is persistent due to continuing segregated residential patterns [87]. Racial segregation has diminished significantly since 1960s but the African-American population still remains the most segregated racial group and is hypersegregated in the largest urban agglomerations (c. f. [88]). Significant spatial segregation can also be found for Hispanics and Asians. Native Americans are segregated but to a much lower extent. In contrast to racial segregation, the separation of income grew during the last decades. More than half of all low-income working families are racial minorities.

6.2.3. Gentrification

Gentrification is a well-discussed issue in the USA, which leads to residential segregation. A recent study in 2015 found that gentrification, as a national problem, is actually less pervasive than perceived. But in urban agglomeration areas, gentrification is a significant issue, where rising housing costs are forcing the low-income classes to relocate to cheaper, less attractive areas. Recent research discusses that trends towards gentrification are only one part of the dynamic. Youthification, an increasing presence of young adults within a given neighbourhood, can be also seen as a trend in North American cities. This trend is uneven distributed, whereas it seems that cities with the highest presence of you adults are those rich in amenities and opportunities but with lower cost of living and dense vibrant urban neighbourhoods (c. f. [89]).

7. Conclusions

The structures, dynamics and general challenges of urban agglomerations in the addressed continents are not that different, unless some phenomena do not occur in all metropolitan regions (e.g. informal housing, ethnic segregation, cultural/religious barriers, etc.). However, the ways to cope with the challenges and the success are different and the agglomerations can learn to some extent from each other.

In all continents, monocentric agglomerations show a more simple structure and more concentric development dynamics and frequently a centralized municipal or regional government, which makes it easier to elaborate integrated development concepts for the agglomeration that consider the region as single entity and addressing functional and relational issues from an internal perspective. Polycentric agglomerations consist of various cores with own municipal administrations, which require intensive collaboration and coordination to make the urban agglomeration a success.

Strategic urban planning is necessary for urban agglomerations to reduce urban sprawl to a minimum and to provide a proper distribution of functions, which allows easy supply of the population with the relevant services through a walkable city or at least a city of short distances. This would reduce traffic and thus congestions and would mitigate air pollution and thus increasing environmental and life quality.

Sufficient housing development is a crucial point as all agglomerations show immigration pressure and growth. Smart social housing is a necessity to avoid the appearance of deprived areas in the future. Segregation can be avoided by a smart mixture of different housing qualities and by a smart distribution of urban functions and infrastructure, serving all social classes in the same way. Participation in planning processes helps to develop common solutions and to achieve better acceptance in decision-making.

Gentrification is a mechanism which can be used actively to improve certain areas, financed by private investors, but negative effects should be mitigated, e.g. by establishment of local social housing estates in the area to accommodate those which must leave their—now renovated and expensive—apartments because of loss of affordability. This also supports the addressed mixture in housing qualities avoiding full segregation.

A social divide must be avoided as far as possible not just through provision of mixed-use areas but also through education and provision of equal opportunities for all, which is easier in agglomerations than in small towns. Skilled people secure economic development in the agglomeration as they attract new companies to establish production sites and thus gain new income in the area.

Traffic is the final big topic in urban agglomerations. The European style star-like major street networks promote the city centre as central place and promote public transport service to provide sufficient accessibility to the relevant services located in the target region, as well as to the housing districts as source regions. Grid-like street networks mitigate the location quality regarding centrality in the core city centres. Central functions can be placed anywhere and need just a road network to provide accessibility, less usable for walking or cycling short distances.

Public transport is a major issue, which makes accessibility of functions, interactions, commuting and thus urban life easier. As the size of the agglomeration increases the risk on congestion, high quality public transport services are the more important, the larger the urban agglomerations are.

Author details

Wolfgang Loibl*, Ghazal Etminan, Ernst Gebetsroither-Geringer, Hans-Martin Neumann and Santiago Sanchez-Guzman

*Address all correspondence to: wolfgang.loibl@ait.ac.at

AIT Austrian Institute of Technology, Vienna, Austria

References

[1] Fang C, Yu D. Urban agglomeration: An evolving concept of an emerging phenomenon. Landscape and Urban Planning. 2017;**162**:126-136. DOI: https://doi.org/10.1016/j.landurbplan.2017.02.014

[2] Knieling J. Metropolitan regions: Definitions, typologies and recommendations for development cooperation. Deutsche Gesellschaft für Zusammenarbeit [Internet]. 2014. Available from: https://www2.giz.de/wbf/4tDx9kw63gma/Metropolitan-Regions_Definitions_Typologies.pdf [Accessed: Sep 10, 2017]

[3] Loibl W, Tötzer T, Köstl M, Steinnocher K. Simulation of polycentric urban growth dynamics through agents – Model concept, application, results and validation. In: Koomen E, Stillwell J, Bakema A, Scholten H, editors. Modelling Land-Use Change – Progress and Applications. Dordrecht: Springer; 2007. pp. 219-235. DOI: 10.1007/978-1-4020-5648-2

[4] Bogue DJ. Principles of Demography. New York: Wiley; 1969. (DLC) 68026847

[5] De Jong G. Expectations, gender, and norms in migration decision-making. Population Studies. 2000;**54**(3):307-319. DOI: 10.1080%2F713779089

[6] Richardson HW. Economies and diseconomies of agglomeration. In: Giersch H, editor. Urban Agglomeration and Economic Growth. Berlin-New York: Springer; 1995. pp. 123-156. DOI: 10.1007/978-3-642-79397-4_6

[7] Sfetcu S. Economies of Agglomeration [Internet]. 2014. Available from: https://www.setthings.com/en/economies-agglomeration [Accessed: Sep 10, 2017]

[8] Straszheim R. Urban agglomeration effects and employment and wage gradients. Journal of Urban Economics. 1984;**16**(2):187-207. Available from: http://www.sciencedirect.com/science/journal/00941190/16/2

[9] Bhatta B. Analysis of Urban Growth and Sprawl from Remote Sensing Data. Berlin-New York: Springer Press; 2010. DOI: 10.1007/978-3-642-05299-6

[10] Bruegmann R. Sprawl: A Compact History. University of Chicago Press. Chicago, IL 2005. ISBN: 9780226076904

[11] Hornstein JM. A Nation of Realtors: A Cultural History of the Twentieth-Century American Middle Class. Durham, NC: Duke University Press. 2005. ISBN: 978-0-8223-3528-3

[12] Harris J. Urban and industrial deconcentration in developing economies: An analytical framework. Working Paper 53, Dep. of Econ. Cambridge: MIT; 1970. Available from: https://dspace.mit.edu/bitstream/handle/1721.1/64336/urbanindustriald00harr.pdf?sequence=1

[13] Kassa F. Growth rates and effects of urban sprawl in developing countries, the case of Addis Ababa Ethiopia International Journal of Area Studies. 2014;**9**(2):135-151. DOI: https://doi.org/10.2478/ijas-2014-0009

[14] UNECE. Informal Settlements in Countries with Economies in Transition in the UNECE Region [Internet]. 2012. Available from: https://www.unece.org/fileadmin/DAM/hlm/documents/Publications/Literature_Review_on_Informal_Settlements.pdf [Accessed: Oct 11, 2017]

[15] Barreto TB. Obstacles to Legalization of Squatter Settlements in Venezuela [Internet]. 2001. Available from: http://www.lincolninst.edu/publications/articles/obstacles-legalization-squatter-settlements-venezuela. [Accessed: Nov 21, 2017]

[16] Alonso W. Location and Land Use: Toward a General Theory of Land Rent. Harvard: Harvard University Press; 1964. ISBN: 9780674730854

[17] Bertaud A. Metropolis: A Measure of the Spatial Organization of 7 Large Cities [Internet]. 2001. Available from: http://alainbertaud.com/wp-content/uploads/2013/06/AB_Metropolis_Spatial_Organization.pdf [Accessed: Sep 15, 2017]

[18] Zetter R, Griffiths D, Signona N, Flynn D, Pasha T, Beynon R. Immigration, Social Cohesion and social capital [Internet]. York: Joseph Rowntree Foundation; 2006. Available from: https://www.jrf.org.uk/sites/default/files/jrf/migrated/files/9781899354440.pdf [Accessed: Sep 15, 2017]

[19] Gordon I, Monastiriotis V. Urban size, spatial segregation and inequality in educational outcomes. Urban Studies 2006;**43**(1):213-236. DOI: 10.1080/00420980500409367

[20] Atkinson R. Measuring gentrification and displacement in greater London. Urban Studies. 2000;**37**(1):149-165. DOI: 10.1080/0042098002339

[21] Glass R. Introduction: Aspects of Urban Change. Centre for Urban Studies, Report 3 [Internet]. London: Mac Gibbon and Kee; 1964. Available from: http://www.worldcat.org/title/london-aspects-of-change/oclc/1618199&referer=brief_results/ [Accessed: Sep 25, 2017]

[22] Peri GM. Young workers, learning, and agglomerations. Journal of Urban Economics. 2002;**52**(3):582-607. DOI: 10.1016/S0094-1190(02)00510-7

[23] Tötzer T, Loibl W. Ansprüche einer alternden Bevölkerung an Wohnen, Wohnumfeld und Mobilität – Zukunftsthemen für die Wiener Stadtpolitik. In: Hanappi-Egger E, Schnedlitz P, editors. Ageing Society, Altern in der Stadt: Aktuelle Trends Und Ihre Bedeutung für Die Stadtentwicklung. Facultas Universitätsverlag Wien. 2009. pp. 606-666. ISBN: 978-3-7089-0357-6

[24] Christaller W. Die zentralen Orte in Süddeutschland. Eine ökonomisch-geographische Untersuchung über die Gesetzmäßigkeit der Verbreitung und Entwicklung der Siedlungen mit städtischen Funktionen. 1933. 340 p. Darmstadt: Reprint by Wissenschaftliche Buchgesellschaft. 2009. ISBN-13: 978-3534197361

[25] Christaller W. Central Places in Southern Germany. Translated by Carlisle W. Baskin. Englewood Cliffs, NJ: Prentice-Hall; 1966. ISBN: 978-0131226302

[26] Cox W. Traffic Congestion in the World: 10 Worst and Best Cities [Internet]. 2014. Available from: http://www.newgeography.com/content/004504-traffic-congestion-world-10-worst-and-best-cities [Accessed: Sep 25, 2017]

[27] Jacobsen S. More Density Does Not Equal More [car] Traffic [Internet]. 2014. Available from: http://transitized.com/2014/01/14/density-equal-car-traffic/ [Accessed: Sep 25, 2017]

[28] Tosics I, Ravetz J. Managing growth. In: Piorr A, Ravetz J, Tosics I, editors. Peri-urbanisation in Europe [Internet]. Frederiksberg, Copenhagen: Academic Books; 2011. pp. 80-87. Available from: http://www.openspace.eca.ed.ac.uk/wp-content/uploads/2015/12/Peri_Urbanisation_in_Europe_printversion.pdf [Accessed: Sep 25, 2017]

[29] Ahrend R, Gamper C, Schumann A. The OECD Metropolitan Governance Survey: A Quantitative Description of Governance Structures in Large Urban Agglomerations, OECD Regional Development Working Papers, 2014/04 [Internet]. Paris: OECD Publishing; Available from: http://www.oecd-ilibrary.org/urban-rural-and-regional-development/the-oecd-metropolitan-governance-survey_5jz43zldh08p-en [Accessed: Oct 10, 2017]

[30] Brunet R. Les villes europeénnes: Rapport pour la DATAR. Montpellier: RECLUS. 42nd European Congress of the Regional Science Association. Dortmund, Germany. [Internet]. 1989. 20 p. Available from: http://www-sre.wu-wien.ac.at/ersa/ersaconfs/ersa02/cd-rom/papers/210.pdf [Accessed: Oct 10, 2017]

[31] Hospers GJ. Beyond he Blue Banana? Structural Change in Europe's Geo-Economy. Dortmund, Germany: 42nd European Congress of the Regional Science Association; 2002. 20pp. Accessible from: http://www-sre.wu-wien.ac.at/ersa/ersaconfs/ersa02/cd-rom/papers/210.pdf [Accessed: 2017-10-10]

[32] Suffolk Community College. Urbanization – When, Where and Why Did Urbanization Occur in Europe? [Internet]. 2012. Available from: http://www2.sunysuffolk.edu/westn/urban.html [Accessed: Oct 10, 2017]

[33] Pevsner N, Bridget C. London I: The Cities of London and Westminster. Rev. edition. London: Penguin; 1973. ISBN-10: 0140710124

[34] Wikipedia. History of London. [Internet]. Available from: https://en.wikipedia.org/wiki/History_of_London [Accessed: Oct 10, 2017]

[35] Wikipedia. Demography of London. [Internet]. Available from: https://en.wikipedia.org/wiki/Demography_of_London [Accessed: Oct 10, 2017]

[36] Wikipedia. Demographics of Paris. [Internet]. Available from: https://en.wikipedia.org/wiki/Demographics_of_Paris [Accessed: Oct 10, 2017]

[37] Wikipedia. Rhine-Ruhr. [Internet]. Available from: https://en.wikipedia.org/wiki/ Rhine-Ruhr [Accessed: Oct 10, 2017]

[38] Batty M, Longley P. Fractal Cities: A Geometry of Form and Function. San Diego, CA and London: Academic Press; 1994. ISBN: 978-0124555709

[39] Zach WD, Evers K. Die Berliner Schnellbahnnetze – ihre Entwicklung und Funktionen. In: Monatsschrift 2003 /07. [Internet]. Available from: http://docplayer.org/29591956-Die-berliner-schnellbahnnetze-ihre-entwicklung-und-ihre-funktionen.html [Accessed: Nov 10, 2017]

[40] Cardiff University: The Growth of Gated Communities. Summary from the European FP7 project GATEPLUG. [Internet]. Available from: http://cordis.europa.eu/result/rcn/148820_en.html [Accessed: Nov 10, 2017]

[41] Adhikar S. 10 Oldest Ancient Civilizations Ever Existed. [Internet]. Available from: https://www.ancienthistorylists.com/ancient-civilizations/10-oldest-ancient-civilizations-ever-existed/ [Accessed: Nov 10, 2017]

[42] Echeverria. A. Urban Change in Iran: Two Millennia of Urban Planning. 2002. [Internet]. Available from: http://thisbigcity.net/urban-change-in-iran-two-millennia-of-urban-planning [Accessed: Nov 10, 2017]

[43] Wikipedia. List of Agglomerations. [Internet]. 2017. Available from: https://en.wikipedia.org/wiki/List_of_agglomerations_by_population [Accessed: Nov 10, 2017]

[44] Ghanbran AH. Iranische Basare im Wandel. [Internet]. 2004. Available from: https://elib.unistuttgart.de/bitstream/11682/44/3/Kap00_Einfuehrung.pdf [Accessed: Nov 11, 2017]

[45] Abdoos M, Norouzi M, The Recognition of Structural-Spatial Development Models in the Immediate Primary Nucleolus Ranges of Iranian Historical Cities and Presenting

Design Approach, Urban Change in Iran. UK: University College London [Internet]. 2014. Available from: https://www.ucl.ac.uk/bartlett/development/sites/bartlett/files/proceedings-10-02-2014.pdf [Accessed: Nov 15, 2017]

[46] The Statistics Portal [Internet]. Available from: https://www.statista.com/statistics/455841/urbanization-in-iran/ [Accessed: Nov 12, 2017]

[47] Zare M, Panagopoulos T, Loures L. Simulating the impacts of future land use change on soil erosion in the Kasilian watershed, Land use Policy. Amsterdam: Elsevier; 2017;**67**:558-572. ISSN: 0264.8377

[48] Banakar R. Driving Culture in Iran: Law and Society on the Roads of the Islamic Republic. Tehran: L.B.Tauris. Sew edition; 2016. 304 p. ISBN-13: 978-1784534486

[49] Financial Tribune. Iran's Housing, Urbanization Policies, Performance Outlined. [Internet]. 2016. Available from: https://financialtribune.com/articles/economy-business-and-markets/51820/iran-s-housing-urbanization-policies-performance [Accessed: Nov 13, 2017]

[50] Zanjani H, Housing in Iran. [Internet]. 2004. Available from http://www.iranicaonline.org/articles/housing-in-iran [Accessed: Nov 14, 2017]

[51] Arjumandnia A. National Strategy for Housing of Youth. Tehran: National Youth Organization; 1988

[52] Rassoolimanesh SM, Jaafar M. Urban planning and management system in Iran: Review and assessment. Middle-East Journal of Scientific Research. 2013;**18**(2):220-229. DOI: 10.5829/idosi.mejsr.2013.18.2.12435

[53] Wikipedia: Ancient Chinese Urban Planning. [Internet]. 2017. Available from: https://en.wikipedia.org/wiki/Ancient_Chinese_urban_planning) [Accessed: Oct 13, 2017]

[54] Wikipedia. Urbanization in China. [Internet]. 2017. Available from: https://en.wikipedia.org/wiki/Urbanization_in_China [Accessed: Nov 13, 2017]

[55] Wikipedia. Beijing City Fortifications. [Internet]. 2017 Available from: https://en.wikipedia.org/wiki/Beijing_city_fortifications [Accessed: Oct 13, 2017]

[56] The World Bank. Data. [Internet]. Available from: https://data.worldbank.org/indicator/NY.GDP.PCAP.CD [Accessed: Oct 13, 2017]

[57] Chan KW. The Chinese Hukou system at 50. Eurasian Geography and Economics. 2009;**50**(2):197-221. DOI: www.tandfonline.com/doi/abs/10.2747/1539-7216.50.2.197

[58] Baculinao E. Jing-Jin-Ji: China Planning Megalopolis the Size of New England [Internet]. 2017. Available from: https://www.nbcnews.com/news/world/jing-jin-ji-china-planning-megalopolis-size-new-england-n734736 [Accessed: Oct 13, 2017]

[59] Shao Y. Current State of China's national new-type urbanization construction and deployment of new reinforcement measures. Mizuo China Monthly. [Internet]. 2014. Available from: https://www.mizuhobank.com/fin_info/cndb/economics/monthly/pdf/R512-0079-XF-0103.pdf. [Accessed: Oct 13, 2017]

[60] UN Habitat. The State of Latin American and Caribbean Cities 2016. Towards a New Urban Transition. United Nations Human Settlements Programme. Lanham: Rowman & Littlefield; 2016. 164 p. ISBN: 987-1-4422-4916-5

[61] Kingstone P. The Political Economy of Latin America. Reflections on Neoliberalism and Development. New York, NY: Routledge; 2011. ISBN-13: 978-0415998277

[62] Inostroza L, Baur R, Csaplovics E. Urban sprawl and fragmentation in Latin America: A dynamic quantification and characterization of spatial patterns. Journal of Environmental Management. 2012;**115**:87-97. DOI: 10.1016/j.jenvman.2012.11.007

[63] CEPAL. Panorama Social de América Latina. Naciones Unidas: Comisión Económica para América Latina y el Caribe; 2010

[64] Hidalgo R. ¿Se acabó el suelo en la gran ciudad? Las nuevas periferias metropolitanas de la vivienda social en Santiago de Chile. EURE; 2007. XXXIII, pp. 57-75. http://dx.doi.org/10.4067/S0250-71612007000100001

[65] Ortiz J, Morales S. Impacto socioespacial de las migracionesintraurbanas en entidades de centro y de. EURE; 2002. XXVIII, pp. 171-185. https://www.eure.cl/index.php/eure/article/view/1249/346 [Accessed: 2017-10-10]

[66] Inostroza L. Informal urban development in Latin American urban peripheries. Spatial assessment in Bogotá, Lima and Santiago de Chile. Landscape and Urban Planning. 2016;**165**:267-279. DOI: 10.1016/j.landurbplan.2016.03.021

[67] Borsdorf A, Hidadgo R, Sanchez R. A new model of urban development in Latin America: The gated communities and fenced cities in the metropolitan areas of Santiago de Chile and Valparaíso. Cities. 2007;**24**(5):365-378. DOI: 10.1016/j.cities.2007.04.002

[68] Coy M. Gated communities and urban fragmentation in Latin America: The Brazilian experience. Geojournal. 2006;**66**:121-132 03432521

[69] Janoschka M. El nuevo modelo de la ciudad latinoamericana: fragmentación y privatización. 2002. EURE, XXVIII. n85

[70] Kanitscheider S. Condominios und fraccionamientos cerrados in Mexiko-Stadt. Sozialräumliche Segregation am Beispiel abgesperrter Wohnviertel. Geographica Helvetica. 2002;**57**(4):253-263 ISSN: 0016-7312

[71] Caldeira TPR. City of Walls, Crime, Segregation, and Citizenship in São Paulo. Berkeley, Los Angeles, CA: University of California Press; 2000. ISBN: 9780520221437

[72] UN Habitat. The State of Latin American and Caribbean Cities 2012. Towards a New Urban Transition. United Nations Human Settlements Programme. Nairobi: Earthscan publications; 2012. 160 p. ISBN: 978-92-1-132468-6

[73] UN Habitat. State of the World's Cities 2010/2011 – Cities for All: Bridging the Urban Divide. United Nations Human Settlements Programme. Nairobi: Earthscan publications; 2011. 220 p. ISBN: 978-1-84971-175-3

[74] LSE Cities. Cities and Social Equity: Inequality, Territory and Urban Form. Detailed Report. Urban Age Programme. [Internet]. 2009. Available from: http://eprints.lse.ac.uk/26754/ [Accessed: Oct 16, 2017]

[75] LSE Cities. Cities Health and Well-being. Urban Age Programme. [Internet]. 2011. Available from: https://lsecities.net/objects/research-projects/cities-health-and-well-being [Accessed: Oct 16, 2017]

[76] Hidalgo R, Borsdorf A. El crecimiento urbano en Europa: conceptos, tendencias y marco comparativo para el área metropolitana de Santiago de Chile. Estudios Geográficos. 2009;**LXX**(266):181-203. ISSN: 0014-1496

[77] Germani G. Estructura Social de Los Argentinos. Raigal: Buenos Aires; 1955

[78] Germani G. Política y sociedad en una época de transición: de la sociedad tradicional a la sociedad de masa. Buenos Aires: Editorial Paidos; 1968

[79] UNDP Human Development Report 2016. Human Development for Everyone. United Nations Development Programme. New York: United Nations; 2016. ISBN: 978-92-1-126413-5

[80] Roy A. Urban Informality, towards an epistemology of planning. In: Journal of American Planning Association. 2005;**71**(2):147-158. DOI: https://doi/abs/10.1080/01944360508976689 [Accessed: 2017-10-08]

[81] Roy A. Slumdog cities: Rethinking subaltern urbanism. International Journal of Urban and Regional Research. 2011;**35**(2):223-238. DOI: http://onlinelibrary.wiley.com/doi/10.1111/j.1468-2427.2011.01051.x

[82] Sen A. Development as Freedom. New York: Anchor Books Edition; 1999

[83] Gallagher D. The Heavy Price of Santiago's Privatised Water. The Guardian, Sustainable Business, Water. [Internet], 2016. Available from: https://www.theguardian.com/sustainable-business/2016/sep/15/chile-santiago-water-supply-drought-climate-change-privatisation-neoliberalism-human-right [Accessed: Sep 20, 2017]

[84] America2050.org. Megaregions. [Internet]. 2017. Available from: http://www.america2050.org/content/megaregions.html [Accessed: Sep 30, 2017]

[85] The Guardian. Cities in Numbers: How Patterns of Urban Growth Change the World [Internet]. 2015. Available from: https://www.theguardian.com/cities/2015/nov/23/cities-in-numbers-how-patterns-of-urban-growth-change-the-world [Accessed: Sep 30, 2017]

[86] Turner MA, Fortuny K. Residential Segregation and Low-Income Working Families. [Internet]. 2016. Available from: https://www.urban.org/research/publication/residential-segregation-and-low-income-working-families [Accessed: Oct 5, 2017]

[87] Lieb E. How Segregated Schools Built Segregated Cities. [Internet]. 2017. Available from: https://www.citylab.com/equity/2017/02/how-segregated-schools-built-segregated-cities/515373/ [Accessed: Oct 5, 2017]

[88] Vigdor JL, Glaeser E. The End of the Segregated Century: Racial Separation in America's Neighborhoods, 1890-2010 [Internet]. 2012. Available from: https://www.manhattan-institute.org/html/end-segregated-century-racial-separation-americas-neighborhoods-1890-2010-5848.html [Accessed: Oct 5, 2017]

[89] Moos M. From gentrification to youthification? The increasing importance of young age in delineating high-density living. Urban Studies. 2016;**53**(14):2903-2920. DOI: https://doi.org/10.1177/0042098015603292

Chapter 3

A Multilevel Approach to Urban Regional Agglomerations: A Swedish Case of Transition Paths toward a "Fossil-Free Society" by 2050

Uno Svedin and Hans Liljenström

Additional information is available at the end of the chapter

http://dx.doi.org/10.5772/intechopen.73104

Abstract

This article has a focus on the changing patterns of connected urban spaces forming large super-regional aggregates made up of cities of various sizes and regional functions as well as the interconnecting space of much smaller municipalities of agricultural or forestry types of character. The multi-scalar level analysis of these connected clusters is pursued from the level of the individual to the regional, national, Nordic and EU levels. The enfolding of the regional pattern also has global connotations in terms of trade connections, but also in the context of bio-geo challenges as climate change, biodiversity depletion or food security considerations. The transition dynamics involves governance, economic, social and cultural aspects. International negotiations, as the Paris agreement on climate change and agreements at the UN level as the 17 "Sustainable Development Goals" (SDG), or agreements at the EU level, provide an international political frame to this process.

Keywords: urbanization, low carbon society, climate, sustainability, resilience, transformation, region, Stockholm, Sweden

1. Introduction

The interest in urban development has drastically increased during the last few decades. The focus has broadened to cover not only cities themselves but also more strongly the changing context within which a broader urbanized space is evolving—in particular reflecting sustainability and resilience systemic considerations and connected conditions. The tendency of such studies has stressed two different simultaneously existing directions. First, there exists a broadening from just the interest about individual cities themselves now moving toward an increased attention to the issue how a number of geographical sites with urban characteristics are linked to form a wider super-regional spatial and functional agglomeration with both urban and less urban features as parts.

In this direction it is the pattern of agglomeration—and its dynamics—that is at the core of interest. This involves technical (e.g., infrastructural considerations related to the transport system), administrative/political governance issues (as those dealing with the planning of a broader labor market) and other economic concerns. It also involves patterns of commuting preference of inhabitants, and quality of the livelihood in the involved communities. Also the spatial distribution of the knowledge centers of higher learning (and the distribution and roles of their competitive competences) as well as the industrial profile features of the region at large is of importance. This involves, e.g., the interplay between private companies and public bodies with regulating or supportive/promoting functions and the concerns of civil society at large.

The earlier analytical dichotomy, sometimes referred to as "cities-and-their hinterlands," has now been transformed into a much more complex concern about patterns in a set of micro-meso-macro cross-going links [1]. However, the framing—either based in a new increased complexity oriented fashion or in a more traditional way—indicates important choices between preferentially a bottom-up or top-down reform political philosophy (or variations of combinations of these). The earlier more "narrow" city focus has thus expanded to investigations over a much broader and more complex techno-bio-social geographically based pattern. In Sweden, the analysis and political considerations concerning the "Stockholm-Lake Mälar region" is such an agglomeration case in quick development.

The area spans some 100.000 km^2—and covers not only the Swedish capital Stockholm but also an array of larger and smaller cities—as well as rural areas in between dominated by agricultural sectoral activities and forestry. The area harbors around 1/3 of the Swedish population (of totally around 10 million persons) and a somewhat larger proportion of the Swedish economy. It is one of the fast growing regions of Europe and has a very strong innovation and research profile.

In Sweden, there are other emergent regions of a similar kind. One is in the south-west part of the country, in and around the city of Gothenburg. Another one is in the south around the city of Malmö, in that case with the connection to the large vital national capital of Denmark, i.e., the Copenhagen region (connected to Malmö with the strategically important Öresund Bridge).

These here indicated regional agglomerations have different features, but also similarities. They are all built on common Nordic history developments over at least 1000 years and the similarities could be seen in terms of cultural, economic and political styles in handling the different regional challenges, especially if comparing these regional developments with a wider set of similar European experiences and—of course—even more in a global comparison.

All these sub-national regions are bound together in a planned regional pattern of connecting infrastructure. The physical connections thus materialize different dimensions of connectivity between a number of cities within a broad geographical space. The patterns of cities (of different kinds and sizes) are embedded in the larger region of diversified space that also includes, in a complex interwoven fashion, less urban (or even non-urban) areas.

This development has a strong interest in a politico-governance-economic way, as well as in a scholarly way. However, that expansion of perspective toward a broader cluster or aggregation of mixed urban and rural is associated with an interest focusing on the characteristics of urban space itself facing transforming factors from outside that impact the respective city system within its own formal "boundaries." These dual interests have to consider a contextual multilevel connotation. It concerns ways to insert solutions (i.e., to respond to various "problems") related to the grand challenges of our time as they create the future framework for a particular urban space. Such "grand challenges" could, e.g., relate to

- climate change (in terms of demands of further "greening" the urban space);
- food security in terms of generating possibilities for growing food within the city limits (e.g., to improve the freshness of the produce, but also drawing its momentum from a social concern to provide a wide range of inhabitants better connection to the food generation process, and—for future problematic situations—to provide a "volume insurance" of availability of deliveries also in turbulent times);
- sustainability and energy efficiency considerations (by compacting the cityscape) and
- social wellbeing consideration (by not only designing the city from a technical-economic "rationality" point of view, but also framing the social setting involving considerations about "the good life" for different age groups—not least children and the aged—but also responding to gender considerations).

Our Swedish experiences are not least drawn on our work within the EU project COMPLEX, which was completed in 2016 [2]. In this context, we have used our investigation area as a sort of "laboratory" for reflections, data collections and various types of modeling. These studies have also provided a wide background for formulating policy advice aimed at a ladder of decision-making levels: municipalities, counties, regional and national bodies. In the end of this article, we will come back to some of these policy considerations.

Our chosen timeframe has been set to 2050. We are well aware that other time frames have been voiced in the political debate about specific solutions dealing with climate change by reducing the carbon emissions to the atmosphere. However, our timeframe is chosen to connect to some of the formal assignments of timeframes in the EU on these matters, as well as time frames expressed at the Swedish national level. This pertains, e.g., to the timeframe within which a net zero-carbon situation ("fossil-free society") should—as a goal—be reached in an overarching sense. However, strong efforts should be mobilized to shorten the time if possible until these goals have been reached. This holds true especially for specific partial goals associated with different sectors (as the transport or the building sectors just to take two examples). In 2017, the Swedish Parliament decided on a set of such targets for these purposes.

2. Agglomerated space and complexity patterns

In our approach dealing with a widened and "mixed" agglomerated regional space, the emergence of patterns and their complexity (as indicated above) comes into the forefront. It is here that the multilevel and multiagent phenomena are found and where the integrating aspects of the wider agglomeration challenge come fully into play.

So what could complexity mean in this "agglomeration" context? As we have addressed elsewhere [1], there are several ways in which complexity can be interpreted and studied, e.g., in terms of structure, dynamics, function and behavior.

2.1. On complex systems

Structural complexity can be seen as an expression of the degree of interplay of many elements in a varied fashion due to a great number of variables and structural constraints. *Algorithmic* complexity refers to the degree of complexity that can be measured in terms of information. ("The more information that is needed to describe a system, the more complex it is"). Another type of complexity is called *effective* complexity, which can be defined as the length of the descriptive code that is needed to describe its regularities. In addition, a study of complex adaptive systems has to separate regularities from randomness.

Yet, there is no obvious – or by all accepted – definition of complexity covering a wide range of contexts. It is measured in praxis by how well we understand causes, behaviors and – in assessment terms – achieved purposes. Hence, issues about large numbers of variables and non-linear relations among them, as well as aspects of the degree of openness of a system are important elements of scrutiny, especially considering the degree to which they present barriers to an improved understanding [3].

Characteristic features of complex systems have to do with the web of (often non-linear) interrelations between variables that may provide thresholds, lags and discontinuities. Feedback and feedforward loops enable amplification, as well as attenuation and control. This often results in unpredictable and non-intuitive behaviors. Complex systems are often self-organizing and adaptive, where spatio-temporal order may emerge out of disorder, and where new qualities may emerge above a certain threshold of complexity. Such systems could have a high degree of redundancy, which makes them less vulnerable to disturbances and malfunctioning parts. Examples of such systems are found among some ecosystems, or in some combined socio-natural systems, as well as those of socio-technical character [4, 5].

2.2. Micro-, meso-, and macroscale relationships

Many complex systems are characterized by a hierarchy of subsystems and different levels, both in time and space. In relative terms, these levels can be viewed as *micro, meso* (i.e., "the in-between level"), and *macro* levels, often equally applicable to temporal, as well as to spatial scales. In fact, often the temporal and spatial scales are correlated, so that spatially "microscopic" systems and structures are characterized by a "microscopic"

http://dx.doi.org/10.5772/intechopen.73104

time scale (i.e., "the smaller the system, the faster it moves"). Similarly, larger systems are typically characterized by longer time scales (i.e., slower motions). For example, the characteristic time scale of molecules is fractions of seconds, whereas the time scales of mesoscopic systems (such as "touchable objects") might be at the time scale of seconds to years, or macroscopic systems (such as planets or galaxies) relate to much larger time scales [1].

In our urban agglomeration spatial case, the time variation could at the level of the neurocognitive processes of individuals (dealing with reactions and decisions) span a frame of some 300 ms to a few seconds (depending on the art of the reaction). At another level the time frame for the decisions and strong transformation impact on the "realities" of a sub-national region is at the level of one to two (human) generations. The year of 2050 as a target year for the transition to a fossil-free society for our case region mirrors this chosen analytical time frame of our approach.

The characterization of agglomerated urban-rural systems requires an attention to the interactions between different spatial and temporal scales, with a focus on the mesoscopic level, between micro- and macroscopic level descriptions. The mesoscopic level can be seen as the level where bottom-up processes meet top-down processes in societal transformations.

Approaching a systems "organization" (as a region) through its micro-, meso-, and macroaspects, applies not only to its technical connotations but also to social structures. That approach is appropriate when discussing the relation between an individual ("microscopic") and the group or population ("mesoscopic"), within which it is a part, and to the entire socio-technical-ecological system ("macroscopic"). In our case, we regard the sub-national regional scale as the mesoscale, in relation to the micro-scale of individuals or households, and to the macro-scales of nations and the world at large.

In this context, it is worth focusing on the connectivity of network structures, whether they are hierarchical or not. Many properties of networks, such as stability, resilience, and flexibility, as well as diffusion of any kind, depend on the strength and types of the network connections. For example, strong connectivity may lead to local resilience, but may also result in fragmentation and social/cultural polarization, while weak connections are more likely to be more efficient for linking different groups to each other, and for creating a more stable relation between micro and macro levels in an organization [6].

Similarly, there seems to be an optimal degree of randomness in social networks, with regard to efficiency in information transmission and social learning, as is described through the concept of "small-world-networks" [7]. Analogue results have been found for models of neural networks [8–10].

Environmental changes occur at several time scales, but it is demanding to analytically describe the interaction between these scales when trying to match the characteristics of the different life processes (not least due to interesting time span overlaps). One attempt is to crudely relate to the average lifespan of an individual or a generation. This becomes obvious when we discuss the relation between individual choices and those policies taken by the society at large (e.g., with relation to a region as our case demonstrates). Recently the interest of inquiry has started to involve much longer time scales (several generations) than what politics normally has been used to do and involving all spatial (and associated time interval) scales up to the planetary one [11].

This also involves "grand" approaches to the issue of the historical period we are living in, seemingly moving into the Anthropocene—the time period in which humans have at least the same level of impact on the planet as all other more "natural" processes demonstrate [5, 12, 13]. This already has strong and increasing implications at lower levels of aggregation—as the regional one—as is shown by the handling of the climate change issue (with global causal connotations and local impacts).

In a social system, e.g., dealing with the structure of decisions, the multilayered structures play more and more important roles [14]. This was one of the important starting points for the development of sustainability science [15]. Globalization phenomena (macro) meet phenomena at a local level (micro). However, it is very difficult to deduce the effects at one level to those generated at another. Of course, global tendencies are built up by ensembles of micro events. But what are the mechanisms of shaping (potentially appreciated) coherence on the way up the ladder? And to which extent do globalization phenomena really frame local events, more than just preconditioning them in certain fairly vague directions?

Analyses about such layered interdependencies [16] need to start with the understanding that no specific actor may claim full knowledge about the total system. In fact, the problem of an inadequate overview of multilayered phenomena from any particular point of view is in itself of importance. This contributes to the complexities in understanding the interplay between different spatial as well as temporal scales. This also explains—to some extent—the difficulties in creating a deepened understanding of such nested systems and to find appropriate operational designs of processes of governance. In itself these concerns call for new and important research efforts—also with regard to agglomeration phenomena.

Social phenomena are often generated in an interplay with phenomena emerging in the natural environment. Thus these connections—operating at a sequence of levels—can be visualized as two inter-related ladders of description: one of socio-economic-cultural character, and one of natural origin—sometimes referred to as the issue of "fit" [17, 18]. The specificities in these two realms, and the logic running them, may be very different. However, when combined these two systems connect—and even become a joint new type of system (e.g., what sometimes has been called a socio-ecological system—currently often adding the technological features to the social frame).

Then, the issue of how to organize the match between these parts becomes paramount. One example is the handling of a watershed geographic area and its management, e.g., connected to a river and its contributing water bodies. Here the natural system of the water flow has to be managed (i.e., providing the relevant "fit") within a socio-economic and cultural context, which comprises the other part of the joint socio-natural system. In our case region the waterbody of the large lake Mälaren (and its watershed area) and all its urban, semi-urban and rural subareas connected to it provides an example of this type of integrated connectedness. The interplay between the levels—and the specific role of the mesoscale—is here of strong importance.

For an earlier study (with relation to one of the authors of this paper—US) about the interplay of humans and the environment in a sustainability and democracy context with relevance for Swedish regions, these issues were exemplified with the two municipalities of Linköping and Åtvidaberg [19]. A number of other studies have been devoted to regional developments of this kind [20–23].

2.3. Dynamics and transitions

Features of systems in terms of their development over time are, of course, highly interesting. In particular in our case, this holds true for social systems, including the changes in cultural patterns. In order for individuals and societies to survive in a complex and changing environment, they need to be able to respond and adapt to environmental events and changes at several time scales. In particular, actors have to learn from experience. At a general level, "learning features" in systems often provide early warning capacities and starting points for solutions. The capacity to transform such signals to adaptive changes of various systems makes them more robust and increases their resilience.

We usually try to distinguish the evolution of a system under "normal conditions" (which often means slowly varying conditions), from those which are due to unexpected external shocks, such as a meteorite plunging the surface of the earth (or an unexpected financial crisis, such as the one in the early 1930s or in 2008). Examples of how the dynamics of a specific complex system can change due to various intrinsic and extrinsic effects are given in [10].

When discussing the long-term evolution of complex systems, we sometimes have to deal with catastrophic events, analyzing them in terms of intrinsic and external influences, respectively. Mostly, catastrophic events (such as a hurricane or a tsunami, mass extinctions of species, or breakdown of a technical or social system) are considered to occur due to external "forces" acting on the system (such as a sudden weather event, an earthquake, the eruption of a volcano or the impact of a meteorite—or due to "external" societal factors, e.g., changes in global trade patterns). However, there is a need also to reflect on the intrinsic factors of the system itself, which either enables the impacts of the catastrophic event to be very severe (flooding in a landscape with insufficient preventive measures or insufficiently reflected zooming spatial policies), or even as being the real source of such a catastrophic events (e.g., long-term developments within a financial system of destructive corruptive practices).

Small fluctuations might inevitably, sooner or later, bring the system across the threshold of instability. This phenomenon is sometimes referred to as "self-organized criticality," and has popularly been illustrated by the growth and collapse of a sand pile [24].

2.4. Resilience and vulnerability

An interesting question in this perspective is when microscopic fluctuations can have effects at a macroscopic level, and become the roots of shocks to the entire system. Many examples can be given from the physical sciences, but different types of systems in the socioeconomic domain express similar features. It is generally considered that the joint action of a large number of stockholders can avalanche into a stock market crash.

An important feature of a combined social and ecological system is its *resilience* [25, 26]. When such a system loses its capacity of resilience, it becomes vulnerable to change in a way that previously could be avoided through risk absorption [27]. Many of these features have to be understood in terms of multiscale interplays.

A closely connected issue deals with the role of diversity—including the layered structure connecting different roles of organisms and their functions. In the societal management of such combined bio-social systems also multilayered governance systems have to be designed and mobilized in which the stratification of the appropriate roles at the various levels and their interplay should be outlined. This often happens as a social nested process within which political will is only one of the components in the causal chain leading to a specific situation (see Svedin *et al.* in [16]).

Are there general features of resilient systems which can recover smoothly after shock treatments? It is very difficult to provide a list of general features making a system resilient. However, some properties that make systems less vulnerable to shocks include: an appropriate network structure with a carefully designed type of high interconnectivity, a high degree of redundancy, heterogeneity, elasticity, and self-organized regulatory mechanisms (i.e., relevant forms of negative feedbacks).

A reduction of resilience creates an increased vulnerability to societies. For example, freshwater systems may have their vulnerability increased with regard to flood events, but also to toxic algal blooms, which originates from intensified fertilizer use, higher densities of animals and poorly performed other types of agricultural practices. (For general surveys about resilience and socio-ecological systems approaches, see e.g., [26, 28].) At a very general level, Hägerstrand [29] traces some of the dangers to our planetary socio-natural systems to the split between the external world and the "projects" within our minds. "It is as if our well-developed capacity to store and hold together systems of ideas makes us unable intuitively to feel the limitations of the external world to accommodate our projects." Indeed, this is an expression of a severe vulnerability of our civilization.

3. A guide to our case region: the Stockholm-Mälar region and its various parts

We now go back to our case region. We are zooming in on the Swedish geography in order to point at the place from which our examples are drawn.

Our studies of the Stockholm-Mälar region within the EU project COMPLEX [2] have given an array of very different insights about this regional system and its potential capacities to transform into a (net) zero-carbon emission region by 2050.

The selected case region has a central place in the history of Sweden. Without going into details it spans in time about more than a millennium of development—and with an even deeper time depth to Viking times and before. It is easily stated that the region is the holder of the

http://dx.doi.org/10.5772/intechopen.73104

combination of central (royal) governmental functions (including the democratic institutions later in history) for the country at large. It has been and still is a key holder of network nodes of trade, the seat of many internationally renowned academic institutions and others of highly qualified character (e.g., at university level from 1477 in Uppsala and later in many other places in the region as – in the present days – world-leading institutions in Stockholm). The region also has a history of carrying the seat of central church offices over at least half a millennium. It has been a vibrant mining (mostly iron-based) area, today spilling over to the high tech industries in the modern world in sectors as manufacturing, forestry, agriculture and financial services.

In short, this has been—and continues to be also in our times—a deeply European core region over a millennium and a carrier today of both long tradition and very advanced technology of contemporary excellence. It is a still expanding global network hub for industrial endeavors. It is also a region balancing the formal role as holder of the capital of Sweden (i.e., Stockholm) with its highly urban features structurally based on a core associated with integrated medium and smaller urban nodes as well as regionally embracing a rural countryside area with highly developed agricultural and forestry activities—and upgrading facilities of these raw material contributions. It is also a region with new types of functions and capacities, from emergent types of consultancies (e.g., in the finance sector) to the sector dealing with a considerable

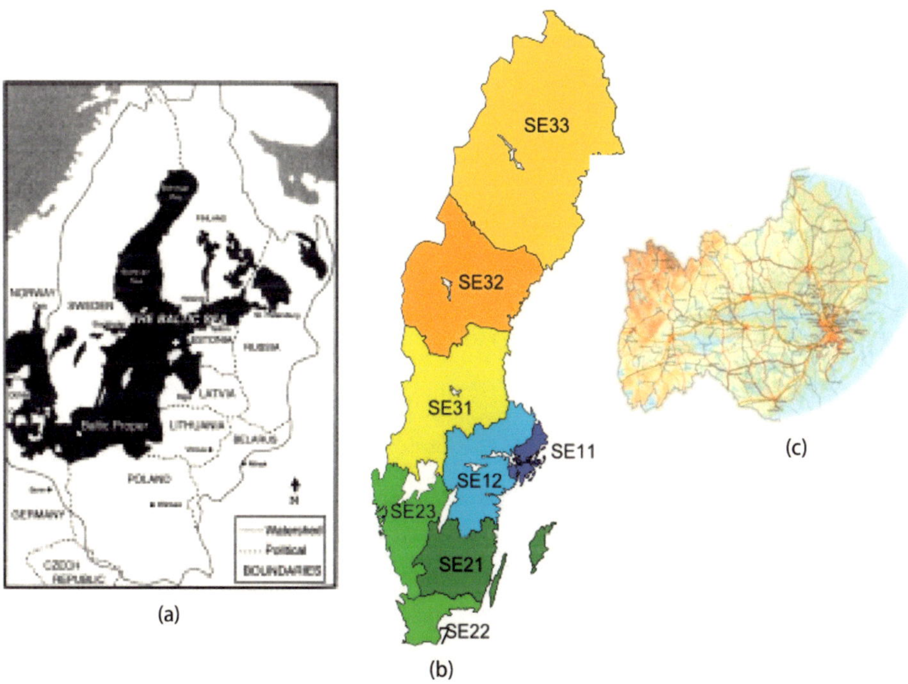

Figure 1. (a) (left) Map over northern Europe. Black part in the center is the Baltic Sea. (b) (middle) Map over Sweden with labels of European regions inserted. (c) (right) Map over the Stockholm-Mälar case region. (Stockholm city is the center in the east where all the communication lines—here the rail lines—converge. The connections over the region correspond to railway links between different medium-sized cities in the region). Our case region relates roughly to the two NUTS regions SE 11 and SE 12 (see **Figure 1(b)**).

capacity to harbor large streams of tourists, not least during the summer period when the large cruiser ships arrive from all over Europe to Stockholm. It is definitely a region with very high innovation orientation. This holds true in many cases of global high tech features as in telecom, electronics, robotics, medical and pharmaceutical specialities, and in other areas of cutting-edge technologies.

Thus, due to its modern societal features and cultural value frame as well as its technological knowledge capacities this region is a part of the world where a successful experimentation leaning toward the "nonfossil society" transition in principle could take place. A great deal of the necessary conditions for this change already exist that are needed for the low carbon trajectory to be possible – or at least are potentially available. The character of the region is thus not only providing means for itself to face the challenges needed now to address the operative requirements. It is also a region with a potential development path that is of strong interest for a wider European (and even global) set of actors in a time when solution structures are searched for in many places of the world in order to manage the climate change challenges within less than a generation.

The Stockholm-Mälar region is the home of some 3 million inhabitants which is about a third of Sweden's total population—and it is quickly expanding in terms of its population. It consists formally (i.e., politically) of the five Swedish counties (län) Stockholm, Uppsala, Örebro, Västmanland and Södermanland with close connections to other counties close by, dealing with similar challenges. And it is one of Europe's approximately 500 regions, in this case with strong expansion and leadership ambitions in the sustainability and other arenas. (In its broader context of the SE11 and SE12 regional areas still other counties—i.e., "län"—are included).

To sum up: the region has several interesting characteristics, such as:

- A great variety of geographic and social features—i.e., it is diverse
- A long historical evolution in a socio-economic cultural frame
- A layered governance structure
- A high cultural profile and with several academic institutions
- A highly qualified "high-tech" type of industry

1. Reasons for regional possibilities to move successfully toward low carbon targets by 2050

What speaks for the capacity of the region to evolve into a low carbon society in a few decades? Main features are of course coupled to the high innovation profile of the current technologically inclined industry. This in turn is embedded in a large, regionally distributed and competent university/university colleges structure with considerable attendance in terms of percentage of the population living in the area. This research and education structure has in addition the last decade advanced strongly in undertaking sustainability-oriented research and educational tasks and developed new types of institutions for promoting such ideas. This holds true not only for the climate-related part of issues but for the broad set of challenges outlined in the set of the UN defined Sustainable Development Goals (the 17 SDGs).

Several of the current industries work within such sectors that are candidates to become lead sectors for the path to a low carbon society (i.e., what could in this sense be labeled as a "fossil free society"). The existing industrial sector also has the potential to expand into new areas of possible innovative solutions with products (and processes) of strong future interest for the world market, i.e., key innovations that might be of crucial importance for the transition in many places. Solutions exist both in the technological domain as well as in the social one—or are born from a combination of both. One example is provided by the municipality of Eskilstuna. The decision-making body has seen to it that the technical solutions of solar panels have been installed on the roof tops of many of the municipality buildings. This in turn has already given encouragement to private house owners to follow this trait in that direction. This is a social structure contribution.

The currently existing professional planning employees in many of the municipalities are already well educated in these matters and there has become a competitive mood among the cities and the municipalities to be on the "leading edge" for these efforts. And a reasonably large fraction of the population of voters are favorably inclined toward such measures. Many individuals are already trying to set up production units for wind power—although still there are a lot of rules and regulations that have to be upgraded in order to make a smooth passage for decentralized extra electric power production (solar or wind as examples) to be fed into the grid from such sources.

Also the formal governance (in the architecture of the current multilayered system) tries to address these issues in terms of concerted efforts outlining a sequence of investigatory innovation projects with such types of aims. Parallel activities are going on, e.g., in the bio-fuel industry and in other parts of industry and civil society in order to promote sustainability-oriented innovations—and to some degree also raise the early investments for such prototype actions. This is obvious, e.g., in the transport sector, not least in the introduction rush for electricity cars (or as an interface in time hybrid versions), including the need to promote and develop the supporting infrastructure for such solutions—be it public or privately owned or a mixture of these.

Thus, indications of alertness in the population at large in promoting such development lines are definitely there, although it is at every time step difficult to fully assess to which extent such support also carries over to acceptance of extra cost implications, e.g., for new solutions for private cars. Here it should be mentioned a rising interest for other forms of ownership of vehicles, e.g., through new type of taxi business models, but also of car pools ("circular economy" business models). Promotion of public transport solutions for large segments of persons—especially in the larger cities—should be noticed, through the provision of the services (prolonging subway lines or added parallel railway track for increased temporal services for regional trains). One example is the line between the university city of Uppsala and Stockholm some 70 km to the south. This is a very intense commuting line with high demand on performance. The connection of this expanded capacity of service for transport has also been made in such a way as to improve the number of connections to the large international airport Arlanda situated between Uppsala and Stockholm. But there are also social innovations coming along as those recent advances in the way how the ticket cost arrangements are done (including monthly based cards for the system in the region at large providing sufficiently inviting fees for the travelers).

Such thought patterns and ambitions within the more formal public sphere of operators are distinctly and rather quickly seen in terms of the number of officials employed to serve such coordinating and planning tasks and the quality of their educational support and thus mounting professional credibility.

But also in society at large it is possible to identify a rising citizen concern and interest with regard to these issues about the direction of development. This provides an important basis for the policy formation for the future in the broader democratic system.

4. The issue of paths toward a fossil-free society

The strong increase all over the world in the tendency for urban spaces (in close connection to non-urban areas) to conglomerate into a vast system of linked patches holds true also for a highly industrialized country like Sweden. And it is clearly demonstrated in our chosen case region, the Stockholm-Mälar region. But what does that mean considering the major grand challenges our world is facing at the moment? A very clear example is the needed build up of a capacity to handle the strategic challenge of climate change. Sweden has as a country (at the level of parliament) recently defined—as a national goal—to aim to be in the forefront of measures to handle these and connected emerging issues. This means that there is a formal aspiration to be at the leading edge among industrialized countries in the world for these purposes. This means in an operative way a deliberate effort to address problems in an array of sectors to reduce carbon emission to the atmosphere in a historically unprecedented speed.

Efforts have to be launched in an array of sectors, such as dealing with the energy production systems for the future in order to eliminate the current fossil fuel component within a few decades. This should be done without jeopardizing the need for Swedish industry—also in the future—to have sufficient access to energy. In technical terms, this means the development or establishment of new sources of energy as solar, wind and bio, which should be done in parallel to the process of facing out the present nuclear component (although this long-term decision is still contested in the Swedish political environment). The build up and expansion of the new energy sources—with a strong electrification tendency on the distribution side—have to go hand in hand with strongly increased energy efficiency efforts both in industry, but also in a broader sense in society as a whole.

All of this has put pressure on the urban segments of society in several ways. One aspect is the push for low (or zero) net energy housing in the building sector. The development has already started and prototype housing and Avant-guard testing sites are already being built in many places (e.g., a student housing complex connected to the campus of the KTH—the Royal Institute of Technology in Stockholm). A possible shift toward wood as building material, also for several level housing is going hand in hand with the low carbon efforts in the energy domain.

At a more aggregated level (i.e., in terms of city planning), the climate concerns—through the push for better energy efficiency—has evoked an increased search for compacting the city space, e.g., creating more housing functions (including offices and other uses) per square meter than has been the case before. In a recent information technology-oriented methods study (led by KTH and Stockholm University) in Upplands Väsby—i.e., one of the fairly new Stockholm suburban agglomeration hubs—people in the street were asked to judge how they wanted the balance to be struck in their urban neighborhood between the compacting tendency versus the "greening" of the city interest (including, e.g., protecting the already existing parks). It is interesting—but not surprising – to note that the answers differed considerably between different types of actors and between people with varying socio-economic types of background. Here the interplay between the level of personal preferences and the level of collective outcome for an entire municipality/small town is demonstrated. A socio-technical rationality in handling the sustainability issue is here competing with a socio-cultural and behavioral type of rationality and connected sets of values.

Also the vaster urban agglomeration transport system is under quick transformation. The connectedness in the core parts of the cityscape (e.g., through the widening of the already existing Stockholm subway net) is further emphasized and given space in an investment perspective. But also the wider frame of sprawl related to the region in its totality calls for increased public transport (rail) development, but with a somewhat other technical sets of means than within more limited patches of urban space with high density. In areas with less compactness of "town connectedness" a system of public transport based on busses might be the solution, often connected to already activated change of the fuel base (e.g., biogas) but also on innovation in logistic handling in terms of new business models. Somewhat further into the future (already heavily discussed and also started to be tested in prototype arrangements) electric battery-based technologies for busses and cars enter the picture. Here, it is the interplay between a sequence of levels: the individual user ("the citizens") preference, the type of local/municipality reach and character, and in the case of the greater city level with its heavy long-term investment capacities expansion of the subway system. At even larger agglomerate regional level over a vast territory (with speed trains in focus) calls for technical, organizational and economic types of solutions. The transport sector concerns do not stop at the regional borders (if such could be considered to really and fully exist, considering the constant inflow and outflow of goods through trade arrangements among conglomerates and into the international hinterlands).

The air traffic component is here of great importance for links between several super regions within the nation (in this case Sweden as a country with quite a wide geographic space of which some parts are more thinly populated than others – especially in the north of the country). But further on—at even higher levels of geographical organization, Nordic constellations are of importance as indicated by interests in binding together Stockholm and Oslo, or Gothenburg with Oslo—or as some futurists have considered long ago, to bind together (probably in another political situation) Stockholm, St. Petersburg, Helsinki, the coastal parts of the Baltic republics, the north part of Polen, north Germany and Denmark in what could be called a "neo-Hansa." Some elements, although still limited, could already be seen in new transport lines across the Baltic Sea, but also in the domain of electricity transfer as well as the Russian driven efforts to build gas lines ("the North stream") on the bottom of the Baltic sea connecting the producer Russia to the EU user markets—with entry in northern Germany. With regard to an even grander agglomeration it could be contemplated how the strategic interests of the Republic of China strongly promoting cross-continental lines from the East coast of China into the markets of Europe under the label of the new "Silk road." In such a perspective the classical trade routes between the continental east and the continental West could be contemplated as a mega agglomeration of "bands of affinity" where the trade routes could be more tightly connected through large infrastructural investments over a long time and with mixtures of super urban nodes and integrated non-urban areas as very long strings of connected urbanized space from Shanghai and Bejing to Frankfurt, London—and why not Stockholm. These possibilities are still to some extent only on the strategically drawing board (especially led by China). But there is already more de facto existing physical infrastructure in place, e.g., in terms of trans continental railway systems for goods, oil pipelines of considerable extension, and similar trade arrangements manifesting sea scape solutions in global connectivity. All these future mega agglomerations with their internal urban nodal points at different hierarchical levels should not be considered—even today—as science fiction. But at the same time such heavy trends for future mega global structures are discussed we also can elaborate with some more precision the ongoing development of more limited (e.g., at regional level as in our Swedish case) agglomerations characterized by strong features of "urbanness"—but intertwined with other types of spatial functions, i.e., a sort of "big cake with raisins everywhere in the bun." Already now, there is a strong interest in mega city connectivity at a global level—i.e., the issue of "teleconnections" [30], which corresponds to the issue of "agglomeration emergence." This type of connectivity is already now applied in more limited nationally existing super regions as those existing in many places in Europe (and also elsewhere in other continents).

Analysis of these emergent structures also has to be devoted to what it means in terms of the aggregated effects on core issues of deep concern for the planet as a whole, as codified in analytical approaches as "planetary boundaries" or development within a safe planetary frame [11]. This also connects to a rising concern about what could constitute a new area of the "Anthropocene" [12, 13], i.e., the current time when humankind has evolved in its global operations to be at least on equal par with the natural phenomena in the capacity to transform the world at a planetary scale [5]—disregarding the qualitative direction by which this change is taking place.

Returning to our more limited regional considerations in the Stockholm-Mälar region, it should not be forgotten that transport of people and transport of goods are not the same thing. At a somewhat limited part of this region in terms of a few collaborating municipalities (as is under development among a set of suburban municipalities in the south part of the greater Stockholm conglomerate) incentives have been struck to generate socio-economic solutions (and not only technical ones) addressing the efficiency of transport of goods. Here, new and thoughtful combined logistics have been mobilized to use the fleet of the lorries to provide schools, houses for elders, etc. with their daily supports. In combination with computerized informatics control systems, distinct advances have been reached to boost efficiency considerably, both with regard to limitations of greenhouse gas emissions (per unit of function), but also in a direct economic way. Here, the innovation profile is on the organizational side with logistic renewal based on support by technologies from the IT sector. It is also an example of a level-depending type of solution.

At another level, an increased load of regional transport of goods is gradually shifting the interest and possible emphasis on to boat/ship transport solutions—especially in a region as the Stockholm-Mälar region with its vast seascape as a core geographic feature of the region.

But transport by sea is not only confined to bulk transport (or heavy products from industrial endeavors) connecting inflows to the region from vaster "hinterlands" to spotlike places in the regional agglomeration. Transport by sea can also take place within a partial cityscape as in the core ("inner") Stockholm City. A new set of lines for commuters within this geographical space is tried out at the moment. Here the level in the geographical hierarchy between the individual and the region is "in between"—i.e., not in any way binding together the region, but making a part of it (those with a reasonably dense cityscape) more connected, and thereby providing solutions for choices by individuals about how to handle (and "smoothing") their calendar relation between housing—jobs—and free time.

An early considerable (and internationally recognized) contribution to such types of space-time thinking was already expressed in the 1960s by the late Swedish cultural geographer Torsten Hägerstrand. Among other topics, he also raised concern for the uncontemplated development of not sufficiently connecting physical structures (and organizational ones as bus line connections over county borders between too limited regions, which together could otherwise have been linked—i.e., been made to agglomerate—into a greater urban-rural conglomerate).

Finally, we should also indicate that not only the energy sector, the building sector, the transport sector, but also the food sector (production, distribution and consumer preferences) has a strong part and role in the transformation into a fossil-free society.

5. The multilevel considerations

As we have just seen, when contemplating various examples from the Stockholm-Mälar region, a key feature is that the various levels are all very important—but somewhat different in character. Thus, any set of solutions has to recognize this insight and use it analytically in a way to bind together the different level qualities, constraints and possibilities. Further, we have been using the case of the transformation of this Swedish "test" society toward a fossil-free fuel society, not only as a tool to find solution responses to the (grand) task to better and more forcefully mitigate (and not only adapt to) the grand challenge of climate change.

We have through this choice of scrutinizing the functional challenge in the region of climate change been provided a sort of "fiber optics" to the more general issue of the agglomeration process of systems challenges. We can thus draw from these types of insights and to some extent generalize the findings. However, this should not be done in an overextended way as many of these processes are contextual in their nature, i.e., different cases have both similar, but also deeply dissimilar characteristics, which should be taken into account when reflecting on any urbanized space in its connection to other linked areas in the larger conglomeration.

In order to give some structure to this multilevel (which also is a multi-actor) issue, let us make a few summing up remarks on this issue

- We have discussed different levels of the following kinds. The level of
- the individual
- the local municipality (Swedish:"kommun," with strong taxation and planning authority in Sweden)
- the county ("län")
- the region (the set of a number of counties collaborating in an organized way on certain issues. In our case region the formal body is "Mälardalsrådet"—an organized body of politicians from the region, but the individuals have been elected to functions at other levels, e.g., the municipality level)

Outside the region and providing the context and sometimes resources for its operations we encounter the following levels and concerns:

- the national
- the EU
- the planet as a whole
- Different actors (the public official bodies, industry/companies, representation of associations covering in terms of NGOs various interests of civil society) have different types of presence at the different levels
- The processes of transformation are both top-down and bottom-up—often in a mixed fashion
- Different actors at different levels have different time frames
- Different actors have different focus of interest (choice of questions, of strategies, of endurance, etc.)
- Different actors engage in varying patterns of alliances in a dynamic fashion
- Different actors have varying political influence and financial resources at varying times

Within this mosaic realm of interests, political and economic power and focus and different time frames the agreement on components and overarching architecture of solutions in a systemic way on "paths toward a fossil-free society within a few decades" has to be dynamic, evolving, explorative. It also has to be served by a process that is democratically considered as legitimate. And it has to be connected to a broad process of participation in the population at large, including strong educational components.

6. Policy considerations aiming at a path toward a fossil-free society at a regional level

In shaping the process that should lay the foundation for a successful transition to a fossil-free society by 2050, the following elements are important to consider, with special regard to the body responsible for our case region (The Stockholm-Mälar region) [2]. (But the points of recommendations also pertain to other levels, as well as other configurations of regional agglomerations in Sweden):

I. The character of the transformation

a. *The transformation to a low carbon society includes all aspects of society.* It relates to different levels of society and types of actors, including all sectors, forms of stakeholder types and has to involve civil society at large. It concerns the living conditions of all citizens of the region and spans the generations. This means that it is not only an issue of change of the technical aspects that is at stake, e.g., of the energy system and related infrastructural mechanisms. The transformation also connects human factors as consumer behavior, and in more general terms issues about where we want to live and work in the future. The overriding perspective has to be how the inhabitants of this region within a few decades would consider what a well-functioning society might entail, especially caring for the different and particular needs of persons of all ages and gender.

b. *The transformation thus requires a mobilization of the entirety of our society.* This means that our democratic processes have to involve a wide array of instruments enabling all citizens to be encouraged to invent and implement changes, i.e., using a deepened planning process with democratic consolidation. Innovations should be encouraged—not only with regard to technology, but also with regard to how society could be changed, e.g., through changes of laws, rules, administrative processes, stimulation to risk taking and renewal in all sectors and by all actors—as well as through the creation of new patterns of collaboration. There will be a need to creatively scrutinize our current patterns of values, facing the new challenges within all strata of society—public official structures, the business community and civil society alike. The further move toward an increased interest in future-oriented activities and openness toward change will be of considerable importance. But this will also put pressure on the "stronger" and more affluent segments of society to responsively constructively relate to the parts in society with more limited capacities and resources.

II. Governance

a. *The transformation is being performed in a societal context within which there are several interplaying levels* (e.g., the level of the individual, of the local municipality, the county, the sub-regional, the region, the national and the EU levels—also influenced by the constantly changing international conditions at larger frames). This means that the interplay between levels has to be given considerable attention and institutional innovative consideration. What once was a reasonable distribution of labor and responsibility might not be the same in the future due to changed conditions. The pressure to move quickly to a fossil-free society thus also puts stress on the governance architecture. Different versions of interplay between "bottom-up" and "top-down" solutions have to be conceived, developed and tested.

b. *The transformation is made within a very large complex system with many partial couplings.* This means that the complexity will have to be orchestrated in partially new ways. This can be prepared through various ambitious experiments both at limited levels and in a variety of actor spaces, but also in large constellations involving the needed investments for such actions. Such transformation experiments must be conducted in line with the goals of a fossil-free society—and be done through strong encouragement, maybe deliberate relaxation of certain rules in combination with appropriate new ones. This must be done by setting up a metric of several partially new and diversified "mirrors" reflecting varied starting points. However, the total overview of the process will never exist at any one time. Thus, the constant upgrading of the vision in relation to path experiences has to be developed in a dynamic interplay over time.

c. *The broad transition has distinct regional operational connotations.* This means that the societal conditions that historically have been developed over long time in our specific case region also in the future must be guarded and cared for—but now in a directed fossil-free context. The new possibilities that probably might be generated should be encouraged. An essential factor for success in this endeavor is a well-spread sense of participation in the change process by large segments of the population in society. This means that all persons in society should be needed in one way or the other—and this should be conceived of in a multi-generational perspective

III. Necessary, doable and emerging possibilities

a. *The change toward a fossil-free society is necessary.* However, it can also provide advantages for other aspects of change in society. This means that a diverse set of solutions developed for the purpose to bring us to a fossil-free society might also be supportive of other changes that are needed. One already very well-known example is that goals related to the handling of climate change may go very well hand in hand with efforts to reduce health effects from harmful components in the atmosphere—not least in heavily urbanized areas. Such synergies have to be better explored and mobilized—much better than what is the case today. This also calls for more cross-sectorial connectivity innovations.

b. *The changes are in most cases very doable.* If the time horizon is creatively used for early investments in what could fit the goals of a longer time frame vision, the costs are more restricted than those emerging in connection with actions taken later under more chaotic and drastic circumstances. But the changes have to be anchored in a transparent and broadly agreed societal process.

c. *The transformation process could not only be used to meet the challenges of change, but might also open up for many new possibilities that may emerge.* This means that although the change is necessary and deep going, it may also provide new competitive means in an international context—given that a change trajectory is chosen that encourages such possibilities. Thus, the region should use the transformation process to serve these purposes—also as input to discussions at national and European levels to demonstrate solutions developed in the societal, technical and ecological domains. In practice, this means to foster "Avant guard" forms of societal competitive ways to operate—as well as supporting and collaboration with other countries with less initial advantages for such performances. This may also be a competitive advantage for Swedish interests abroad.

As a closing line we consider these recommendations in many cases to be of a broader character than just pertaining to the target case of "transformation of society toward a fossil-free society by 2050." Indirectly, these particular points also say something about the decision-making challenges related to territories of agglomerations with strong urban connotation in our time. This holds true in our case for the level of a region in a European historical and political context. But it also has wider relevance for linked urban and nonurban agglomerations at even wider geographical and institutional space.

Acknowledgements

The research leading to these results has received funding from the European Union Seventh Framework Programme (FP7 2012-2016) under grant agreement n° 308601. We acknowledge the influence of ideas not least in our own institutional embedding at both Stockholm University (especially the Stockholm Resilience Center), and at SLU. This includes the researchers and external discussants that were involved in the now finished EU project COMPLEX.

Author details

Uno Svedin[1]* and Hans Liljenström[2]

*Address all correspondence to: uno.svedin@gmail.com

1 Stockholm Resilience Center (SRC), Stockholm University, Stockholm, Sweden

2 Department of Energy and Technology, Swedish University of Agricultural Sciences (SLU), Sweden

References

[1] Liljenström H, Svedin U, editors. Micro Meso Macro: Addressing Complex Systems Couplings. Singapore: World Scientific Publ. Co; 2005

[2] Liljenström H, Svedin U, editors. Towards a Fossil-Free Society—in the Stockholm-Mälar Region. COMPLEX EU Project 2012-2016, n°: 308601. WP4 Final Scientific Report. Sigtuna: The Sigtuna Foundation, Sweden, Human Nature Series, 2016. ISBN: 978-91-976048-4-0. DOI: 10.13140/RG.2.2.36656.97284

[3] Holling CS. The resilience of terrestrial ecosystems, local surprise and global change. In: Clark W C, Munn R E, editors. Sustainable Development of the Biosphere. Cambridge: Cambridge University Press; 1986. p. 292-317

[4] Liljenström H. Cognition and the efficiency of neural processes. In: Århem P, Liljenström H, Svedin U, editors. Matter Matters?—On the Material Basis of the Cognitive Activity of Mind. Heidelberg: Springer Verlag; 1997

[5] Steffen W, Persson Å, Deutsch L, Zalasiewicz J, Williams M, Richardson K, Crumley C, Crutzen P, Folke C, Gordon L, Molina M, RamanathanV, Rockström J, Scheffer M,

Schellnhuber J, Svedin U. The Anthropocene: From global change to planetary stewardship. In: Contribution, background report nr. 2 to the 3rd Nobel laureate symposium on global sustainability, Stockholm, Ambio 2011. DOI:10.1007/s13280-011-0185-x

[6] Granovetter MS. The strength of weak ties. American Journal of Sociology. 1973; **78**:1360-1380

[7] Watts DJ. Small Worlds: The Dynamics of Networks between Order and Randomness. Princeton, USA: Princeton University Press; 1999

[8] Liljenström H. Neural stability and flexibility: A computational approach. Neuropsychopharmacology. 2003;**28**:S64-S73

[9] Liljenström H. Network effects of synaptic modifications. Pharmacopsychiatry. 2010; **43**:S67-S81

[10] Liljenström H. Inducing transitions in mesoscopic brain dynamics. In: Steyn-Ross DA, Steyn-Ross M L, editors. Modeling Phase Transitions in the Brain. New York: Springer; 2010. p. 147-175

[11] Rockström J, Steffen W, Noone K, Persson Å, Chapin F S, Lambin E F, Lenton T M, Scheffer M, Folke C, Schellnhuber H J, Nykvist B, de Wit C A, Hughes T, van der Leeuw S, Rodhe H, Sörlin S, Snyder P K, Costanza R, Svedin U, Falkenmark M, Karlberg L, Corell R W, Fabry V J, Hansen J, Walker B, Liverman D, Richardson K, Crutzen P, Foley J. A safe operating space for humanity. Nature. 2009;**461**:472-475

[12] Crutzen PJ. Geology of mankind: The Anthropocene. Nature. 2002;**415**:23

[13] Steffen W, Crutzen PJ, McNeill JR. The Anthropocene. Are humans now overwhelming the great forces of nature. Ambio. 2007;**36**:614-621

[14] Rolén M, Sjöberg S, Svedin U, editors. International Governance on Environmental Issues. Dordrecht: Kluwer Academic Publishers; 1997

[15] Kates RW, Clark WC, Corell R, Hall JM, Jaeger CC, Lowe I, McCarthy JJ, Schellnhuber HJ, Bolin B, Dickson NM, Faucheux S, Gallopin GC, Grubler A, Huntley B, Jäger J, Jodha NS, Kasperson RE, Mabugunje A, Matson P, Mooney H, Moore III B, O'Riordan T, Svedin U. Sustainability science. Science. 2001;**292**:64-65

[16] Svedin U, O'Riordan T, Jordan A. Multilevel Governance for the Sustainability Transition. In: O'Riordan T. Editor. Globalism, Localism and Identity. Fresh Perspectives on the Transition to Sustainability. London: Earthscan; 2001

[17] Folke C, Pritchard L, Berkes F, Colding J, Svedin U. The problem of fit between ecosystems and institutions: Ten years later. Ecology and Society. 2007;**12**:30

[18] Galaz V, Olsson P, Hahn T, Folke C, Svedin U. The problem of fit among biophysical systems, environmental regimes and broader governance systems: Insights and emerging challenges. In: Young O, Schroeder H, King L, editors. Institutions and environmental change: Principal findings, applications, and research Frontiers. Cambridge, MA: MIT press; 2008. p. 147-186

[19] Svedin U, Hägerhäll Aniansson B, editors. Sustainability, Local Democracy and the Future: The Swedish Model. Dordrecht: Kluwer Academic Publishers; 2002

[20] Lafferty W, Eckerberg K, editors. From Global Summit to Local Forum. Studies of Local Agenda 21 in Europe. Oslo: ProSus (program for Research and Documentation for a Sustainable Society); 1997

[21] Eckerberg K, Forsberg B. Implementing agenda 21 in local government: The Swedish experience. Local Environment. 1998;**3**:333-347

[22] Lafferty W. Editor. Implementing LA 21 in Europe. New Initiatives for Sustainable Communities. ProSus: Oslo; 1999

[23] Coenen F, Eckerberg K, Lafferty WM. Implementation of Local Agenda 21 in twelve European countries: A comparative analysis. In: Häkinnen L, editor. Regions: Corner Stones for Sustainable Development. Helsinki, Finland: Publications of the Academy of Finland 8/99; 2000

[24] Bak P. How Nature Works. Oxford, UK: Oxford University Press; 1998

[25] Gunderson LH, Pritchard Jr. L, editors. Resilience and the Behaviour of Large-Scale Systems, Scope 60. Washington: Island Press; 2002

[26] Folke C. Resilience: The emergence of a perspective for social-ecological systems analyses. Global Environmental Change. 2006;**16**:253-267

[27] Kasperson JX, Kasperson RE, Editors. Global environmental risk. London: United Nations University Press/Earthscan

[28] Carpenter SR, Arrow KJ, Barrett S, Biggs R, Brock WA, Crépin A-S, Engström G, Folke C, Hughes TP, Kautsky N, Li C-Z, McCarney G, Meng K, Mäler K-G, Polasky S, Scheffer M, Shogren J, Sterner T, Vincent JR, Walker B, Xepapadeas A, Zeeuw AD. General resilience to cope with extreme events. Sustainability. 2012;**4**:3248-3259

[29] Hägerstrand T. Focus on the corporeality of man, society and environment. In: The Science and praxis of Complexity. Tokyo: The United Nations University; 1985, pp. 193-216. (Also in French translation in "Science et pratique de la complexité. La Documentation Francaise". Paris 1986, pp. 225-250

[30] Seitzinger SP, Svedin U, Crumley C, Arif S, Alfsen C, Broadgate W, Biermann F, Bondre N, Dearing J, Deutsch L, Dhakal S, Farahbakhshazad N, Gaffney O, Haberl H, Lavorel S, Mbow C, McMichael AJ, Morais J, Olsson P, Pinho P, Seto K, Sinclair P, Stafford-Smith M, Steffen W, Sugar L. Planetary stewardship in an urbanising world: Beyond city limits. Ambio. 2012;**41**:787-794

Chapter 4

The Multivariated Effect of City Cooperation in Land Use Planning and Decision-Making Processes: A European Analysis

Luís Carlos Loures, Rui Alexandre Castanho,
José Manuel Naranjo Gómez, Ana Vulevic,
José Cabezas and Luis Fernández-Pozo

Additional information is available at the end of the chapter

http://dx.doi.org/10.5772/intechopen.72191

Abstract

Spatial and urban planning processes regarding border cooperation have reached unprecedented levels in recent decades, not only due to their potential for territorial integration, i.e., infrastructure construction and planning activities worldwide. Bearing in mind the European project, for a united and strong network of nations, this scenario is more evident in European territories. In this regard, through multivariated analyzes of city cooperation on European border areas, it is possible to identify the factors that influence the territorial success and also a sustainable regional development and even their effects over the urban agglomerations. From the identified factors, the study pointed out one that is common to all cases: connectivity-movement between cities.

Keywords: border territories, cooperation, planning, sustainability, urban planning

1. Introduction

Recent changes in the European landscape introduced by geopolitical, socioeconomic and/or cultural issues have been continuously contributing to strengthen the magnetism of urban areas, increasing their capacity to change land use, thus fostering important transformations not only at the socioeconomic level but also in terms of urban morphology all over Europe [1–7]. Even though sometimes positive, these changes have reinforced the clear tendency of depopulation of rural territories—confirmed on the last decades—consequently increasing the development of progressively larger urban agglomerations [8–10].

Even if these issues are currently seen as part of countries' evolutionary processes, it urges to develop specific strategies that might tackle this scenario, contributing to revert this situation. In this regard, it is crucial to study new ways of cooperation between smaller cities, highlighting the ways in which this type of synergic projects and strategies related to cross-border and city cooperation might revert this process contributing to city sustainability.

Considering the aforementioned opportunity and taking into account that very often the cooperation between cities occur among sovereign nations, such issues gain more emphasis not only because one-third of European population live in border areas [11, 12] but also because this fact takes the discussion to another level, associated with intricate bureaucratic procedures related to the lack of standardization practices and policies between these territories, which generally lead to a reckless attitude by the main actors/decision-makers toward city cooperation and development, facts which have a high influence not only in the development capacity of these regions but also on the overall quality of life of people leaving in these areas [7, 13].

The analysis of city cooperation multivariated effect in land use planning and decision-making processes is, in this respect, seen as a pivotal procedure that, throughout the implementation of specific planning principles and frameworks, might contribute to strengthen the development of these areas, enabling Europe to achieve a more balanced polynucleated territory, less dependent on few large metropolitan areas.

In this regard, in order to identify a set of planning principles and city cooperation frameworks, a group of the best practice case studies will be analyzed, using both Case Study Research (CSR) method [14, 15] and Territorial Impact Assessment (TIA) analysis, in turn identifying the impacts produced by this type of cooperation [16–18] and highlighting how these processes might contribute to reverse the impacts promoted by current mononucleated tendencies.

This study is considered a fundamental basis that will enable the identification of precise factors, which influence spatial planning procedures, leading to the definition of new methods and approaches to one of the main urban problems affecting Europe during the last decades, in line with the scope of the book and considering several relevant issues related to spatial planning, sustainable growth, and urban development approaches.

2. European planning framework: a brief review

The European Union (EU) has no direct mandate and no clear institutional and political framework for spatial planning: according to the principle of "subsidiarity," territorial planning is a responsibility primarily of the member states. While this is so, it is equally clear that the EU's indirect role in spatial planning is steadily increasing, mainly through sector policies particularly in the areas of regional policy, rural development, environment, and transport. Another way in which the role of the EU is indirectly increasing in spatial planning is the principle of territorial "cohesion" [19].

In this regard, European territorial cooperation is the third objective of the EU Cohesion Policy for 2007–2013, as well as being the new umbrella under the European Spatial Development Perspective (ESDP) of 1999, such as European Spatial Planning Observation Network (ESPON), and European Territorial Cooperation (INTERREG) [20, 21].

In fact, the documents of EU from 1999 regarding spatial policy foster to promote a sustainable urban development. European Territorial Cooperation, and the EU INTERREG Initiative in particular, is one of the five main means of application identified by the European Spatial Development Perspective (ESDP). ESDP is a document approved by the Informal Council of Ministers of Spatial Planning of European Commission in Potsdam in 1999 and forming a policy framework with 60 policy options for all tiers of administration with a planning responsibility.

An entire chapter of the ESDP is in fact devoted to considering the application of its concepts and ideas. The five main means of application comprise (1) application at the European Community level, (2) application via transnational cooperation between member states, (3) application via cross-border and interregional cooperation, (4) application of the ESDP in member states, and (5) application via pan-European and international cooperation. The influence and application of the ESDP clearly have an important transnational learning dimension. As a result, the ESDP remains the most significant policy document guiding the EU's determinations in the territorial development sphere [22].

Even though the document does not have the power of a law, it has indeed been successful in establishing a framework [20]. Once that the strategic aim achieved a balanced and sustainable spatial development strategy. In the late 1990s, the ESDP represented a "new dimension of European policy" since for the first time the EU was starting to pay explicit attention to territorial planning as an instrument to achieve broader social and economic goals [19]. A key objective of the ESDP was to facilitate better coordination of the territorial impacts of European policy: horizontally across different sectors, vertically among different levels of government, and geographically across administrative boundaries [23].

The ESDP was created to achieve three central aims of European policy: socioeconomic cohesion, conservation of natural resources and cultural heritage, and a more balanced competitiveness of the European territory. In order to achieve these goals, the ESDP highlighted three crucial spatial development policy objectives [19–21, 24]:

1. Development of a polycentric and balanced urban system and strengthening of the partnership between urban and rural areas. This implicates overcoming the outdated contrast between city and countryside.

2. Promotion of unified transport and communication concepts, which support the polycentric development of the EU territory and are an important precondition for enabling European cities and regions to pursue their integration into the Economic and Monetary Union. Equivalence of access to infrastructure and knowledge should be realized gradually. Regionally, adapted solutions must be brought into being for this.

3. Development and conservation of natural and cultural heritage through wise management. This contributes both to the preservation and deepening of regional identities and the maintenance of the natural and cultural diversity of the regions and cities of the EU in the age of globalization.

A significant outcome of the ESDP process was the establishment of the European Spatial Planning Observation Network [21]. The mission of ESPON is to support policy development and build a European scientific community in the field of the European territorial development [19]. The main aim is to increase the general body of knowledge about territorial structures, trends, perspectives, and policy impacts in the enlarging European Union [25].

One of the concrete offspring of the ESDP is the ESPON established in 2001 for providing data and information about spatial trends and developments in the EU. ESPON's aim was to support spatial policy making for all territorial levels [26]. For planners, the value added of ESPON is that it supplies the technical and scientific knowledge needed to help implement the policy options in the ESDP and translates them into appropriate legal and financial instruments [27, 28].

However, the planning and execution of the specific political objectives of the ESDP should take into account the specific economic, social, and environmental situation of each area. These policy objectives for agriculture and rural development, infrastructure, and transport, as well as the environment, in a synthesized way are policy aims and options for ensuring productive and diverse rural areas, policy aims and options for promoting accessibility to transport and sustainable infrastructure, and policy aims and options aimed at preservation and development of the natural heritage [21, 29].

On the other hand, the policy priorities of the ESDP have been addressed by means of cofinancing of spatial planning projects involving partners in different countries through the INTERREG Initiative [19]. In this regard, INTERREG is the EU's primary instrument to support cooperation across national borders, and it is financed by the European Regional Development Fund. INTERREG was launched in 1990 to overcome the disadvantages presented by administrative boundaries of adjacent regions in the emerging common market [30]. INTERREG has included three phases with spatial planning agenda: Phase II (1994–1999), Phase III (2000–2006), and Phase IV (2007–2013). There are three territorial levels: (1) cross-border projects, involving geographically contiguous border regions; (2) transnational initiatives, across large multinational spaces; and (3) interregional initiatives, among noncontiguous regions across the whole territory of the EU.

In the latest phase (2007–2013), INTERREG has been incorporated into the EU's territorial cohesion agenda under the objective of territorial cooperation [19]. This implies cooperation on regional and spatial planning together with other economic growth issues more generally related to regional policy [30]. As for territorial cohesion in 2005, the Commission explained that territorial cohesion becomes a key element of promoting stronger integration of the territory of the Union in all its dimensions, and cohesion policy supports the balanced and sustainable development of the territory of the Union at the level of its macro-regions and reduces any barrier effects through cross-border cooperation (CBC) and the exchange of best practices [31].

The latter objective (strengthening territorial cooperation) is closely tied to the notion of European spatial planning (in INTERREG Phase III). The strong relation between cohesion policy and planning in the EU is reinforced by the work of DG Regio, the Commission Directorate-General for Regional (Cohesion) Policy which has taken the lead, with member states, on European spatial planning initiatives [30].

On the one hand, in European urban areas, the focus is on improving competitiveness through clustering, networking, and achieving more balanced development between the economically strongest cities and the rest of the urban network. Priorities are promoting entrepreneurship, local employment, and community development and measures to rehabilitate the physical environment, redevelop brownfield sites, and preserve and develop historical and cultural heritage [19, 31].

On the other hand, in European rural areas, the member states should support economic regeneration by ensuring a minimum level of access to "services of general economic interest," with a view to improving conditions in rural areas and limiting outmigration. Priorities include building connectivity to the main national and European networks; developing an integrated approach to tourism development; investing in development poles in rural areas (e.g., in small- and medium-sized towns); and developing economic clusters based on local assets combined with the use of new information technologies [19, 31].

3. City cooperation (CC) effect in land use planning

City-to-city (C2C) cooperation is not a novel phenomenon, through a scientific study on the matter is quite new. C2C was started, and evolved, in Europe—in wider terms: local governments in developed countries tend to determine the content of it. The first international relations between local governments in Europe were recognized after the Second World War, especially in the 1950s. The Council of European Municipalities and Regions, established in 1951, took a strong position to encourage these international contacts at the local level. The idea was to build a united Europe. Strict regulations were determined for these relations by the Council of European Municipalities and Regions and by the French-oriented Fédération Mondiale des Villes Jumelées—Cités Unies [32].

In the Western world of attention for development, cooperation was high in the 1970s, and characteristic of this development was the total absence of formal regulations. Then, a new tendency of international relations of local governments became apparent in the 1980s. In the United Kingdom, the United States, and the Netherlands—among others—local authorities united themselves against the apartheid in South Africa [32, 33].

As for Asia, China has undergone economic and political restructuring in the post-socialist era under the background of globalization. At the same time, its provincial governments have mobilized various forms of booming North-South city-to-city (C2C) cooperation within their respective jurisdictions during the last 10 years [34]. As for cross-border cooperation (CBC), during the recent decades, border areas increased great importance on the international scene concerning their potential and integrative functions such as demonstrated along the unification of Europe [12, 35–37].

The experiences of CBC, assumed not only in Europe but also all over the world, as is the case of several CBC projects between the United States and Mexico, China-India (Asia), Argentina-Chile, or Brazil-Bolivia (South America), among many other examples through the globe [38, 39], fostered the creation of a global network of relationships among people and states, which enabled the achievement of several political, economic, environmental, and sociocultural win-win situations [40, 41]. These networks have been increasingly recognized by urban planners, landscape architects, and other urban development specialists, as crucial elements which enable the introduction of recent urban development challenges and paradigms [13, 42–47] into future planning activities [10].

As for spatial planning system, an ensemble of territorial governance arrangements that seek to shape patterns of spatial development in particular places is considered [48, 49]. In this regard, the set of systems and policies of the planning of the EU expanded planning system increasing criteria such as the extent of the planning system, the extension, and the type of planning at the national and regional level [31, 50, 51]. Also, it has redefined the role of the public and the private, maturity and integrity of the system and the distance between the intended objectives and the results actually obtained [52]. And, it divides traditional planning mainly into four types, including regional economic planning, urban planning, comprehensive planning, and land use planning [49].

In fact, the efficacy of the CBC at different scales of cooperation in planning the land use has been under discussion, through the review of CBC's initiatives in developing countries [10, 53]. On a governmental scale normally stands as the promise of the project from the management and land use, in several cases, it contributes to overshadow the limitations of land use. For example, the extraction of natural resources and the massive agricultural exploitation. Secondly, at the regional level, the design and implementation of the CBC have overlooked often asymmetries of power within a community. This effect has been gone unnoticed at the individual level. Thus, there has been unequal access to the design, decision-making, and the intended results. It also highlights at the community, local, and national bargaining power that has taken the elites, because this has been instrumental in shaping the governance of the CBC and even in the promotion of the CBC by external agents. In fact, the adaptation of the CBC to new scenarios depends largely on the process of empowerment of the community and the construction of optimal networks with external agents [53]. In this way, as will be achieved through CBC, cross-border regions can reach (1) comparative advantages and (2) economies of scale.

4. The impact of city cooperation in decision-making processes

Planning potentially influences and connects a wide range of issues, behind which are most diverse and conflicting interest [49]. The traditional special planning focuses on the position, intensity, form, quantity, and coordination of the development of the land in different spaces. However, the issues and challenges faced by the local areas need to be addressed by a process of socio-spatial integration through which occurs a vision, coherent actions, and means of implementation for shaping and structuring making it a place and what this place might be in the future position [54, 55]. In this sense, cooperation between cross-border cities is directed to the solution of problems arising from the dysfunctions caused by the existence of the border [56].

As for sovereign and cooperation problems, nation-states are under pressure to find innovative ways to redefine their relationships with space. The traditional understanding of the state, as the ultimate repository of sovereignty over a bounded portion of the Earth's surface and the society that inhabits it, is at odds with the current world of cross-border flows of capital, goods, people, and ideas [57]. At the same time, there is a reterritorialization of economic and political activity that transcends the spatial framework of the nation-state [58].

Border, cross-border regions, and CBC studies address state territorial restructuring at the subnational level. They have primarily examined the reterritorialization of state power and institutions across borders, documenting the emergence of cross-border governance networks and power relations [57, 59–65]. In this regard, CBC should be placed in the context of the emergence of multilevel and participative governance, which requires an active involvement on behalf of, on the one hand, different tiers of government (from the EU institutions to local governments) and, on the other, civil society and private actors working alongside public authorities [66]. The diversity of arrangements existing in the field of CBC includes the fact that Euroregions may be established according to either private or public law [67]. The geopolitics of Euroregions suggest that cross-border reterritorialization across the latest EU borders is driven by a scalar conflict of territorial logics [57]. The border-induced territorial logic of the nation-state conflicts with the border-bridging territorial logic of CBC [68].

In political and institutional terms, the emergence of CBC and the setting-up of cross-border frameworks such as Euroregion and European Grouping of Territorial Cooperation (EGTC) can be seen as an expression of broader developments in the field of European governance, including the subsidiarity principle and its adaptation to a borderless Europe, the increasing centrality of regions as spaces combining a political, economic, social, and cultural dimension [69].

A "territoire de project" amounts to a bottom-up process, which in turn can contribute to moving from a vertical, interlocked approach to regional development and multilevel governance to the one characterized by shared responsibilities and by horizontal, interdependent relationships among different regions and tiers of government [70]. Besides, the added value of EGTCs lies in their ability to fulfill cross-border tasks by common decisions on the regional/local level and to reinforce the ability of local and regional authorities to contribute to bottom-up regional development [71]. But public authorities on the regional/local level need an EGTC; otherwise, a bottom-up approach in territorial cooperation is very difficult [66]. Also, in comparative terms, the adoption of the EGTC Regulation also serves to highlight the relevant role played by CBC within the European integration process, since the sovereignty of the state ends at its borders, but the differences and problems of these borders continue to exist and require sustainable solutions [66]. Against this background, they cannot allow their borderland to follow special rules without compromising the theoretical model of the territorial container that the nation-state follows [72]. Euroregions for their part need exemptions from national regulation in order to be able to function meaningfully across state borders [57]. Indeed, the degree of cross-border integration can be a result of political will and the benefit perceived by the communities involved in the collaboration, but it can also be interpreted as resulting from the prevailing tension between the aim to supersede traditional borders and the ongoing weight of national traditions and structures [66].

In this regard, research has gained additional momentum considering current developments, such as the recent situation in Crimea (Ukraine) with Russia, the Greek-German frictions, the rise of terrorism in European continent (leading to an increasingly fractious debate about free movement in Europe as well as the resurgence of nationalist and extremist sentiments among European citizens), the economic fallouts in Europe, or even the recent Brexit scenario [3, 7, 73–75]. Due to the fact, other factors to ensure the sustainability of experiences include political will, i.e., political commitment and transparency, and the integration of a "cross-border awareness" or "cross-border culture" in the standard design and management of policies and legislation, i.e., common objectives and Master's Plans [10, 66].

5. City cooperation in Europe: assessing the past, envisioning the future

In the inception of the twenty-first century, especially in a European and Western context, it is almost given that borders are just lines drawn on a map. For this, the establishment of the European Union has contributed as a catalyst [76], almost vanishing borders and promoting CBC and city border cooperation, i.e., as is the case of the establishment of the Euro-cities. Still, as abovementioned, the recent developments going on European territory are threatening such relationships among European nations and even with other continents, leading to the necessity to rethink some of the EU policies.

In this regard, four European CBC case studies (**Figure 1**) focused from different aspects and perspectives will be exposed, analyzed, and assessed, enabling a multivariated effect analysis of city cooperation in land use planning and decision-making processes. Thus, the case studies need to meet the following criteria:

- Cities must present a historic of city cooperation.
- Countries should present CBC projects toward the integration of environmental, sociocultural, and economic development goals.
- The distance between cities should be no longer than 100 km.

Figure 1. Selected case studies. (A) Vienna-Bratislava-Brno-Györ, (B) Copenhagen-Malmö, (C) Oradea-Debrecen, and (D) Ruse-Giurgiu.

5.1. Case study analysis

5.1.1. Vienna-Bratislava-Brno-Györ: the labor force as a catalyst "4" city cooperation

The institutional relations between the cities of Bratislava, Vienna, Brno, and Györ are driven, mainly, by the desire to promote economic development. With the end of the transitional period—limiting the right of Slovak workers to enter the labor market in the European Union in 2011—the cooperation levels between the cities have considerably increased. Nowadays, the leading entity, regarding CBC, is the Centrope Strategy 2013—which includes territories of Hungary, the Czech Republic, Austria, and Slovakia. In this regard, one of the main goals of such strategy is to foster a coordinated approach toward a sustainable regional planning to improve connectivity and the movement between cities, in this large region, regarding accessibility and transportation infrastructure and services [77]. All the regions present great results; however, it should be highlighted that the city of Bratislava once is perhaps one of the best examples of urban growth in the globalization *Era* [1, 7]. These great results may be explained mainly by its privileged geographical situation, at the heart of a rapid European development, presenting interesting future indicators for their citizens, along with new opportunities and horizons perfectly possible to reach.

Identified critical factors: common objectives and Master's Plans, connectivity-movement between cities, and stronger economy.

5.1.2. Copenhagen-Malmö: a strong cooperation through a bridge below Öresund

The cooperation around the Øresund Strait mainly focuses on cross-border economic development based on knowledge and innovation—several clusters in life sciences and clean technologies. Projects as the Øresund Bridge, the train or the freeway, are just some example that have enabled an increase on accessibility standards and improving border functional integration among regions, which have led to the development and construction of new urban districts, i.e., Ørestad, in Copenhagen, and Hyllie, in Malmö, through this transboundary linkage. Being peripheral cities normally seen as a handicap, however, in this case, it has turned into an advantage thanks to the great relationships among these territories and the commitment demonstrated by the main actors of the common regional planning processes and the decision-makers.

Identified critical factors: common objectives and Master's Plans, connectivity-movement between cities, and stronger economy.

5.1.3. Oradea-Debrecen: CrossTrans toward the EU standards

If the cooperation between these two cities is still young and also poorly developed, the possibilities for growth through new cooperations and synergies are considerably high, once these two cities perform one of the major urban agglomerations of the region—leading to a significant amount of human resources. By the other hand, the existence of a Hungarian minority in the Romanian side contributes to a stronger regional cohesion. Still, both cities are inserted in the Euroregion Hajdu-Bihar Bihor, where one of the pre-established goals of the Euroregion is fostering a better integration of the projects carried out by the public regarding: health, culture, education, and economic development. The distance between these two cities, about 65–70 km, should also be taken into account. So, their territorial success depends on well-developed connectivity between cities; so, the *CrossTrans* project, as well as the establishment of other infrastructures, has been critical to achieving the so desired success. Besides, it also should be considered that the region is not Schengen Area, or a member of the common currency (Euro), leading to more handicaps for the territories. Nevertheless, and against all the low odds, the region through CBC strategies shows significant positive performances which conduct an approach to European standards.

Identified critical factors: access to European funds, connectivity-movement between cities, and increasing life's standards.

5.1.4. Ruse-Giurgiu: the importance of being inserted in a Euroregion

The cities of Ruse and Giurgiu constitute the largest border urban agglomeration between Bulgaria and Romania, separated by the Danube River, which is in itself a reason for cooperation since a long time, i.e., common urban planning due to extreme flood phenomena. Along with the previously identified reason for cooperating, also exists the need for the rehabilitation of accessibility infrastructures promoting a transboundary integration avowing infrastructural duplication in CBC [7]. Also, in this case, the area is inserted within a Euroregion—*Danubius Euroregion*. Thus, it is possible to verify, once more, the importance of being inserted in a Euroregion, mainly, for cities and regions that denote an estrangement from the European standards—which, unfortunately in many cases, is more evident. So, their integration in larger common regional development programs, i.e., the Euroregions, conducts that the performances presented by these "Euroregionalized regions" start to reverse the estrangement tendency and consequently a closeness to the EU standards.

Identified critical factors: access to European funds and connectivity-movement between cities.

5.2. Outcomes

In this regard, general settings including statistical data for the population of the cities and the corresponding region or influenced area (**Table 1**), as well as data for the distance-time between the cities of these border areas (**Table 2**), were analyzed. Through that analysis, it is possible to understand their spatial configuration, i.e., different spatial configurations and different urban patterns and dynamics, concurring with previous researches and studies as the one conducted by ESPON regarding Urban Functions [78], among many other studies and works developed in the same context.

Also, an indicator which allows that to analyze the economic dynamics of the territories, i.e., GDP per capita in terms of purchasing power parity, is presented in **Table 3**, covering both sides of the border. Analyzing **Table 3**, it is possible to understand the socioeconomic dynamics of the urban areas under study, i.e., GDP per capita for each case study region for the 3 years 2000, 2006, and 2011, where in all cases, the GDP per capita has increased in absolute terms but with very different rates. For an easy reading, in the last column, the information is synthesized using signs which allow seeing whether GDP has increased at rates, which are:

Case study	Country	Area (km²)	Spatial structure	Population
Vienna-Bratislava-Brno-Györ (Centrope region)	AT SK CZ HU	48,200	Cross-border polycentric metropolitan region	6,500,000
Copenhagen-Malmö	DK SE	21,800	Cross-border polycentric metropolitan region	3,800,000
Oradea-Debrecen	RO HU	13,600	Cross-border polycentric metropolitan region	1,134,255
Ruse-Giurgiu	BG RO	195	Cross-border agglomeration	204,297

AT, Austria; SK, Slovakia; CZ, Czech Republic; HU, Hungary; DK, Denmark; SE, Sweden; RO, Romania; and BG, Bulgaria.

Table 1. General settings and spatial structure (source: [77, 79]).

- Still below the average increase for European regions in NUTS 3 (−), which was 5500 €.
- Between 5000 € and 10,000 € above the EU average growth (+).
- Very above (++), over 10,000 € elation to the population.

It is a well-documented fact that labor market can play a critical role through the impact on cross-border integration [81]; so, throughout **Table 4**, it is possible to analyze the differences in unemployment levels and where cross-border cooperation can potentially allow a higher fluidity of the labor market for the benefit of both sides of urban regions [80, 81].

Table 5 shows the identified factors for territorial success on the case studies.

Case study	Country	Between main cities	Travel time (min)		
			By public transport	By car	By bus
Vienna-Bratislava-Brno-Györ (Centrope region)	AT	Vienna-Bratislava	83	54	/
	SK	Vienna-Györ	116	81	
		Vienna-Brno	137	103	
	CZ	Bratislava-Györ	91	55	
		Bratislava-Brno	87	78	
	HU	Brno-Györ	248	122	
Copenhagen-Malmö	DK SE	Copenhagen-Malmö	34	48	/
Oradea-Debrecen	RO HU	Oradea-Debrecen	54	69	/
Ruse-Giurgiu	BG RO	Ruse-Giurgiu	/	15	15

AT, Austria; SK, Slovakia; CZ, Czech Republic; HU, Hungary; DK, Denmark; SE, Sweden; RO, Romania; and BG, Bulgaria.

Table 2. Connectivity and accessibility-movement between cities (source: [80]).

Case study	Country	GDP per capita			Evolution (2000/2011)
		2000	2006	2013	
Vienna-Bratislava-Brno-Györ	AT	30,263	34,547	36,562	+
	SK	12,400	17,200	19,000	+
	CZ	12,803	16,139	18,652	+
	HU	15,652	26,855	34,357	++
Copenhagen-Malmö	DK	22,100	25,700	27,100	–
	SE	29,161	33,419	36,667	+
Oradea-Debrecen	RO	7600	10,900	12,500	–
	HU	4700	9300	10,100	–
Ruse-Giurgiu	BG	4700	7100	8700	–
	RO	2700	4800	9000	+

Notes: (–) means that the evolution of the GDP per capita is below 5.500 € (EU NUTS 3 average growth) between 2000 and 2011; (+) means that the evolution of the GDP per capita is between 5.500 and 10.000 € between 2000 and 2011; (++) means that the evolution of the GDP per capita is higher than 10.000 € between 2000 and 2011; The threshold values have been calculated based on the average and standard deviation in all the NUTS 3 regions in the EU. AT, Austria; SK, Slovakia; CZ, Czech Republic; HU, Hungary; DK, Denmark; SE, Sweden; RO, Romania; and BG, Bulgaria.

Table 3. GDP per capita dynamics (source: [77, 79]).

Case study	Country	NUTS2*	Active population 2006/2013 (%)	Unemployment rate (%)			Employment 2006/2013 (%)
				2006	2013	Difference (%)	
Vienna-Bratislava-Brno-Györ		Burgenland	3.5	5	4.7	−1	4.7
	AT	Niederösterreich	6.2	4	5.6	0.5	5.6
	SK	Wien	7.4	8.8	7.8	−0.4	7.8
	CZ	Jihovýchod	3.7	7.1	4.1	−0.3	4.1
	HU	Nyugat-Dunántúl	−0.7	5.7	−2.7	2.1	−2.7
		Bratislavský kraj	0.3	4.6	−1.7	1.8	−1.7
Copenhagen-Malmö	DK	Sydsverige	9	8.2	6.8	1.7	6.8
	SE	Hovedstaden	8.8	5	3.8	2.4	3.8
		Sjælland	−2	4	−5.6	2.8	−5.6
Oradea-Debrecen	RO	Észak-Alföld	5.7	11	1.7	3.4	1.7
	HU	Nord-Vest	0.6	5.9	2.7	−1.8	2.7
Ruse-Giurgiu	BG	Severen Tsentralen	−9.1	13.5	−11.1	1.8	−11.1
	RO	Sud-Muntenia	−8.8	9.4	−9	0.5	−9

AT, Austria; SK, Slovakia; CZ, Czech Republic; HU, Hungary; DK, Denmark; SE, Sweden; RO, Romania; and BG, Bulgaria.*Totally or partly incorporated in cross-border urban areas.

Table 4. Labor market dynamics (2006/2013) (source: [77, 79]).

Identified factors	Case studies			
	Vienna-Bratislava-Brno-Györ	Copenhagen-Malmö	Oradea-Debrecen	Ruse-Giurgiu
Access to European funds			x	x
Common objectives and Master's Plans	x	x		
Connectivity-Movement between cities	x	x	x	x
Increasing life's standards			x	
Stronger economy	x	x		

Table 5. Identified factors.

6. Discussion and conclusions

The performed multivariated analysis of the case studies enabled us to identify critical factors for territorial success through CBC projects. Nevertheless, similar studies have already been developed, i.e., [7, 10, 32, 66], among many others, however, not through a multivariated analysis focusing on the addressed case studies. Thus, the present study allowed to define specific factors for these cases as well as to establish a correlation with land use, urban planning, city cooperation, and CBC.

So, from an individualized perspective, two rhythms of development and consequently two groups of factors and objectives can be defined; in order words, through the analysis of the case studies, it is verified that the cases located in Central and North Europe, Bratislava-Vienna-Brno-Györ, and Copenhagen-Malmö, have similar objectives, i.e., common objectives, Master's plans, and stronger economy, while in the cases of Eastern Europe, Oradea-Debrecen, and Ruse-Giurgiu, the factors and objectives are access to the EU funds and increase in life's standards, demonstrating significant disparities in the development levels within the European continent. However, one of the identified factors is common to all case studies: connectivity-movement between cities, showing unequivocally its relevance to achieve territorial success as well as the so desired sustainable development. In this regard, to achieve a sustainable, well-developed and abiding CBC project, all the critical factors, even the ones that have been assigned only for some of the cases, should be considered and not be underestimated [6]; such statement is valid not only for the planners but mainly for the decision-makers.

The historical and social evolution that European territories have felt through time should also be focused; along with the land use, changes as well as territorial landscapes, urban and rural, are the outcome of policies and administrative actions leading to strengthen urban agglomerations giving them the consistency that they present as a result of the application of regional strategies, i.e., the plurality of Europe [41]. In fact, the reality of the twenty-first century is increasingly the change tendency in the urban landscapes of these "new times," i.e., technological, and socioeconomically, where city cooperation is not an exception.

In fact, throughout history, territories have always sought an approach to the more developed/avant-garde nations of their time. Nowadays, and based on the results of the present study, economies and countries of Central and Northern Europe, along with the United States, continue to be references to good practice, formatting the urban agglomerations and their landscapes in the demand for such standards.

Acknowledgements

The authors would like to acknowledge the financial support given by the Junta de Extremadura/FEDER for the support to the research Group ARAM - Environmental Resources Analysis Research Group (GR15149).

Author details

Luís Carlos Loures[1]*, Rui Alexandre Castanho[2,5], José Manuel Naranjo Gómez[3], Ana Vulevic[4], José Cabezas[2] and Luis Fernández-Pozo[2]

*Address all correspondence to: lcloures@gmail.com

1 Polytechnic Institute of Portalegre, Portugal and Research Centre for Spatial and Organizational Dynamics (CIEO), University of Algarve, Portugal

2 Environmental Resources Analysis Research Group (ARAM), University of Extremadura, Badajoz, Spain

3 Polytechnic School, University of Extremadura, Caceres, Spain

4 Department of Architecture and Urban Planning, Institute of Transportation – CIP, Belgrade, Serbia

5 Department of Landscape, Environment and Planning; School of Science and Technology, University of Évora, Portugal

References

[1] Špirková D, Ivanička K. Dynamics of Bratislava urban and housing development. In: Panagopoulos T, editor. New Models for Innovative Management and Urban Dynamics. Faro: University of Algarve; 2009. pp. 37-53

[2] Martins S. Labirintos Mundiais: As Revoluções Pós-Modernas e os Caminos da Incerteza Global. Lisbon, Portugal: Sílabo; 2015. 248 p

[3] Holmes DR. Fascism 2. Anthropology Today. 2016;**32**(2):1-3. DOI: 10.1111/1467-8322.12235

[4] Scott JW. Rebordering Central Europe: Observations on cohesion and cross-border cooperation. In: Scott JW, editor. Cross-Border Review Yearbook 2016 European Crisis and

its Consequences for Borders and Cooperation. Budapest: Central European Service for Cross-border Initiatives (CESCI), European Institute of Cross-Border Studies; 2016. pp. 9-29

[5] Castanho R. Sustainable Urban Planning in Transboundary Areas—Analysis of Critical Factors for Territorial Success. [PhD Thesis]. Badajoz, Spain: University of Extremadura; 2017

[6] Castanho R, Cabezas J, Loures L, Fernández-Pozo L, Nunes J, Gama J. Processes to achieve territorial cohesion and sustainable development. Proceedings of the International Congress on Interdisciplinarity in Social and Human Sciences. Faro-Algarve, Portugal; 2017

[7] Castanho R, Loures L, Cabezas J, Fernández-Pozo L. Cross-border cooperation (CBC) in southern Europe—An Iberian case study. The Eurocity Elvas-Badajoz Sustainability. 2017;**9**(3):360. DOI: 10.3390/su9030360

[8] McCann E, Ward K. Policy assemblages, mobilities and mutations: Toward a multidisciplinary conversation. Political Studies Review. 2012;**10**(3):325-332. DOI: 10.1111/j.1478-9302.2012.00276.x

[9] Batista T, Caballero C, Ceballos F, Carriço C, Mateus J, Lopes H, et al. IDE-OTALEX C. The first crossborder SDI between Portugal and Spain: Background and development. Journal of Earth Science and Engineering. 2013;**3**(6):393. DOI: 10.1553/giscience2013s6

[10] Castanho R, Loures L, Fernández J, Pozo L. Identifying critical factors for success in cross border cooperation (CBC) development projects. Habitat International. 2016; (in press). DOI: 10.1016/j.habitatint.2016.10.004

[11] Smallbone D, Welter F. Cross-border entrepreneurship. Entrepreneurship & Regional Development. 2012;**24**(3-4):95-104. DOI: 10.1080/08985626.2012.670907

[12] Martín G. Border Space as a Vehicle for Territorial Cooperation in the New Programming Period 2014-2020. Badajoz, Spain: OTALEX-C. Direccíon General de Transportes, Ordenación del Territorio e Urbanismo. Consejeria de Fomento, Vivienda, Ordenación del Territorio y Turismo. Junta de Extyremadura; 2013

[13] Loures L, Panagopoulos T, Burley JB. Assessing user preferences on post-industrial redevelopment. Environment and Planning B: Planning and Design. 2016;**43**(5):871-892. DOI: 10.1177/0265813515599981

[14] Yin RK. Case Study Research: Design and Methods. London: Sage Publications; 1994

[15] Levy JS. Case studies: Types, designs, and logics of inference. Conflict Management and Peace Science. 2008;**25**(1):1-18. DOI: 10.1080/07388940701860318

[16] Medeiros E. Efeito barreira e cooperação transfronteiriça na Raia Ibérica. Guarda/Lisboa: Centro de Estudos Ibéricos/Áncora Editora; 2011

[17] Medeiros E. Assessing territorial impacts of the EU cohesion policy: The Portuguese case. European Planning Studies. 2013;**22**(9):1960-1988. DOI: 10.1080/09654313.2013.813910

[18] Medeiros E. Territorial impact assessment and cross-border cooperation. Regional Studies, Regional Science. 2015;**2**(1):97-115. DOI: 10.1080/21681376.2014.999108

[19] Rabé P, Toto R, Dhima S. Analyzing the Likely Impact of European Union Policies on Territorial Planning in Albania. Rotterdam, Netherlands: Netherlands government (MATRA); 2013

[20] Faludi A. European Territorial Cooperation and Learning. disP: The Planning Review. 2008;**44**(172):3-10. DOI: 10.1080/02513625.2008.10556998

[21] Gualini E. Political economy of scale in European spatial policy. In: Rethinking European Spatial Policy as a Hologram: Actions, Institutions, Discourses. London, United Kingdom: Routledge; 2017

[22] Stead D, Waterhout B. Learning from the application of the ESDP: Influences on European Territorial Governance. disP: The Planning Review. 2008;**44**(172):21-34. DOI: 10.1080/02513625.2008.10557000

[23] Faludi A. The application of the European spatial development perspective: Introduction to the special issue. Town Planning Review. 2003;**74**(1):1-9. DOI: 10.3828/tpr.74.1.1

[24] Commission of the European Communities. European Spatial Development Perspective: Towards Balanced and Sustainable Development of the Territory of the European Union. Luxemburg: Office for Official Publications of the European Communities; 1999

[25] ESPON. Study on Urban Functions, No. 1.4.3. Luxembourg: The ESPON Monitoring Committee; 2007

[26] Vogelij J. Does ESPON support planning practice? Planning Theory & Practice. 2014; **15**(1):139-143. DOI: 10.1080/14649357.2013.873232

[27] Ravesteijn N, Evers DVH, Middleton D. Unseen Europe: A Survey of EU Politics and its Impact on Spatial Development in the Netherlands. Rotterdam, Netherlands: NAi Publishers; 2004

[28] Knieling J, Othengrafen F. Planning culture—A concept to explain the evolution of planning policies and processes in Europe? European Planning Studies. 2015;**23**(11):2133-2147. DOI: 10.1080/09654313.2015.1018404

[29] Vulevic A. Accessibility concepts and indicators in transportation strategic planning issues: Theoretical framework and literature review. Logistics & Sustainable Transport. 2016;**7**(1):58-67. DOI: 10.1515/jlst-2016-0006

[30] Dühr S, Stead D, Zonneveld W. The Europeanization of spatial planning through territorial cooperation. Planning, Practice & Research. 2007;**22**(3):291-307. DOI: 10.1080/02697450701688245

[31] Commision of the European Communities. Cohesion Policy in Support of Growth and Jobs: Community Strategic Guidelines, 2007-2013. Luxemburg: Office for Official Publications of the European Communities; 2005

[32] Buis H. The role of local government associations in increasing the effectiveness of city-to-city cooperation. Habitat International. 2009;**33**(2):190-194. DOI: 10.1016/j.habitatint.2008.10.017

[33] Musch A, van der Valk C, Sizoo A, Tajbakhsh K. City Diplomacy. Hague, Netherlands: VNG International; 2008

[34] Xian S, Chan RCK, Qi Z. Booming provincial-led North–South City-to-city cooperation in China: A case study of Suzhou-Suqian Industrial Park of Jiangsu Province. Cities. 2015;**46**(Supplement C):44-54. DOI: 10.1016/j.cities.2015.04.006

[35] Fadigas L. Urbanismo e natureza: os desa os. Lisbon, Portugal: Edições Sílabo; 2010

[36] Dominguez Castro L, Varela AE. Building Europolis on the basis of local "informal" cooperation in European cross-border spaces: The case of RIET. Geopolítica(s) Journal. 2015;**6**:225-246

[37] Trillo Santamaria J, González R, Paül V. Border-crossing cities: A critical analysis on the Eurocidade Chaves-Verin project. Journal of Geography. 2015;**54**:160-185

[38] Heredia C, Lawson C. Managing the United States–Mexico border: Cooperative solutions to common challenges. In: Report of the Binational Task Force on the United States–Mexico Border. Washington DC., USA: Pacific Council for International Policy (PCIP) and the Mexican Council on Foreign Relations (COMEXI); 2012

[39] Lee S, Na S. E-waste recycling systems and sound circulative economies in East Asia: A comparative analysis of systems in Japan, South Korea, China and Taiwan. Sustainability. 2010;**2**(6):1632-1644. DOI: 10.3390/su2061632

[40] Bacova A, Puskar B, Vrablova E. New Housing Models—Case Studies. Albena, Bulgaria: SGEM; 2015. DOI: 10.5593/ sgem2015/b62/s27.066. Pub. SGEM. Albena, Bulgaria

[41] Fadigas L. Urbanismo e território: as políticas públicas. Lisbon, Portugal: Edições Sílabo; 2015

[42] Slotterback CS, Runck B, Pitt DG, Kne L, Jordan NR, Mulla DJ, et al. Collaborative Geodesign to advance multifunctional landscapes. Landscape and Urban Planning. 2016; **156**:71-80. DOI: 10.1016/j.landurbplan.2016.05.011

[43] Follmann A. Urban mega-projects for a 'world-class' riverfront–the interplay of informality, flexibility and exceptionality along the Yamuna in Delhi, India. Habitat International. 2015;**45**:213-222. DOI: 10.1016/j.habitatint.2014.02.007

[44] Joseph M, Wang F, Wang L. GIS-based assessment of urban environmental quality in port-au-prince, Haiti. Habitat International. 2014;**41**:33-40. DOI: 10.1016/j.habitatint.2013.06.009

[45] Yu L. Low carbon eco-city: New approach for Chinese urbanisation. Habitat International. 2014;**44**(Supplement C):102-110. DOI: 10.1016/j.habitatint.2014.05.004

[46] Norman B. Principles for an intergovernmental agreement for coastal planning and climate change in Australia. Habitat International. 2009;**33**(3):293-299. DOI: 10.1016/j.habitatint.2008.10.002

[47] Saavedra C, Budd WW. Climate change and environmental planning: Working to build community resilience and adaptive capacity in Washington State, USA. Habitat International. 2009;**33**(3):246-252. DOI: 10.1016/j.habitatint.2008.10.004

[48] Nadin V, Stead D. European Spatial Planning Systems, Social Models and Learning. disP: The Planning Review. 2008;**44**(172):35-47. DOI: 10.1080/02513625.2008.10557001

[49] Zhou X, Lu X, Lian H, Chen Y, Wu Y. Construction of a Spatial Planning system at city-level: Case study of "integration of multi-planning" in Yulin City, China. Habitat International. 2017;**65**(Supplement C):32-48. DOI: 10.1016/j.habitatint.2017.04.015

[50] Commision of the European Communities. The European Compendium of Spatial Planning Systems and Policies. Luxembourg: Office for Official Publications of the European Communities; 1997

[51] Williams M, Williams R. European Union Spatial Policy and Planning. Thousand Oaks, CA, USA: Sage; 1996

[52] Munteanu M, Servillo L. Romanian spatial planning system: Post-communist dynamics of change and Europeanization processes. European Planning Studies. 2014;**22**(11):2248-2267. DOI: 10.1080/09654313.2013.830696

[53] Calfucura E. Governance, Land and Distribution: A Discussion on the Political Economy of Community-Based Conservation. Ecological Economics. 2018;**145**:18-26. DOI: 10.1016/j.ecolecon.2017.05.012

[54] Albrechts L. Strategic (spatial) planning reexamined. Environment and Planning B: Planning and Design. 2004;**31**(5):743-758. DOI: 10.1068/b3065

[55] Albrechts L. Shifts in strategic spatial planning? Some evidence from Europe and Australia. Environment and Planning A. 2006;**38**(6):1149-1170. DOI: 10.1068/a37304

[56] González XO, Guimerà AD, Perkmann M. Las regiones transfronterizas: balance de la regionalización de la cooperación transfronteriza en Europa (1958-2007). Documents d'anàlisi Geogràfica. 2010;**56**(1):21-40

[57] Popescu G. The conflicting logics of cross-border reterritorialization: Geopolitics of Euroregions in Eastern Europe. Political Geography. 2008;**27**(4):418-438. DOI: 10.1016/j.polgeo.2008.03.002

[58] Sassen S. Territory, Authority, Rights: From Medieval to Global Assemblages. New Jersey. USA: Princeton University Press; 2008. DOI: 10.1515/9781400828593.473

[59] Anderson J, O'Dowd L, Wilson TM. New Borders for a Changing Europe: Cross-Border Cooperation and Governance. London, United Kingdom: Psychology Press; 2003. DOI: 10.4324/9780203508435

[60] Hooper B, Kramsch O. Cross-Border Governance in the European Union. London, United Kingdom: Routledge; 2004. DOI: 10.4324/9780203563380

[61] Paasi A, Newman D. Boundaries, Territory and Postmodernity. Abingdon: Taylor and Francis; 1999. 206 pp

[62] O'Dowd L. The changing significance of European borders. Regional & Federal Studies. 2002;**12**(4):13-36. DOI: 10.1080/714004774

[63] Perkmann M. Euroregions: Institutional entrepreneurship in the European Union. In: Globalization, Regionalization and Cross-border Regions. Springer; 2002. pp. 103-124. DOI: 10.1057/9780230596092_5

[64] Scott JW. European and North American contexts for cross-border regionalism. Regional Studies. 1999;**33**(7):605-617. DOI: 10.1080/00343409950078657

[65] Sparke M. Between post-colonialism and cross-border regionalism. Space and Polity. 2002;**6**(2):203-213. DOI: 10.1080/1356257022000003635

[66] Baltà PJ. Cross-Border Cooperation and Cultural Communities in Europe. Brussels, Belgium: Centre Maurits Coppieters; 2015. DOI: 10.3726/978-3-0352-6452-4/13

[67] Commision of the European Communities. Practical Guide to Cross-border Cooperation. Brussels, Belgium: EU; 2000

[68] Browning CS. The internal/external security paradox and the reconstruction of boundaries in the Baltic: The case of Kaliningrad. Alternatives. 2003;**28**(5):545-581. DOI: 10.1177/030437540302800502

[69] Föhl PS. Regionale Kooperationen im Kulturbereich. Begri e und Systematisierungen. In: Föhl PS, Neisener I, editors. Regionale Kooperationen im Kulturbereich Theo-retische Grundlagen und Praxisbeispiele Bielefeld. Wetzlar, Germany: Majuskel Medienproduktion GmbH; 2009. pp. 15-46. DOI: 10.14361/97838394 10509-001

[70] Perrin T. Culture et eurorégions: enjeux institutionnels de l'action culturelle eurorégionale. Grenoble: École Doctorale Sciences de l'Homme, du Politique et du Territoire; 2010

[71] Gabbe J. General Secretary. Association of the European Border Regions (AEBR), Gronau, Germany; 2010. p. 17

[72] Taylor PJ. The state as container: Territoriality in the modern world-system. Progress in Human Geography. 1994;**18**(2):151-162. DOI: 10.1177/030913259401800202

[73] Perrin AJ. Political microcultures: Linking civic life and democratic discourse. Social Forces. 2005;**84**(2):1049-1082. DOI: 10.1353/sof.2006.0028

[74] Kornyeyeva L, Boehnke K. The role of self-acceptance in authoritarian personality formation: Reintroducing a psychodynamic perspective into authoritarianism research. Psychoanalytic Psychology. 2013;**30**(2):232. DOI: 10.1037/a0029879

[75] Dale G. Karl Polanyi: A Life on the Left. New York, USA: Columbia University Press; 2016. DOI: 10.7312/ dale17608

[76] Treaty on European Union. Maastricht Treaty. Luxembourg: Publications of the European Communities; 1992

[77] LISER. Opportunities of Cross-Border Cooperation between Small and Medium Cities in Europe. Luxembourg: Institute of Socio-Economic Research; 2015

[78] ESPON. Scenarios on the Territorial Future of Europe. ESPON Coordination Unit: Luxembourg; 2007

[79] Eurostat. Eurostat – Statistics Explained. Luxembourg: European Commision; 2015

[80] ESPON. Cross-Border Polycentric Metropolitan Regions. Targeted Analysis. Luxembourg: The ESPON Monitoring Committee; 2013

[81] Decoville A, Durand F, Feltgen V. Opportunities of cross-border cooperation between small and medium cities in Europe. In: Report Written in the Frame of the Spatial Development Observatory. Luxembourg: On Behalf of the Department of Spatial Planning and Development, Ministry of Sustainable Development and Infrastructure; 2015. pp. 1-55

Chapter 5

The City as an Experimental Space: The Interface between Public Satisfaction and Effects on Urban Planning Resulting from Kampala City's Sprawl

Fred Bidandi

Additional information is available at the end of the chapter

http://dx.doi.org/10.5772/intechopen.72186

Abstract

This paper analyses the interface between public satisfaction and effects on urban planning in the context of Kampala City, Uganda. The interface is significant because it provides an understanding of the effects of urban sprawl service delivery in relation to planning. It further discusses the underlying forces responsible for the city's urbanisation process. The objective of the study is to establish what defines Kampala's public satisfaction with urban changes resulting from the city's sprawl. The study involves focus group discussion interviews, which were used to collect the qualitative data from a group of respondents simultaneously. Moreover, purposive sampling was used to select the respondents interviewed. The analysis indicates that public dissatisfaction with poor urban environment has resulted in urban changes, which are officially sanctioned to take place in their residential areas, and the desire to access services easily that explains urban dwellers' decisions.

Keywords: urban sprawl, urban planning, public satisfaction, Kampala, Uganda

1. Research problem

A careful examination of studies on urban planning and sprawl reveals that many of them neglect public satisfaction with the urban changes that result from urbanisation process. Yet the satisfaction of these processes is necessary to understand because they explain the manner in which cities like Kampala urbanise especially a role played in determining how the public make decisions concerning the location of their residences, the design of urban environment, water, infrastructure, transport, physical investments and which place to work in. In support of the argument above, Kährik et al. [1] observe that public satisfaction is significantly related to decisions urban dwellers make pertaining to location of a residence in relation to the nature of the neighbourhood and affordability of available services. Public satisfaction is also critical to private investment decisions and to employment choices made by individuals in relation to the location of the employing organisation [2–6]. It also determines how urban population reacts to changes resulting from urbanisation process [7]. These decisions and reactions subsequently determine how a city urbanises residentially and in terms of attracted physical investments and workers. This is why such satisfaction is necessary to investigate in order to understand the planning implications of urbanisation cities like Kampala.

Aside from the existing scholarly work, Kampala's urban planners and policy implementers are also concerned about the current urban sprawl, but they are not providing an elaborate understanding of the changes underlying the city's urbanisation [8, 9]. This study seeks to answer this question by analysing public satisfaction in terms of the flaws that cause city government in Kampala fail to guide the city to urbanise in a planned manner.

2. Introduction to urbanisation

Urbanisation is not a phenomenon limited to Kampala or Uganda but a global, regional and national problem especially in sub-Saharan Africa. Globally, urbanisation is increasingly becoming a problem since almost half of the world's population (3.9 billion) now live in cities [10]. There were only 16 cities with the population of 1 million people at the beginning of the twentieth century globally, and many of them were in industrially developed economies such as Great Britain and France [11]. It is however noted that there are now more than 400 cities around the world that contain over a million residents, and about three-quarters of these are in developing regions such as Latin America and sub-Saharan African countries, to name but two [11].

In the 1960s, for example, Johannesburg was the only city in sub-Saharan Africa with the population exceeding over 1 million residents. By 2010, Africa had 33 cities with the population exceeding 1 million, including Kampala [10]. These trends of urbanisation seem to be cutting across all nations of the world, including Uganda.

As cities across the globe continue to attract people for better employment, education, health care and culture, they unduly contribute to national and urban economies. However, often rapid urbanisation in this case is associated with poverty, environmental degradation and population demands that outstrip service capacity. These conditions create unpleasant urban environment leading to numerous unsatisfactory outcomes such as poor housing, overcrowding, air pollution, transportation, insufficient or contaminated drinking water, inadequate sanitation and solid waste disposal services, industrial waste, increased motor vehicle traffic, stress associated with poor implementation of urban planning programmes and unemployment, among others.

http://dx.doi.org/10.5772/intechopen.72186

Urbanisation in sub-Saharan Africa has translated into rising slum establishments, increased poverty and inequality. Most of the cities in this regard are characterised by insufficient basic infrastructure, increasing poverty, poor service delivery, inequality and rising slum formations [12]. While sub-Saharan countries such as Uganda are experiencing unprecedented rate of urbanisation, the rates are higher than the resources, and these urban centres can offer given the population demands. This scenario provides an understanding of how urbanisation process in itself impacts negatively on urban service delivery and therefore the need to investigate its associated effects on urban planning and public satisfaction in terms of service delivery [13].

3. Theoretical framework

This study is underpinned by demographic transition theory, general theory, theory of capitalist urbanisation and modernisation theory, respectively: sociologists to describe how urbanisation results from the transition of high birth and death rates to low birth and death rates [15] propounded the demographic transition theory. This theory breaks the transition into four stages, namely, the pre-industrial stage, the industrial revolution, post-industrial revolution and stabilisation [16]. The theory contends that during the pre-industrial stage, societies were characterised by high birth and death rates, and because both rates are high, population grows slowly, and the rate of urbanisation tends to be low [17]. The industrial revolution is characterised by industrialisation that causes death rates to drop rapidly by making improvements in food production, health and sanitation. Food production is improved through more efficient agricultural practices and better transportation and food distribution, which collectively prevent death that would have resulted from starvation and lack of water. Health is improved through medical progress and advanced sanitation methods such as water supply, sewerage, food handling and general personal hygiene [18]. Accordingly, the industrial revolution is characterised by rapid urbanisation.

According to Fox [19], the post-industrial stage is typified by falling birth rates and lowering death rates. The falling birth rates result from better access to contraception, higher wages, commercialisation of agriculture and greater parental investment in the education of children, increasing female literacy and employment which lower opportunities for childbearing and motherhood and higher levels of investment in fixed assets such as housing, shopping malls, arcades and other physical infrastructure which increase urbanisation. During the fourth stage, population growth stabilises as birth rates fall into line with death rates, leading to stability in development and urbanisation [20].

In short, the demographic transition theory indicates that changes in human populations caused by the historical economic development of society are direct effect that explains the way urbanisation occurs. This theory is therefore used in helping to analyse whether Kampala's current urbanisation is a result of natural changes in its population or not.

Meanwhile, according to Puig [21], the general theory identifies five interrelated forms of dynamics that account for urbanisation of cities. These include 'the technical, administrative, political, legal and economic changes'. The theory posits that all these dynamics work together to promote urbanisation. According to Miller [22], this theory first posited that it is technical changes that are most critical to how cities urbanise because of their influence in determining the plans and policies that guide this process.

However, based on the more detailed analysis of the forces that determine cities' urbanisation, the theory shifted the emphasis to administration. This shift was based on the fact that it is urban administration that determines the effectiveness of all other changes because they are the ones that are responsible for implementing all enacted acts and designed urban programmes and policies, including technically designed city or town plans [22]. It is urban administration that not only guides urbanisation economically, socially and environmentally but also counters urban processes that are environmentally and politically undesirable, technically unapproved and legally prohibited, especially when they are acting ethically and impartially [23]. Politics tend to operate in much the same way in that it can counter undesirable urbanisation, but it can also cause this kind of urbanisation, depending on the political interests at play [22]. The political and administrative discourses can be official or unofficial [23]. This is important as a basis for analysing the nature of the political and administrative discourses as they operate in Kampala's urbanisation.

Another version of the general theory of urbanisation is referred to as the human ecology perspective, which was developed by Robert Park to explain the ways in which the population of urban areas expands or declines [24]. Therefore, this version focuses on how urban populations change as a result of the interplay of the five elements mentioned earlier. It contends that the manner in which each of these elements occurs affects the political, administrative, demographic, spatial, environmental and socioeconomic structure of a city [22, 25, 26]. This theory assumes that urbanisation should provide greater access to jobs, basic services and social safety nets [27]. Therefore, as Kasibante [28] points out, the two versions provide good grounds for analysing this nature as it applies to Kampala City. The rationale of the versions of the general theory is particularly important to the analysis of the elements characterising Kampala's urbanisation.

The theory of capitalist urbanisation offers important grounds for understanding the factors accounting for Kampala's urbanisation. Harvey [29, 30] to explain the effects and challenges of urbanisation developed this theory, especially in capitalist societies characterised by tendencies of intentional political and economic forces, especially those pertaining to capital investment. Harvey [29] developed this theory to offer a definitive Marxist interpretation of the urban process under capitalism. Harvey [29] believes strongly that 'capitalism has to urbanise to reproduce itself'. Capitalism can only survive if in addition to the conventional path of purchase, production of profit and distribution for consumption in a cash economy it also promotes the secondary path of circulation of fixed capital. This is well summarised by Christophers [31] that, 'A capitalist society generates surplus value and hence profit, must invest not only directly in the production process (e.g. in labour and machinery) but also in the built environment that houses companies, the state institutions that regulate them, and the employees that work for them'.

It is the secondary path that translates into built environments or urbanisation; as a result, Harvey [29] maintains that capital accumulation and the production of urbanisation have to go hand in hand for capitalism to survive. He bases this argument on the notions of capital over accumulation, also called surplus capital and capital switching. Harvey [30] considers surplus capital as that which has no value in the conventional production and distribution process but can gain value when it is not switched from this process and absorbed into a built-up environment (urbanisation). The switching takes the form of using this capital to construct factory buildings, administration offices, warehouses, employee residences, sewers, schools and hospitals and shops and other fixed developments and infrastructure such as roads, canals, docks and harbours and so on. According to Christophers [31], the gained value of surplus capital takes different forms such as increased consumption (when shopping malls are constructed) and social reproduction (housing when the capital is converted into construction of workers' residences) and easier access to labour (people attracted to live in the built housing).

Harvey [32] warns, however, that the conversion of surplus capital into urbanisation tends to deny poor people their right to the city, which was first recognised by Henri Lefebvre as the right to access urban resources. Harvey [32] expanded the meaning of this right by describing it as 'a right to change ourselves by changing the city and it is a common rather than an individual right since this transformation inevitably depends upon the exercise of a collective power to reshape the processes of urbanisation'. Harvey [32] observes that absorption of surplus capital into urbanisation sometimes takes place in the form of urban restructuring through 'creative destruction'. He argues that this restructuring nearly always has a class dimension since it is the poor, the underprivileged and those marginalised from political power that usually suffer from this process. Harvey [33] observes that the restructuring does not pay attention to the poor and, in most cases, destroys the housing and business structures that belong to the poor, causing homelessness, unemployment or redundant labour, undesirable neighbourhoods and other consequences, all of which tend to create dissatisfaction to the affected populations. The dissatisfaction tends to translate into urban revolutions to which capitalists react by expanding the built environment in form of constructing more housing, arcades and other infrastructure that can accommodate the revolting poor ([33, 34] Harvey, 2009).

Generally, the theory of capitalist urbanisation indicates that capital switching is one of the elements that can explain a city's urbanisation. The type of urbanisation this theory stresses is that which occurs when capitalists (investors) construct or restructure built environments as a way of absorbing surplus capital. The importance of this theory to this study is that its rationale helps to investigate whether one of the factors explaining Kampala's urbanisation relates to absorption of surplus capital or not and how is this absorption, if indeed, an effect on Kampala's urban poor's right to the city.

Last but not certainly least, modernisation theory, sometimes called the development doctrine [35], explains the process of countries' systematic transformation or progressive transition from premodern or traditional subsistence economies to modern industrialised economies (abid). This school of thought maintains that subsistence economies develop and urbanise as they adopt more modern industrial, technological, communication and cultural practices [36]. Indeed, Tettey [37] observed that urbanisation varies in line with the development pace of a country and for any country to urbanise, there is need to foster development through adoption of technology and industrialisation. A number of scholars endorse this connection by indicating that the phenomenon and process of urbanisation are irreversible features of modernization and development [38, 39].

The modernisation school of thought posits when their internal productive factors are reinforced by external assistance, predominantly subsistence economies can be developed and urbanised in the same way industrialised countries have developed and urbanised [19, 40]. This theory stresses using processes that bring about socioeconomic change and permit responses to this change [41, 42]. In so doing, it helps identify internal factors that contribute to social progress and development and how these factors can be boosted with external assistance to propel the processes of social evolution, including urbanisation, as desired [43–45].

The internal factors the theory identifies include the nature of politics, the development and urbanisation policy pursued by government and demographic factors [46, 47]. Other internal factors include levels of people's participation in productive activities, regulatory institutions and nature of available markets [48, 49]. Others are the available employment opportunities, level of infrastructural development and quality of the available human capital [37]. The forms of external assistance the theory identifies to lead to improvements in urban planning capacity and to unlock and realise the huge development potential of urbanisation include development aid and direct foreign aid inflows, among others [50].

The modernisation theory recognises that internal factors are not always enough to propel development and urbanisation at the desired pace [37]. Therefore, these factors need to be reinforced by foreign aid [43, 51, 52]. The assistance should be utilised to engage in massive investment in infrastructure, industry, technology and social services needed to propel socioeconomic progress and subsequent urbanisation [37, 53, 54]. Mungai [55] noted that massive investment increases population because of its potential to attract workers and subsequent rise in housing and social infrastructure. It is, however, also associated with negative consequences such as increased congestion and crime [55].

The modernisation school of thought has been criticised in that its prescribed foreign assistance encourages the dependency syndrome [52]. Some scholars even claim that its rationale does not apply to developing countries due to their levels of economic growth [37]. The theory is criticised for failing to prescribe governance values and norms which should be followed in order to bring about desired development and urbanisation [56, 57]. Notwithstanding these criticisms, the rationale of this theory offers the principles upon which the development and urbanisation model pursued in Kampala, Uganda, is based [58]. As a matter of fact, the official development and urbanisation programme pursued in Kampala under the PEAP umbrella was developed based on the modernisation theory [59]. Consequently, the manner in which the government of Uganda is promoting urbanisation depends on how it utilises both internal factors and how it solicits foreign assistance [60–63]. This suggests that the modernisation theory recognises the role of government as an element responsible for this country's urbanisation.

In addition, the modernisation theory permits responding to factors that cause transformation, including urbanisation [41, 42]. This offers another theoretical foundation for analysing

Kampala's urbanisation. The transformation that the modernisation theory advocates is not that which takes place for its own sake. It is transformation that should be felt by the people among or for whom it occurs [43, 51]. This argument is used in this study to analyse and understand Kampala city dwellers' satisfaction as a form of factors responsible for the city's urbanisation.

In summary, the reviewed theories indicate that each theory specifies factors, which explain urbanisation of different cities. They therefore show that the factors that are responsible for the manner in which a city such us Kampala urbanises are multifaceted. A close examination of the theories reveals that no single theory exhausts all the factors related to urbanisation. This implies that a study seeking to analyse urbanisation of a city such as Kampala is rationally safe when its theoretical grounding is hinged on the combined rationale of all the theories. This is therefore the rationale used to develop themes of this study as explained in the next section.

4. Understanding urban sprawl

Though urban expansion in cities like Kampala has benefits, such as increased city revenue and development, it has many negative outcomes for residents, particularly on the environment, water, sanitation, air pollution, increased traffic jams and loss of agricultural land. Polidoro et al. [64] describe urban sprawl as leapfrogging of development characterised by unrestricted expansion, which in this case occurs in considerably unplanned or poor residential settings. Moreover, Polidoro et al. argue that urban sprawl is characterised by the spreading of urban livelihood, which rapidly extends beyond the consolidated city centre. In other words, urban sprawl typically occurs outside the centre of services and available jobs, thus separating the places where people shop and work, and even where they study, from the place where they live.

However, some of the indictors of urban sprawl are encroachment on public land due to population growth, difficulties in access to public services (such as health, education, water, sanitation, etc.), creation of new urbanisation (usually in form of slums), decentralisation of public lands and inability for city governments to control real estate expansion for fiscal use. As Hasse and Lathrop [65] argue, urban expansion in cities like Kampala enhances social inequalities since the taxing of property in such areas does not exactly reflect the spatial and economic condition of the population.

5. Challenges of urban sprawl in the context of Kampala

Urban expansion is an important instrument, which in theory can contribute to any city's development and planning in new areas. As cities expand, they encroach into rural areas and small towns whereby having the opportunity to make the best possible use of land for better infrastructure and services in general becomes difficult, since the already occupied areas may either not be gazetted in the city plans or are simply people's private land on which they can carry out any activity without necessarily considering challenges that lie ahead especially given the population growth as well as services needed. However, urban expansion brings with it scandalous land grab and creation of urban gaps for the valuation of land [66]. Thus, this has become one of the major producers of value and accumulation of capital in cities. In Uganda, it is uncommon to use the expansion of urban districts to allocate social interest housing and low-cost housing projects in locations far distant from the consolidated city centre. Consequently, the infrastructure developed in certain areas serves as a factor for land valuation, while the city outskirts suffer for the lack of or poor-quality infrastructure, as well as difficulties in transportation due to the precarious system of public transport to areas where jobs are concentrated [67]. Moreover, households on the city periphery consider agriculture to be the single most important source of their livelihood, and therefore land is a critical resource for the 42% of households that earn a living from subsistence farming as well as providing the population employment in agriculture [66]. However, Uganda having a dual land tenure facilitates poor service delivery and conflict as land owners would sell to anybody they want without considering consequences resulting from urban sprawl. Jones et al. [68] argue that urban expansion means population demand for services but land scarcity is an important dimension of urban environments that influences the available space for service infrastructure and economic activities of land use planning. Moreover, heightening competition over land ownership is quite a challenge in Kampala.

Studies conducted about Kampala's urbanisation indicate that the process is characterised by lack of proper zoning of economic activities and construction of informal and formal physical infrastructure without regard to the subsequent spatial quality and environmental conservation [8, 69–72]. Kampala's urbanisation depicts sharp differences in residential standards where expensive housing coexist with shantytowns and informal settlements, with about 60% of the city's population living in unplanned residences and using very dirty, largely potholed and narrow roads with no street lights [73]. Other studies show that Kampala's urbanisation is typified by deteriorating environmental health characterised by air and noise pollution [74–76].

The city's drainage channels are silted and contaminated by organic and inorganic waste dumped by city dwellers and workers, causing the channels to get blocked, thereby flooding during rainy seasons [77–79]. It is very common to find similar challenges in peripheral areas where construction in swamps and green belts are closely interspaced with muddy huts in slums [8]. It is also not rare to find office buildings whose access roads are so narrow that even a fire brigade vehicle finds it difficult to access them when need arises. The result has been the development of different types of slums.

Kampala's urbanisation is also characterised by rising unemployment resulting from the demand for jobs far outstripping their supply [77, 80]. The city's unemployment is estimated to be between 60 and 80% and is cited among the critical causes of crime and violence increasing in the city [9, 81]. Growing population pressure on social services and menacing

http://dx.doi.org/10.5772/intechopen.72186

traffic congestion are the order of the day [8, 9, 82–85]. Furthermore, as Kampala continues to expand, services such as water, sanitation, solid waste management and planning remain a challenge. City dwellers establish themselves in areas too difficult to reach in terms of transport and city government which makes planning complex.

6. Urban sprawl and effects on planning

Kampala's urbanisation appears a replica of the general urban situation in Africa. There is a lack of a comprehensive understanding of the effects of urban sprawl on planning. This dates back when Kampala was declared the capital city of Uganda during which time the city was planned for the population of 300,000 without considering future urban changes [28, 87]. This understanding was neglected because like many other cities in the world in their infancy, Kampala was urbanising at a slow pace [88]. Such a pace gave Kampala City authorities time to plan and provide the public services that the slowly increasing urban population needed. This could be done without any need to first understand the underlying dynamics. This understanding was further neglected when the pace of Kampala's urbanisation declined in the 1970s because of the economic embargos, mismanagement of Uganda's economy and an atmosphere of insecurity that characterised the late President Amin's dictatorship [59].

The pace of Kampala's urbanisation started rising from 1990, but little was done to understand the underlying effects inclusively. Yet the rising rate was signalling a clear need for analysing the effects as a basis for containing the negative ones while encouraging those that promote organised urbanisation. The focus was instead on implementing the socioeconomic development agenda, which the government of Uganda had adopted following the Poverty Eradication Action Programme (PEAP) [89, 90]. Since over 90% of Ugandans were at that time engaged in rural subsistence farming, the main aim of the PEAP was to transform the country from being a predominantly rural subsistence economy to a modern economy [91]. In this programme, Kampala was viewed as a nucleus of socioeconomic transformation through industrialisation and commercialisation [59], and this further accelerated its urbanisation.

At the moment, Kampala is urbanising at a rapid rate estimated to be between 5.2 and 16% per annum [9, 92]. The latest statistics indicate that Uganda's urban areas claim 20% of her estimated 35 million people [93], but over 40% of these people are resident in Kampala City [94]. In fact, Kampala's population size grows to over 60% when the transitory population is factored in. Kampala is, however, urbanising in an unplanned manner [95–97] but without a clear picture of the underlying effects on planning.

Nonetheless, a review of the National Physical Planning Standards and Guidelines, 2011, revealed that they were developed by the MLHUD [98] with intent to provide criteria for determining the scale, location and site requirements of various land uses and facilities. The planning standards affect the allocation of scarce land and financial resources. They should, therefore, be applied with a degree of flexibility. Trade-offs may be necessary so that the community at large could benefit most from the development. They were to realise this purpose by ensuring equitable and balanced spatial distribution of development; orderly, efficient and coordinated spatial socioeconomic development; facilitating equitable distribution of services; integration of the functions of rural and urban settlements; and the optimum use of land for agriculture, forestry, industry, human settlements, infrastructure and other competing land uses.

The review of the 1998 Land Act reveals that one of the objectives of this Act is to ensure proper planning and well-coordinated development of urban areas. However, it protects security of land tenure rather than facilitating land development. This makes it weak in promoting planned urbanisation. Indeed, many of the urban development projects proposed by KCCA meet stiff opposition from private land owners, including Buganda Kingdom [99].

Kampala's land tenure system legally provides for both private and public ownerships of land, with private land ownership being far greater than public ownership. The Constitution of the Republic of Uganda also recognises and protects private rights in land, even in the midst of Kampala City. Since land ownership plays a significant role in urbanisation, this type of land tenure provides clues as to negative or positive effects responsible for the way Kampala is urbanising. As Jones et al. [68]. Interrelated factors such as ones mentioned above can intensify the externalities connected to urban services, heighten political aspects of service delivery and create intensified opportunities for rent-seeking of various kinds. The authors (ibid) postulate that urban populations may also be temporary, particularly in informal settlements where tenure is absent or insecure, and can hinder collective action to demand better services such as sanitation. Thus, social and economic polarity, which is more common among urban populations, can also limit collective action in terms of service provision and generally planning. The features of urban population may increase the attention of demand, increase accountability for service provision and expand the diversity of providers; but these advantages have no guarantees in anyway, therefore leading to challenges in public satisfaction.

7. Public satisfaction with urban changes in cities

The review of literature on public satisfaction with urban changes in cities shows that different scholars have shown interest in understanding what defines public satisfaction resulting from urbanisation. A careful review of these scholars' studies reveals the interest is motivated by the fact that public satisfaction plays an important role in determining the development and spatial distribution of settlements, investments, choice of employment and city dwellers' reaction to urban changes resulting from services provided by city government [7, 100, 101]. This way, this satisfaction is one of the good indicators that need to be considered when developing urban policy or legislations intended to ensure that a city urbanises in a manner that satisfies those who live in it.

Explicitly, Kährik et al. [1] observes that public satisfaction is significantly related to decisions regarding not only where to establish a residence but also neighbourhood preferences.

This satisfaction also determines private investment decisions and choices of employment [2, 6, 102]. In other words, investors establish businesses, and individuals choose where to reside and to work depending on their satisfaction with the location and its neighbourhood. Locations and neighbourhoods that induce high levels of public satisfaction are those that are well planned spatially and in terms of zoning of their socioeconomic activities; locations and neighbourhoods that do not satisfy the public are usually those associated with unplanned settlements, chaotic socioeconomic activities, noise and air pollution and poor environmental health [2–4, 103]. City dwellers react to locations and neighbourhoods associated with high levels of satisfaction by getting attracted to them in terms of residing, working or investing there. The public react to locations and neighbourhoods associated with low or no satisfaction by doing the opposite [96, 103].

Evidently, understanding public satisfaction with urban changes taking place in their locations helps urban physical planners, policy makers and implementers to improve the locations, thereby promoting urbanisation that meets people's expectations. This is why understanding such satisfaction is necessary to investigate and understand in Kampala. Literature reveals the specific indicators that can be used to establish this satisfaction in particular; Akaateba and Yakubu [104] indicate that the level of public satisfaction with services provided by city authorities reveals the quality of these services, and this enables authorities to improve where necessary. Akaateba and Yakubu are, however, focusing on public satisfaction with only provided solid waste collection services. Moreover, their study was conducted in Wa in Ghana, not in Kampala, Uganda.

Research has also shown that city dwellers' satisfaction is also measured in terms of urban people's gratification with spatial quality, zoning of commercial activities, availability of job opportunities, adequacy of provided social services, accessibility of residences as well as quality of neighbourhoods [105–108]. The study of Yizhao et al. [109] indicates that public satisfaction can also be measured in terms of these dwellers' contentment with housing supply, housing quality and available housing access options. This study was conducted in China, and its findings show that housing that is satisfactory to residents is that which is either affordable in terms of rent or self-constructed. Such housing is satisfactory to both high- and low-income residents. In contrast, housing which is expensive in terms of rent and constructed by government is less satisfactory to city residents. These findings suggest that if a city is to urbanise in a satisfactory manner, especially in terms of housing, authorities have to promote planned self-constructed residents or to encourage low-cost rentals.

The studies of Hipp [110]; Shieh et al. [111]; Bonnes et al. [112]; and Kahlmeier et al. [113] reveal that urban environments that are free from noise and air pollution are satisfactory to city residents. These studies indicate further that the public satisfaction declines as the health of their environments deteriorates. These studies suggest that proper urbanisation has to take place in a manner that ensures that satisfactory environmental health is maintained. However, none of these studies was carried out in Kampala. Therefore, their findings need to be validated in the case of Kampala City.

A number of studies have shown that urban or city dwellers' satisfaction with the spatial quality increases when proper changes result into improved orderliness of city land uses and activities [90, 91, 114]. The more organised or zoned the land uses and activities are perceived to be, the more satisfaction they yield to urban residents and vice versa [115]. These land uses and activities include commercial business activities, physical infrastructure (roads, power supply and telecommunications lines), settlements and social service provision facilities such as educational centres, health centres and administration blocks [116–118]. These observations suggest that understanding city dweller's satisfaction necessitates finding out how these inhabitants are satisfied with the land uses sanctioned to take place in their neighbourhoods. The observations, however, do not delve further to show how this understanding can be used as a basis for developing an urban policy required to ensure that inhabitants are satisfied with their neighbourhood.

8. Research methodology

Focus group discussion (FGD) interviews were used to collect qualitative data from a group of respondents simultaneously [119]. This method was used in this study to collect more qualitative data from the selected KCCA councillors. It was used because these councillors were all met and asked to provide required data after their morning plenary session. Instead of being interviewed separately, they preferred to provide their responses in a collective discussion session. The suggestion of councillors was adopted. As Hennink [120] observed, FGDs can also facilitate collection of detailed data about the variables of the study in a free, interactive and participative environment characterised by free exchange of views and comments. In fact, the held FGD facilitated a deeper understanding of the nature of factors about Kampala's urbanisation. In all, the FGD and interviews were held with respondents shown in **Table 1**.

Table 1 indicates that 19 respondents were interviewed and 5 of them participated in the held focus group discussion. Therefore, respondents from whom qualitative data was collected were 24 altogether.

All key respondents were reached at their offices after making prior appointments with each one of them. Only KCCA councillors were reached and data collected from them after their plenary session.

Purposive sampling was used to select respondents who were interviewed and those who participated in focus group discussions (FGDs). Purposive sampling is a non-probability sampling technique applied to select respondents in a biased manner justified by the fact that respondents are considered as key informants in the study [122]. In this study, respondents who qualified in this criterion included technocrats from the technical wing of KCCA, leaders from the political wing of KCCA, national service delivery agency officials serving in Kampala and central government political leaders serving Kampala.

Each of the respondent categories outlined above was considered for a reason. In particular, technocrats from the technical wing of KCCA, who included policy developers, implementers and controllers, were selected to provide data needed to answer all the research questions of the study from the administrative perspective. Leaders from KCCA's political wing were

Position	Number of respondents		
	Interviewees	FGD Participants	Total
KCCA divisional mayors	4		4
KCCA deputy divisional mayors	1		1
KCCA councillors		5	5
KCCA divisional town clerks	4		4
NEEMA monitoring official	1		1
NW&SC officials	1		1
KCCA public health and environment official	2		2
Land use officer (MLH&UD)	1		1
Inspector physical planning (MLH&UD)	1		1
Commissioner (Uganda Communications Commission)	1		1
Uganda Roads Authority official	1		1
Uganda Electrical Regulatory officials	1		1
Official from Office of the President (Kampala affairs)	1		1
Total	19	5	24

Source: [121]

Table 1. Number of interviewees and FGD participants.

selected to provide data that were required to answer the research questions from the political perspective. Those specifically selected consisted of mayors and councillors of the five divisions of Kampala City and officials in the Ministry of the Presidency who are in charge of Kampala. Officials from the national service delivery agencies who serve in Kampala City were selected to provide data required to answer the research questions from a service delivery perspective. These officials included personnel from national service delivery agencies such as National Water and Sewerage Corporation (NW&SC), MLHUD, Uganda Roads Authority, NEEMA and Uganda Electricity Regulatory Authority.

9. Research results

As shown in **Table 1**, the key informants included technocrats and political leaders of KCCA, officials from national service delivery agencies serving Kampala and central government political leaders in charge of Kampala. As a way of corroborating the findings obtained from the selected city residents, the key informants were each asked to mention the informal forces that accounted for the urban expansion of Kampala City from 1990 to 2013. Triangulated thematic and descriptive analysis of their responses led to findings shown in **Figures 1–3**.

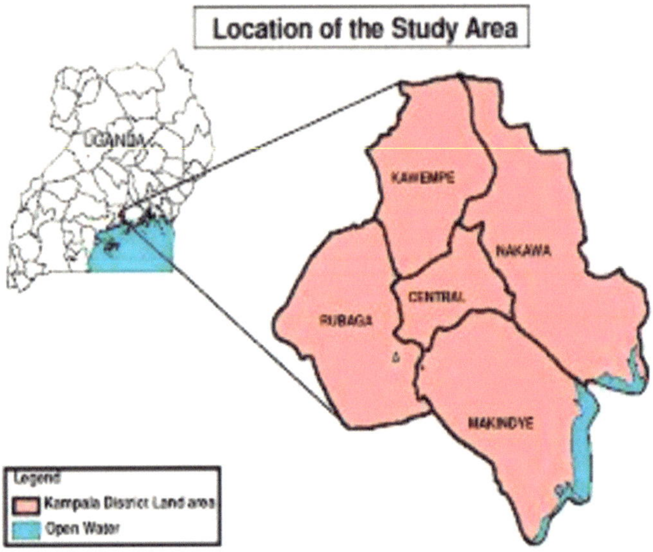

Figure 1. Location of the study. Source: [14].

Figure 2. Some of the challenges of urban sprawl in Kampala. Source: [86].

Figure 1 summarises the various forces or factors which, according to key informants, explained Kampala's urbanisation between 1990 and 2013. The percentage distribution suggests that each informant suggested more than one form of factor. A comparative analysis of proportions reveals that with the exception of natural population increase, each form of factor/forces was identified by at least 50% of these respondents. This indicates that most of the key informants revealed similar factors. Specifically, all key informants (100%) indicated Kampala's urbanisation was due to its attractiveness to jobseekers and job makers. In addition, majority of these respondents (95.8%) identified government's modernisation agenda

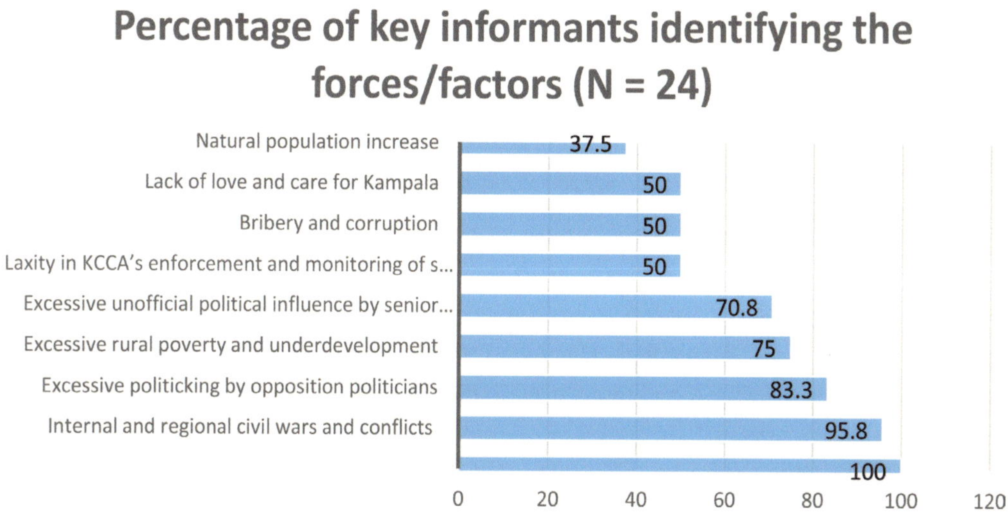

Figure 3. Frequency distribution of informal forces or factors explaining Kampala's urbanisation from 1990 to 2013, as reported by key informants. **Source:** [121].

as another major factor. Other factors specified by most of the key informants included internal and regional civil wars and conflicts (95.8%), excessive politicking by opposition politicians (83.3%), weak urban policy (75%), excessive rural poverty and underdevelopment (75%) and excessive political influence by high-ranking government officials (70.8%). Other specified factors included government tactical intervention (58.3%), private ownership of land and laxity in KCCA monitoring (50%), bribery and corruption (50%) and Lack of love and care for Kampala (50%). In essence, these findings largely substantiate many of the factors revealed by urban dwellers and reveal other factors as well. The findings further confirm Owusu [27] who identify the attractiveness of cities as a factor that accounts for their urbanisation. They also further confirm MLHUD [80] and Mabode [123] since these also specify population increase, politics and administrative forces as factors explaining cities' urbanisation.

However, while the authors cited above specified the forces without explaining how they exactly occur, this study delved further by analysing their nature. This nature was established by asking the key informants to substantiate why the forces they had mentioned were indeed responsible for Kampala's urbanisation.

10. Discussion of results

Key informants were asked about whether they were resident in Kampala or not. Twelve of them (50%) responded affirmatively, and the other twelve (50%) replied negatively. This effectively implied that those who could provide valid data were only the 12 key informants who answered positively. These were, therefore, the respondents who were further asked to indicate how their satisfaction with the urban changes officially introduced in Kampala influenced their personal decisions relating to putting up any development in the city. One of them had this to say:

> *KCCA authorised the establishing of a taxi park in the neighbourhood of where I used to stay at Nateete. I had to sell the place and relocate to Ntinda where I now reside. I could not stand the sudden noise pollution, increase in traffic flow and in the number of people who started passing through my compound to go to the taxi park.*

The finding above indicates that the respondent shifted to another location and promoted Kampala's urbanisation residentially because of dissatisfaction with the noise pollution that the new urban change KCCA had introduced in the neighbourhood of his former residential area.

Another key informant had this to say:

> *In 2009, KCCA authorised my neighbour to establish a maize mill just a few metres away from my residence. When this mill started operating, it made a lot of noise at night. I first tolerated it for some time because I did not have any immediate solution. However, my family members started complaining of failure to sleep and headache resulting from the noise that the mill produced every night. I went and talked the neighbour, but he told me that he had invested a lot of money in the mill. He however, told me that if I was going through such a bad experience, I could look for another place and sell this one to him. He also offered to partly finance establishment of my new home somewhere else. I had to accept. That is how I shifted from Kamwokya to Bweyogerere.*

The preceding findings reveal that the key informant shifted again as a result of dissatisfaction with noise pollution introduced by the maize mill that city authorities permitted in his former neighbourhood. Another key informant noted:

> *I lived on Kawempe Hill from 1987 till 2009. Around September 2008, I saw people putting up a telecommunications mast in the compound of my neighbour. On inquiring to find out why they were doing so, I found out that my neighbour had been approached by MTN and had accepted to sell to them a part of his land. In December 2009, I saw another similar mast being put up in another neighbour's garden. My enquiries made me realise that it was Uganda Telecom putting up the mast. I had tolerated the noise that the generator that powered the MTN mast would make whenever there was load-shading of electricity supplied by UMEME. This time it was going to be double doze. I could not stand it anymore. As I looked for a solution, shifting of course, I was also approached by WARID (Airtel today). I immediately accepted to sell them my land and run away from the noise. I went and bought land in Makindye where I built the residence where I stay now.*

The above narrative further confirms that city residents in Kampala are shifting because of dissatisfaction with noise pollution created by the telecommunications masts erected in their former place of abode. Another key informant said:

> *I cannot stay near a factory and this is the reason why I shifted from Namuwongo where Mukwano (U) Ltd. and House of Plastics Ltd. made extension of their soap and plastics factories, respectively. Not only was this factory causing noise. It also polluted the air.*

Clearly, noise and air pollution caused the preceding respondent to shift.

Another key informant replied:

> *In 1994, I shifted from Naalya and started renting in Naggulu because my former place did not have markets and health centres where I could easily go shopping and for health services my family and I needed. I always had to drive to supermarkets and health centres located in Kampala Central and Nakawa Division. However, the situation changed in 2005. KCCA authorized investors to establish Shoprite and other supermarkets in Naalya as well as Naalya Health Centre. I had to shift back to my own home, and even renovated it to suit the new developments in the area.*

The foregoing narrative indicates that satisfaction with access to services promoted by urban authorities accounted for the respondent's residential decisions, the latter of which contributed to how Kampala urbanised in terms of housing development.

An overview of the findings from the key informants reveals that it was mainly dissatisfaction with the poor environment health (noise and air pollution) that resulted from urban changes officially sanctioned to take place in their residential areas and the desire to access services easily that explained their urbanising decisions. The findings are therefore consistent with what [90, 91] described as organised or zoned land uses.

To recap on the findings, the study indicates that public satisfaction strongly influenced the manner in which Kampala urbanised including dissatisfaction with poor environmental health caused in terms of noise and air pollution by the factories, telecommunications masts and other activities that Kampala city authorities sanctioned. Another form of these factors was satisfaction that residents derived from easy access to needed social services. These factors caused city residents to residentially urbanise Kampala not because they wanted, but because they either wanted to have easy access to needed social services or were dissatisfied with the air and noise pollution caused by the factories, mills, telecommunications masts and taxi parks that had been authorized to operate in their former locations. Two urban policy implications can be derived from these findings.

The first implication is that it is not good to mix noise-making and/or air-polluting activities with residences. This alludes to the need to zone economic activities in Kampala in a manner that ensures that those that make noise or pollute the air are far separated from residential areas. The only way proper urbanisation can take place depends on the manner in which the urban authorities maintain satisfactory health environment as advocated in the studies of Hipp [110], Shieh et al. [111] and Bonnes et al. [112]. The second implication is that it is important to ensure that essential services are provided in a manner that brings them nearer to residential areas so that people do not have to shift in order to have easy access to them. Accordingly, a policy required to ensure that Kampala urbanises in a planned manner needs to consider these two implications.

11. Summary

The paper has argued that as cities continue to attract people for better employment, education, health care and culture, they unduly contribute to national and urban economies.

Nevertheless, often rapid urbanisation is associated with poverty, environmental degradation and population demands that outstrip service capacity. These conditions without proper planning create unpleasant urban environment leading to numerous unsatisfactory outcomes such as poor housing, overcrowding, air pollution, transportation, insufficient or contaminated drinking water, inadequate sanitation and solid waste disposal services, industrial waste, increased motor vehicle traffic, stress associated with poor implementation of urban planning programmes and unemployment, among others. Moreover, the paper discussed public satisfaction in the context of services provided in relation to Kampala's urban sprawl.

Author details

Fred Bidandi

Address all correspondence to: 2257858@myuwc.ac.za

School of Government, Robert Sobukwe Road, University of the Western Cape, Belleville, Cape Town, South Africa

References

[1] Kährik A, Leetmaa K, Tammaru T. Residential decision-making and satisfaction among new suburbanites in the Tallinn urban region, Estonia. Cities. 2012;**29**(1):49-58

[2] Hillman B. The causes and consequences of rapid urbanisation in an ethnically diverse region: Case study of a county town in Yunnan. China Perspectives. 2013;**3**:25

[3] Lixing L. The Incentive Role of Creating Cities in China. Peking University: China Centre for Economic Research; 2008a

[4] Lixing L. The Incentive Role of Creating Cities in China. Peking University: China Centre for Economic Research; 2008b

[5] Sartori D, Catalano G. Infrastructure Investment Long Term Contribution: Economic Development And Wellbeing. 2013(No. 201301)

[6] Sartori D, Gelsomina C. Infrastructure Investment Long Term Contribution: Economic Development And Wellbeing. Centre for Industrial Studies. [Online]. 2013. Available at: http://www.regionalstudies.org/uploads/Infrastructure_investment_long_term_con tribution.pdf [Accessed 18-11-2014]

[7] Aulia DN, Ismail AM. Residential satisfaction of middle income population: Medan city. Procedia-Social and Behavioral Sciences. 2013;**105**:674-683

[8] Kampala Capital City Authority. Kampala City Council Five-Year Development Plan, 2010-2015. Kampala: KCCA; 2010a

[9] Ministry of Lands, Housing, and Urban Development. Kampala, Uganda National Urban Policy. 2013a. (Draft 1)

[10] UN-Habitat. State of the World's Cities 2008-2009: Harmonious Cities. 2010. Earthscan

[11] UN-habitat. State of the World's Cities 2010/2011: Bridging the Urban Divide. 2008. Earthscan

[12] De Blij HJ, Muller PO. Realms of Regions and Concepts. New York: John Wiley and Sons; 2000

[13] Nyakaana JB. The Urbanisation of Kampala city and its Effect on the Natural Environment. Kampala: Makerere University; 2013

[14] Lwasa S, Nyakaana JB, Senyendo H. Population, Urban Development and the Environment in Uganda: The Case of Kampala City and its Environs 2007

[15] Cohen B. Urban growth in developing countries: A review of current trends and a caution regarding existing forecasts. World Development. 2004;**32**(1):23-51

[16] Bocquier, P. World Urbanization Prospects: An Alternative to the UN Model of Projection Compatible with Urban Transition Theory. [Online]. 2004. Available at: http://www.eldis.org/vfile/upload/1/document/0708/DOC20089.pdf [Accessed 18-11-2014]

[17] Hugo G, Champion A. New Forms of Urbanisation. Aldershot: Ashgate; 2003

[18] Njoh AJ. Urbanization and development in Sub-Saharan Africa. Cities. 2003;**20**(3):167-174

[19] Fox S. Urbanisation as a global historical process: Theory and evidence from sub-Saharan Africa. Population and Development Review. 2012;**38**(2):285-310

[20] Wyly E. Theories of Urban System Development. 2012. Introduction to Urban Geography

[21] Puig SY. General theory of urbanisation. The Town Planning Review, Liverpool University Press. 1995;**66**(1):15-39

[22] Miller HT. The Administrative State: A Study of Political Theory of American Public Administration: Dwight Waldo with New Introduction. New Jersey: Transaction Publishers; 2007

[23] Alliance C. Guide to City Development Strategies: Improving Urban Performance. Washington DC: Cities Alliance, UNEP and ICLEI; 2006

[24] Orum A. Urbanisation. Encyclopaedia of Social Theory. Sage Publications. [Online]. Available at: http://www.sagepub.com/oswmedia3e/study/chapters/handbooks/handbook15.2.pdf; 2011 [Accessed 14-09-2014]

[25] Beires L. A Study of Urbanisation in the Province of Kwa Zulu Natal. [Online]. 2010. Available at:http://www.kznded.gov.za/Portals/0/OUTLINE%20OF%20URBANISATION%20DEDT.pdf [Accessed 14-09-2014]

[26] Orum AM, Xiangming C. The world of cities. In: Malden, MA, & Oxford. UK: Blackwell; 2003

[27] Owusu F. Urban Planning: The Way Forward for Africa's Cities? [Online]. 2011. Available at: http://thinkafricapress.com/population-matters/impact-demographic-change-resources-and-urban-planning-africa [Accessed 14-09-2014]

[28] Kasibante J. Kampala was Neglected: The New Vision. [Online]. 2011. Available at: http://www.newvision.co.ug (Accessed 7 October, 2011)

[29] Harvey D. The Urbanization of Capital: Studies in the History and Theory of Capitalist Urbanization. Oxford: Blackwell; 1985a

[30] Harvey D. Consciousness and the Urban Experience. Oxford: Blackwell; 1985b

[31] Christophers B. Revisiting the urbanization of capital. Annals of the Association of American Geographers. 2011;**101**(6):1347-1364

[32] Harvey D. Their Crisis, Our Challenge. Interview with Marco Berlinguer and Hilary Wainwright. [Online]. 2008. Available at: http://www.redpepper.org.uk/Their-crisis-our-challenge [Accessed 19-11-2014]

[33] Harvey D. Rebel Cities: From the Right to the City to the Urban Revolution. London: Virso. [Online]. Available at: http://abahlali.org/files/Harvey_Rebel_cities.pdf; 2012 [Accessed 18-11-2014]

[34] Purcell M. Excavating Lefebvre: The right to the city and its urban politics of the inhabitant. GeoJournal. 2003;**58**:99-108. [Online]. Available at http://faculty.washington.edu/mpurcell/geojournal.pdf [Accessed 18-11-2014]

[35] Sheri, B. Modernization in historical perspective: The case of imperial Germany. World Politics. 2001;**53**(3):431-462

[36] Milkovich GT. Organization and management in the midst of societal transformation: The People's Republic of China. Organization Science. 2004;**15**(2):133-144

[37] Tettey C. Urbanisation in Africa in Relation Social-Economic Development: A Multi-faceted Quantitative Analysis. University of Akron; 2005 (Unpublished PhD thesis)

[38] Sano JC. Urban Infrastructure in Kigali city, Rwanda: Challenges and Opportunities for Modernized Decentralized Sanitation System in Poor Neighbourhoods. Wageningen University; 2007 (unpublished MSc thesis)

[39] Usman, B Confronting the Haphazard Urbanization Challenges in Anambra State of Nigeria through Urban Structure Planning/Re-Planning And Implementation. 2011. Settlements Discussion Paper Series No.4

[40] Goldstone JA. The new population bomb: The four megatrends that will change the world. Foreign Affairs. 2010;**89**:31-39

[41] Chengdan Q. Constructing a new disciplinary framework of modern world history around the theme of modernization. Chinese Studies in History. 2009;**42**(3):7-24

[42] Harding R. Urban-Rural Differences in Support for Incumbents Across Africa. 2010. Afrobarometer Working Paper No. 120

[43] Fox SR. The Political Economy of Urbanisation and Development in sub-Saharan Africa. Department of International Development of the London School of Economics (PhD thesis). framework (PhD thesis). Cape Town: University of the Western Cape; 2013

[44] Herbst J. Population change, urbanisation, and political consolidation. In: Goodwin RE, Tilly CC, editors. The Oxford Handbook of Contextual Political Analysis. Oxford: Oxford University Press; 2009. pp. 649-663

[45] Levy F, Murnane R. The new division of labour. Massachusetts Benchmarks. 2004;7(1)

[46] Kimeria C. Why Rapid Urbanisation in Africa Can be Good News, After All. Nairobi: Dalberg Global Development Advisers; 2011

[47] Wucherpfennig J. Modernization and Democracy: Theories and Evidence Revisited. [Online]. 2009. Available at: http://democracy.livingreviews.org/index.php/lrd/article/viewarticle/lrd-2009-4/13 [Accessed 30-09-2014]

[48] Mungai C. East Africa in Rapid Urbanisation. The East African. [Online]. 2010. Available at: http://www.theeastafrican.co.ke. (Accessed 06-12-2010)

[49] Ssempijja D. Lack of Regulation Dogs in the Housing Sector. [Online]. 2010. Available at: http://www.newvision.co.ug. (Accessed 06-10-2010)

[50] MFPED (Ministry of Finance, Planning and Economic Development). Background to the Budget. Kampala: Ministry of Finance, Planning and Economic Development; 2012

[51] Beinestein H. Modernisation Theory and the Sociological Study of Development. [Online]. 2009. Available at: http://www.artsrn.ualberta.ca/courses/PoliticalScience/670A1/documents/HenryBernstein-ModernizationSociogyofDev.pdf [Accessed 30-09-2014]

[52] Joshi S.Theories of Development: Modernisation Versus Dependency. [Online]. 2005. Available at: http://infochangeindia.org/defining-development/theories-of-development-modernisation-vs-dependency.html [Accessed 30-09-2014]

[53] Kendal D. Sociology in Our Times. Portland: Wadsworth Publications Co.; 2007

[54] Tacoli C. Rural-Urban Linkages and Pro-Poor Agricultural Growth: An Overview. [Online]. 2004. Available at: http://www.oecd.org/development/povertyreduction/36562896.pdf [Accessed 23-09-2014]

[55] Mungai C. Migration to Urban Areas Holds Key to Kenya's Middle Income Status, Nairobi, The East African. [Online]. 2011 Available at: http://www.theeastafrican.co.ke. (Accessed 05-06-2011)

[56] Andorka R. The socialist system and its collapse in Hungary: An interpretation in terms of modernisation theory. Journal of International Sociology. 1993;8(3):317-337

[57] Sergey G. Modernisation of the Empire. Social and Cultural Aspects of Modernisation Process in Russia. Thousand Oaks: Sage Publications; 2004

[58] Nkusu M, Sayek S.Local Financial Development and the Aid-Growth Relationship. 2004. IMF Working Paper 04/238. [Online]. Available at: http://www.imf.org/external/pubs/ft/wp/2004/wp04238.pdf [Accessed 14-09-2014]

[59] Kasekende LA, Atingi-Ego M. Economic Reform Policies in Uganda. Oxford: University of Oxford, Centre for Study of African Economies (CSAE); 2003

[60] Kiiza L. The Relationship Between Aid And Economic Growth: What aid Strategy Should The Government of Uganda promote? [Online]. 2008. Available at: http://www.grips.ac.jp/forum/DCDA/Chapter10.pdf [Accessed 30-09-2014]

[61] World Bank. Uganda Public Expenditure Review 2007. Washington, D.C.: World Bank; 2007

[62] Pearce DW, Atkinson G, Mourato S. Organisation for Economic Co-operation and Development. Paris: Organisation for Economic Co-operation and Development; 2006. Cost-benefit analysis and the environment: recent developments

[63] Rajan RG, Subramanian A. Aid and Growth: What Does The Cross Country Evidence Really Show? IMF Working Paper 05/127. IMF: Washington D.C; 2005

[64] Polidoro M, De Lollo JA, Barros MVF. Urban Sprawl and the Challenges for Urban Planning. Journal of Environmental Protection. 2012;**3**(09):1010

[65] Hasse JE, Lathrop RG. Land resource impact indicators of urban sprawl. Applied geography. 2003;**23**(2):159-175

[66] Kaggwa R, Hogan R, Gowa E. Land Use Change, and Degradation and Human Welfare: Lessons Learned from the Lake Kyoga Catchment Area. Policy Brief. Nairobi: UNDP-UNEP Poverty-Environment Initiative; 2004

[67] Uganda Bureau of Statistics. Statistical Abstract. 2011. Kampala

[68] Jones H, Cummings C, Nixon H. Services in the City: Governance and Political Economy in Urban Service Delivery. 2014

[69] Hepworth N, Goulden M. Climate Change in Uganda: Understanding the Implications and Appraising the Response. Edinburgh: LTS International; 2008

[70] Rwakakamba MT. Examining the Nexus Between Depletion of Water Catchments Systems and the Deepening Water Crisis in Uganda. Kampala: Uganda Advocates Coalition for Development and Environment [ACODE]; 2008

[71] Rwakakamba MT. How effective are Uganda's environmental policies? A case study of water resources in 4 districts, with recommendations on how to do better. Mountain Research and Development. 2009;**29**(2):121-127

[72] Somik L. Planning for Uganda's Urbanisation. Inclusive Growth Policy Note 4. [Online]. 2012. Available at: http://www.urbanknowledge.org/ur/docs/Uganda_Policy%20Note.pdf [Accessed 17-11-2014]

[73] Lwasa S. Managing African urbanisation in the context of environmental change. Interdisciplina. 2014;**2**(2):263-280

[74] Kashaka U. KCCA Overstretched. [Online]. 2014. Available at: http://www.newvision.co.ug. (Accessed 1st June 2014)

[75] National Environment Management Authority (NEMA). Wetlands, River Banks, and Lakeshore Management Regulations. Kampala: NEMA; 2012

[76] Watuwa T. Behind Kampala's Sanitation Woes. The New Vision. [Online]. 2013 Available at: http://www.newvision.co.ug (Accessed 30 May, 2013)

[77] Ogwang J. Slums Present a Fierce Challenge to Uganda's Urbanisation Effort. [Online]. 2013. Available at: http://www.newvision.co.ug/news/640621-slums-present-a-fierce-challenge-to-uganda-s-urbanisation-effort.html [Accessed 12-09-2014]

[78] National Water and Sewerage Corporation. Annual Report, 2011-2012. 2011. Kampala: NWSC

[79] Lwasa S. Adapting Urban Areas in Africa to Climate Change: The Case of Kampala. [Online]. 2010. Available at: http://web.mit.edu/jcarmin/Public/For%20Nina/Lwasa-Kampala.pdf [Accessed 12-09-2014]

[80] Ministry of Lands, Housing and Urban Development. Issues Paper for the Development of the National Urban Policy. Kampala: CME Consult Group Ltd.; 2013b

[81] Mazige J. World Bank Tips Policy Makers on Urban Violence: The Monitor Uganda. [Online]. 2011 Available at: http://www.monitor.co.ug (Accessed 22 April, 2011)

[82] Kampala Capital City Authority. Kampala City Council Five-Year Development Plan, 2010-2015. Kampala: KCCA; 2010b

[83] Kasozi J. Kampala City Hall standoff: Crisis or Opportunity? [Online]. 2013 Available at: http://www.theeastafrican.co.ke (Accessed 07-09-2013)

[84] Nyende M. *Public expenditure tracking infrastructure in Uganda: Kampala*. [Online]. 2010. Available at: http://www.eprc.or.ug/pdf_files/research_report3.pdf [Accessed 7 November 2014]

[85] Kato J. Local government failing on urbanisation. [Online]. 2009. Available at: http://www.newvision.co.ug. [Accessed 21 January 2015]

[86] Elliott L. Scenes of City Life: Urbanisation on Our Cities. Slum Dwellers International: Kampala; 2011

[87] Omolo-Okalebo F. Planning of Kampala city 1903-1962: The planning ideas, values, and their physical expression. Journal of Planning History. 2010;**9**(3):151-169. [Online]. Available at: http://jph.sagepub.com/content/9/3/151.refs [Accessed 14-09-2014]

[88] Byaruhanga A, Ssozi L. The Impact of Population Growth on the Ecosystems and Biodiversity of Kampala: Lessons for Sustainable Urban Development. Sustainable Futures: Architecture and Urbanism in the Global South Kampala, Uganda, June. [Online]. 2012. Available at: http://sfc2012.org/byaruhanga_ssozi.pdf [Accessed 16-09-2014]

[89] Thomson A. Independent Evaluation of Uganda's Poverty Eradication Action Plan (PEAP) 1997-2007. Volume 1 of Synthesis Report. [Online]. 2008. Available at: http://opm.go.ug/manage/pdfs/1304016366.pdf [Accessed 17-09-2014]

[90] United Nations Development Assistance Framework for Uganda. United Nations Development Assistance Framework, 2010 to 2014. [Online]. 2009. Available at: http://countryoffice.unfpa.org/uganda/drive/UNDAF.pdf [Accessed 14-09-2014]

[91] Gollin D, Rogerson R. Agriculture, Roads, and Economic Development in Uganda. 2010. NBER Working Paper No. 15863

[92] Amayo J. Presentation of MRI Pilot Project-Kampala Water Lake Victoria Water and Sanitation Project–(KW-LVWATSAN Project). [Online]. 2013. Available at: http://www.eib.org/attachments/general/events/210113_mri_johnson_en.pdf [Accessed 26-10-2014]

[93] World Bank. Uganda - data & statistics. [Online]. 2013. Available at: http://www.worldbank.org/en/country/uganda [Accessed 14 September 2014]

[94] Statistics U. B. O. Statistical Abstract. Kampala: Uganda Bureau of Statistics; 2012

[95] Rossi-Hansberg E, Wright M. Urban structure and growth. Review of Economic Studies. 2007a;74(2):597-624

[96] Rossi-Hansberg E, Wright M. Urban structure and growth. Review of Economic Studies. 2007b;74(2):597-624

[97] Storeygard A. Farther on Down the Road: Transport Costs, Trade and Urban Growth in Sub-Saharan Africa. Department of Economics Recruiting Seminar Series. Providence, RI: Brown University; 2012

[98] Ministry of Lands, Housing and Urban Development. National Physical Planning Standards and Guidelines, 2011. [Online]. 2011a. Available at: http://www.mlhud.go.ug/documents/National%20Physical%20Planning%20Standards%20and%20Guidelines,%202011.pdf [Accessed 25 October, 2014]

[99] Rugadya M. Land Reform: The Ugandan Experience. [Online]. 1999. Available at: http://www.mokoro.co.uk/files/13/file/lria/land_reform_ugandan_experience.pdf [Accessed on October 2013]

[100] Mossin BB. Internal migration, migration and development Buzz, Rethinking the Migration Development Nexus and Policies, Geneva, International Organisation for Migration. International Migration Journal. 2012;**50**(3):1-97

[101] Wakabi M. New model for Uganda's erratic towns. The East African. [Online] Available at: http://www.theesatafrican.co.ke/business/-/2560/604856. 2009 (Accessed 01-06-2009)

[102] Yasuhiro S, Zenou Y.How Urbanization Affect Employment and Social Interactions. [Online]. 2013. Available at: http://people.su.se/~yvze0888/Sato_Zenou_6.pdf [Accessed 18-11-2014]

[103] You-Tien H. The Great Urban Transformation: Politics of Land and Property in China. Oxford: Oxford University Press; 2010

[104] Akaateba MA, Yakubu I. Householders' Satisfaction towards solid waste collection services of Zoomlion Ghana ltd in Wa, Ghana. European Scientific Journal, ESJ. 2013;**9**(32)

[105] Gidlöf-Gunnarsson A, Ohrström E. Noise and well-being in urban residential environments: The potential role of perceived availability to nearby green areas. Landscape and Urban Planning. 2007;**83**:115-126

[106] Patterson PK, Chapman NJ. Urban form and older residents' service use, walking, driving, quality of life, and neighbourhood satisfaction. American Journal Health Promotion. 2004;**19**(1):45-52

[107] Pisman A. Urban and Suburban Lifestyles and Residential Preferences in a Highly Urbanised Society: Experiences from a Case Study in Ghent (Flanders, Belgium). [Online]. 2011. Available at: http://belgeo.revues.org/6394 [Accessed 17-09-2014]

[108] van den Berg AE, Hartig T, Staats H. Preference for nature in urbanised societies: Stress, restoration, and the pursuit of sustainability. Journal of Social Issues. 2007;**63**(1):79-96. [Online]. Available at: http://agnesvandenberg.nl/jsi.pdf [Accessed 17-09-2014]

[109] Yizhao Y, Wenzhong Z, Zhilin L, Yao L. Does Market-based Housing Offer Higher Housing Satisfaction to Urban Residents than Other Housing Access in China? Evidence from the 2005 Beijing Livable City Evaluation Survey. [Online]. 2013. Available at: lincolninst.edu/pubs/2269_Does-Market-Based-Housing-Offer-Higher-Housing-Satisfaction-to-Urban-Residents-than-Other-Housing-Access-in-China- [Accessed 4th October 2014]

[110] Hipp J. What is the 'Neighborhood' in Neighborhood Satisfaction? Comparing the Effects of Structural Characteristics Measured at the Micro-Neighborhood and Tract Levels. [Online]. 2014. Available at: http://usj.sagepub.com/content/47/12/2517 [Accessed 4th October 2014]

[111] Shieh E, Sharifi A, Rafieian M. Identification of Factors that Assure Quality of Residential Environments, Using Environmental Assessment Indices: A comparative study of Two of Tehran's neighborhoods (Zafaranieh &Khaniabad). [Online]. 2011. Available at: http://ijaup.iust.ac.ir/files/site1/user_files_23s792/admin-A-11-1-31-90f9ca9.pdf [Accessed 4th October 2014]

[112] Bonnes M, Uzzell D, Carrus G, Kelay T.Inhabitants' and Experts' Assessments of Environmental quality for urban sustainability. Running Head: Environmental Quality for Urban Sustainability. [Online]. 2001. Available at: http://epubs.surrey.ac.uk/29734/2/Bonnes-Uzzell-JSI-0701-4.pdf [Accessed 4th October 2014]

[113] Kahlmeier S, Schindler C, Grize L, Braun-Fahrlander C.Perceived Environmental Housing Quality and Wellbeing of Movers. [Online]. 2001. Available at: http://www.ncbi.nlm.nih.gov/pmc/articles/PMC1731791/ [Accessed 4th October 2014]

[114] Alem Y. Life Satisfaction in Urban Ethiopia. Environment for Development Discussion Paper Series. [Online]. 2014. Available at: http://www.rff.org/RFF/Documents/EfD-DP-14-04.pdf [Accessed 4th October 2014]

[115] Fujita M, Henderson JV, Yoshitsugu K, Tomoya M. Spatial distribution of economic activities in China and Japan. In: Henderson JV, Thisse JF, editors. Handbook of Regional and Urban Economics Amsterdam. North-Holland Ltd; 2004

[116] Akten M, Çetinkaya DM.The Analysis of User Satisfaction in Urban Renewal Projects: Kayser İyildirim Beyazit District. [Online]. 2014. Available at: http://www.ijlass.org/data/frontImages/gallery/Vol._2_No._7/15.pdf [Accessed 4th October 2014]

[117] James P, Holden M, Lewin M, Neilson N, Oakley C, Truter A, Wilmoth D. Managing Metropolises by Negotiating Urban Growth. [Online]. 2013. Available at: http://www.academia.edu/7207756/Managing_Metropolises_by_Negotiating_Mega-Urban_Growth [Accessed 15-09-2014]

[118] Kemp RL, Stephani CJ. Cities Going Green: A Handbook of Best Practices. London: McFarland and Co., Inc.; 2011

[119] Morgan DL. Focus Group as a Qualitative Research. London: Sage Publications; 2006

[120] Hennink MM. Focus Group Discussions. Oxford University Press; 2013

[121] Bidandi F. The Dynamics of Urbanisation in Kampala. Uganda: Towards a possible alternative urban policy; 2015

[122] Creswell J. Educational Research: Planning, Conducting, and Evaluating Quantitative and Qualitative Research. 4thed. ed. Upper Saddle River, NJ: Pearson Education; 2012

[123] Mabode S. Energising urban South Africa: Poverty, sustainability and future cities. A paper presented at a conference at the urbanisation and rural-urban migration in sub-Saharan Africa held at Nairobi on 26th–27th November [Online]. 2012. Available at: http://migratingoutofpoverty.dfid.gov.uk/research/migrationandurbanisation/urbanisation_conference/abstracts [Accessed 29-11-2014]

Chapter 6

Scaling in Urban Complex Systems: Mexico City Metabolism

Aida Huerta-Barrientos

Additional information is available at the end of the chapter

http://dx.doi.org/10.5772/intechopen.71660

Abstract

Large cities are usually wealthier, denser in terms of population, more expensive, more congested but also more productive culturally and technologically. Mexico City is one of the most dynamic cities of the global economy but also presents the highest crime and congestion levels on the road network. The socio-metabolic approach interprets cities as a socio-metabolic system that interacts with systems in the natural environment. Although considerable progress has been made in studying cities as complex adaptive systems using such approach, many important issues such as social and innovation dimensions remain unexplored, mainly in Mexico City context. The principal purpose of this study is to analyze the metabolic scaling of socio-cultural and technological aspects in the context of Mexico City in order to predict the energy necessary to maintain the city socially connected and to estimate the impact of such social connections on the socio-economic-environmental indicators. We take into account the total population, the cultural infrastructure, the social cohesion, the traffic congestion level, the cost of fuel car, the minimum income, and the number of patents in the analysis as the agglomeration effects in Mexico City. We consider this study can support the design of public policies using the metabolic approach.

Keywords: urban complex systems, urban economy, urban population, scale

1. Introduction

Large cities are usually wealthier, denser in terms of population, more expensive, more congested but also more productive culturally and technologically. Mexico City is one of the most dynamic cities of the global economy. It is Latin America's financial center and Mexico's political, economic, and cultural capital. Mexico City also provides the most important access to markets and clients. One of the main reasons to invest in Mexico City is the capital's enormous potential to be an active and leading competitor in the new global market. According to [1], Mexico City recorded 4.1 million workers during 2016, mainly in commerce and professional, financial, and corporate services. Following [1], in respect to the national level, the state represented 8.5% of the workers in commerce and 18.0% in professional, financial, and corporate services; the gross domestic product (GDP) of Mexico City was 2.9 trillion pesos in 2015, contributing with 16.7% of national GDP; tertiary sector, which includes trade and real estate services, contributed with 89% of the state's GDP in 2015. Mexico City has 281 km of railways. It also has an international airport. Out of 115,695 students with completed studies from engineering, manufacturing, and construction in Mexico, 13,796 are from Mexico City. Mexico recorded 27,186 researchers at January 2017; 31.6% of them were in Mexico City. Most of the researchers in the state focused on the areas of humanities, as well as medicine and health sciences [1].

In contrast, Mexico City presents the highest congestion level on the road network at global level, causing more than 90% extra travel time for citizens during busy hours [2]. The traffic congestion affects directly on the quality of life. But citizens prefer to use private transportation instead of the public transport network because it offers a poor coverage and a lack of modal transfer centers. Additionally, the general crime rate in Mexico City is above the U.S. national average, and crime varies widely. Armed robberies, kidnappings, car thefts, credit card fraud, and various forms of residential/street crime are daily concerns. Thefts of the vehicle's operating computer and sound systems are also common crimes.

Due to their inherent complexity, Mexico City problematic cannot be studied using traditional approaches of science based on the Physics reductionism perspective because this approach breaks down the problems in their parts, analyzes them taking into account only the most relevant variables and then extrapolates the results to understand their overall behavior. Certainly in the past, in very specific contexts, the Physics reductionist approach was very successful and allowed for great achievements and the scientific and technological advance that we know today. However, the reductionist approach is no longer enough to face the challenges of the 21st century. The decision-making of Mexico City government every day face unpredictable situations that change over time, formed by complex systems of problems. Therefore, it is necessary to use modern science tools based on complex systems paradigm whose methodological framework is complexity science. The complex systems perspective considers open subsystems interacting with their environment, non-reversible energy, and information processes, dynamic states of entities and subsystems far from equilibrium as well as teleological analysis, in the case of human and social subsystems [3]. This approach indicates that phenomena are not the result of a cause-effect chain in which the result can be proportional and predictable only with the knowledge of the parts, but rather, that the systems behave in a non-linear way [4]. Thus, the complex systems approach supports the development of our skills needed to deal with problematic and complex situations so that we are able to change them holistically. In the application of the complex systems approach, the most important feature of study is the emergent properties of complex systems [5] due to a process of self-organization. Such emerging properties cannot be determined by analytical methods or models [6], but rather by using simulation, an experimental tool essential in modern science.

Although it is true that a large number of definitions of a complex system have been proposed from different fields of science such as Biology, Physics, Sociology, and Economics, there is no precise formal definition accepted by the scientific community. Broadly speaking, such proposals co-exist in defining a complex system as a system composed by interrelated elements that generate information and whose operations are based on simple rules. From the interrelations of its elements, new properties or collective behaviors are generated at different scales. To the above, we add that a complex system is increasingly complex as human actions are explicitly considered.

On the other hand, it is observed that the majority of urban science has not treated cities as complex systems [7]; however, methods and models from the complexity science are likely to yield insight into urban structure and dynamics. From the complexity perspective, cities are complex adaptive systems characterized by heterogeneity, interconnectivity, scaling, circular—causality, and development as follows [8]:

- Heterogeneity: diversity of people and organizations.
- Interconnectivity: everything is connected in networks.
- Scaling: cities of different sizes have different problems.
- Circular—causality: cause and effect are mixed.
- Development: cities change in open-ended ways.

As discussed by [9], cities are the first large social networks. It means cities are not just large collections of people instead they are agglomerations of social links. Following [9], space, time, and infrastructure play a fundamental role in enabling social interactions in allowing them to become open-ended in terms of increased connectivity, and sustainable from the point of view of energy use, such interactions imply cost of people, material, energy, and information flows through decentralized networks of infrastructure that are built gradually as the city grows. In this direction, cities, to realize the full socioeconomic potential, need to expand connectivity per person and of social inclusions [9]. Rapport [10] states that the bio-physical approach of studying and quantifying urban material and energy flows, which draws on approaches in the field of industrial ecology, is the predominant interpretation of urban metabolism that has strong connections to the field of industrial ecology which look for ways to optimize the metabolism of industrial systems through industrial symbiosis. Following [10], studies of urban material and energy flow are often used to identify ways to improve the environmental and economic performance of an urban area. The socio-metabolic approach interprets the society as a socio-metabolic system that interacts with systems in the natural environment. Although considerable progress has been made in studying cities as complex adaptive systems using the metabolic approach, many important issues such as social and innovation dimensions remain unexplored, mainly in Mexico City context. The principal purpose of this study is to analyze the metabolic scaling of socio-cultural and technological aspects in the context of Mexico City in order to predict the energy necessary to maintain the city socially connected and to estimate the impact of such social connections on the socio-economic-environmental indicators.

The book chapter is divided into five main sections. In Section 2, the socio-metabolic approach applied to the study of urban complex systems is reviewed. In Section 3, the analysis of the metabolic scaling of cultural and technological issues in the context of Mexico City is presented. The energy necessary to maintain the Mexico City connected and its impact on the socio-economic-environmental indicators is evaluated in Section 4. The concluding remarks are drawn in Section 5.

2. The socio-metabolic approach

Human society has been maintained culturally and biophysically. In the first case by a flow of self-referential communication, organized in subsystems of society each with its own codes, while in the second case by a continuous flow of energy and materials from/to natural environment (social metabolism), and by deliberate interventions into the environment [11]. Geddes [12] explained that from the biology perspective, a city can be conceptualized as an organism, opposed to a mechanical system, made of complex and intricate processes that keep it alive, in a regional balance where the relation between the city and the surrounding agricultural economy is symbiotic [13].

The social metabolism is a concept from biology and transferred to the world of relations between society and nature [14]. The concept of social metabolism has not only proven a useful metaphor to stress the biophysical foundation of social systems and their economy, but it has emerged as a key analytical concept in sustainability science [15]. Associated with the concept of social metabolism, we find the notion of socio-metabolic transition to describe fundamental changes in socio-economic energy and material use that occur in the course of human history; in some cases, such transitions have implied the multiplication of metabolic rates, for instance, the transition from agrarian toward industrial society, and in other cases, such transitions have implied the emancipation of the energy systems from land use, for example, the transition from a solar energy system tapping into renewable flows of biomass toward a fossil fuel powered energy systems based on the exploitation of large stocks of energy resources [15]. One interesting study about the potential of the social metabolism approach to study the industrialization of the agriculture is presented in [14]. It provides information about how the socio-ecological transition took place in agriculture. In this context, Bringezu et al. [16] make the point that in industrial ecology and ecological economics interdisciplinary fields, sophisticated methods and tools have been developed to study material and energy flows in socio-economic systems in order to contribute to the design and implementation of more sustainable types of industrial metabolism.

The concept of urban metabolism was first introduced by Abel Wolman in the middle sixties to compare and organism and a city. In his seminal study, Wolman used national data on water, food, and fuel use, along with production rates of sewage, waste, and air pollutants to determine per capita inflow and outflow rates for a hypothetical American city. Actually, urban metabolism is fundamental to developing sustainable cities and communities. In this direction, cities, like organisms, need energy and resources as inputs to sustain life, which are processes and ultimately released to the environment as wastes.

Urban metabolism may be defined as the sum total of the technical and socio-economic processes that occur in cities, resulting in growth, production of energy, and elimination of waste [17]. In [18], an extended concept or urban metabolism is presented where the metabolic inflows and outflows are its central dimension, but urban quality the biophysical processes determining the environmental impacts associated with environmental sources and sinks as well as urban drivers, patterns, and lifestyles are conceptually interlinked in a holistic approach while economic and social inflows and outflows such as information, cultural goods, and employment are also covered. In this context, urban metabolism framework is the process of material exchanges between the city and its natural environment, whose use involves focusing on material flows within the city as a potential way to uncover environmental, social, economic, and political aspects of resource distribution in a holistic way [19].

According to Hyman [20], material flows are the key intervention point to achieve urban sustainability. Castán Broto et al. [21] note that there is considerable optimism about the opportunities that urban metabolic analysis opens for more sustainable and more just cities.

3. Analysis of the metabolic scaling of cultural and technological issues

In this section, we present the analysis of the metabolic scaling of socio-cultural and technological aspects in the context of Mexico City. We use per capita indicators, which on one hand, conflate general effects of urbanization [22], and on the other hand, are ubiquitous in official statistics and policy documents compiled by Mexican governmental agencies and international bodies worldwide. Mexico City's society has been maintained culturally and biophysically along many hundreds of years. In the first case by a flow of self-referential social communication (information) organized in subsystems of society each with its own codes; while, in the second case by a continuous flow of energy and matters from/to natural environment, and by deliberate interventions into the environment to uncover environmental, social, economic, and political aspects of resource distribution (see **Figure 1**).

According to the INEGI [23], based on the surveys conducted from 1900 to 2010 as well as on the Intercensal Survey in 2015, **Table 1** and **Figure 2** show the population growth in Mexico City from 1900 to 2015. It is important to note that the exponential population growth started in 1950 and continues in 1970.

Mexico City has a 1000-year-old cultural richness in which various expressions give it its own cultural identity. In this direction, the Mexican city is an intercultural mosaic of ethnic varieties in which indigenous communities, indigenous peoples, and communities from different origin interact all together every day. These groups and their worldview through different cultural practices conform the identity of citizens [24]. The cultural heritage of Mexico City

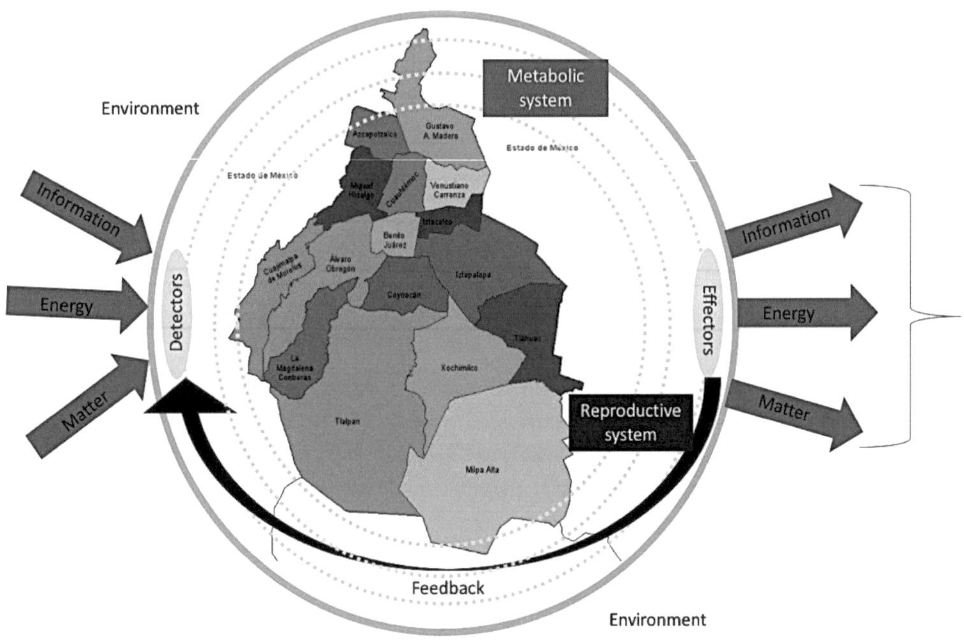

Figure 1. Conceptual model of Mexico City as a complex adaptive system with metabolic and reproductive subsystems.

Millions of inhabitants	Year
0.7	1900
1.2	1930
3.1	1950
6.9	1970
8.2	1990
8.6	2000
8.8	2010
8.9	2015

Table 1. Total population in Mexico City from 1900 to 2015.

is reflected in the diversity of archeological, historical and artistic monuments, expressions of popular culture, gastronomy, festivities, traditions, and normative systems that coexist in the same geographical space [24]. According to the Atlas of Mexico's cultural infrastructure 2010 [25], in this year, there were 310 archeological sites in this city, 5 of them open to the public, and an estimated universe of 7000 historical monuments from the 16th to the 19th centuries, and 11,071 which were characterized by the National Institute of Fine Arts as artistic monuments of the 20th and 21st centuries. As for the natural heritage, in Mexico City there are geographically located 23 natural areas protected by local and national laws and 33 gullies declared as areas of environmental value [24]. **Table 2** shows the cultural infrastructure per capita in Mexico City in 2010.

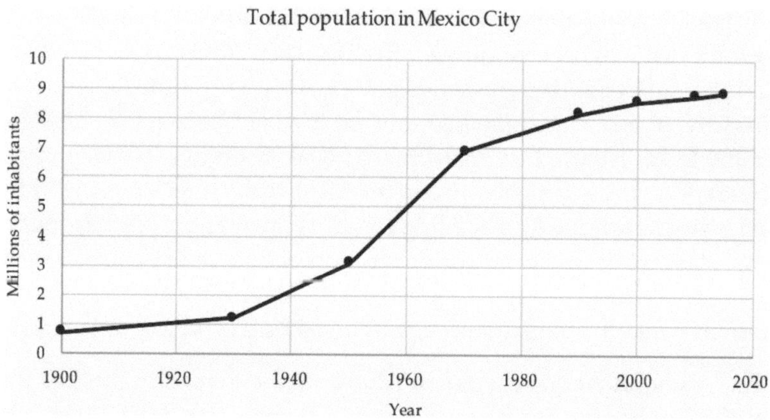

Figure 2. Total population in Mexico City between 1900 and 2015.

Cultural infrastructure	Per capita in 2010
Archeological sites	0.0000352
Historical monuments from the 16th to the 19th centuries	0.0008
Historical monuments from the 20th to the 21st centuries	0.0013

Table 2. Cultural infrastructure per capita in Mexico City, 2010.

On the other hand, the measurement of social cohesion estimated by the CONEVAL [26] incorporates indicators that help to know the level of economic and social inequality of the population at national, state, and municipal levels, as well as indicators of support networks and social exchange at the state level. This allows us to approach the level of equity and solidarity that exists in a society. To measure the degree of social cohesion, CONEVAL uses Gini coefficient that measures the economic inequality of a society, by exploring the level of concentration that exists in the distribution of income among the population. The Gini coefficient takes values between 0 and 1; a value that tends to 1 reflects greater inequality in the distribution of income. On the contrary, if the value tends to zero, there are greater conditions of equity in the distribution of income. **Table 3** shows the Mexico City social cohesion values estimated by CONEVAL for three years: 1990, 2000, and 2010. We observe that the social cohesion in Mexico City has decreased from the last 27 years, with the consequences associated.

Agglomeration effects in Mexico City are manifested, on one hand, as an increase in overall travel times when compared to a free flow situation, and on the other hand, as the increase

Social cohesion	Year
0.536	1990
0.505	2000
0.439	2010

Table 3. Social cohesion in Mexico City [26].

in the cost of public transportation. For the first case, **Figure 3** shows the evolution of the congestion level in Mexico City considering the TomTom Traffic Index [27]. For example, a congestion level of 57% in 2010 corresponds to 57% extra travel time for any trip, anywhere in the Mexico City, at any time compared to what it would be in a free flow situation. The TomTom Traffic Index is based on speed measurements from TomTom's historical traffic database. These speed measurements are used to calculate the travel times on individual road segments and entire networks. By weighting based on the number of measurements, busier, and more important roads in the network have more influence than quieter, less important roads. **Figure 4** shows the evolution of the congestion level versus the population growth in Mexico City. In this case, the congestion level suddenly rises as well as population along time.

Analyzing the cost of public transportation relative to the minimum income, we can see in **Figure 5** that the first has suddenly risen from 2005. In consequence, citizens have not been able to stay socially connected, due to the cost of transportation, leading to expected decreases

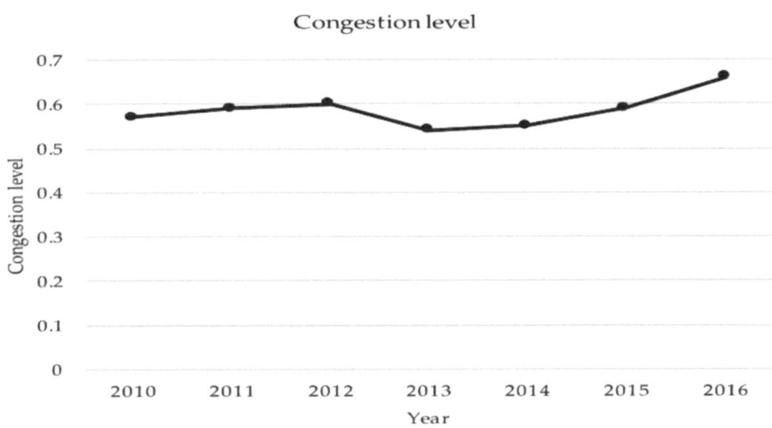

Figure 3. Congestion level evolution in Mexico City.

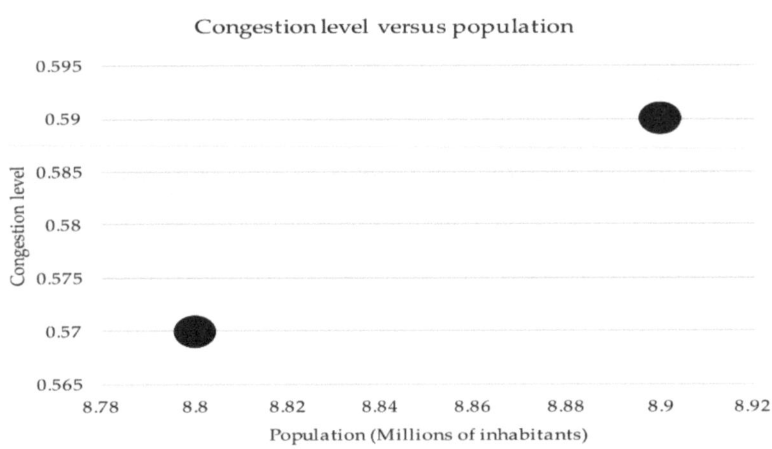

Figure 4. Congestion level versus population in Mexico City.

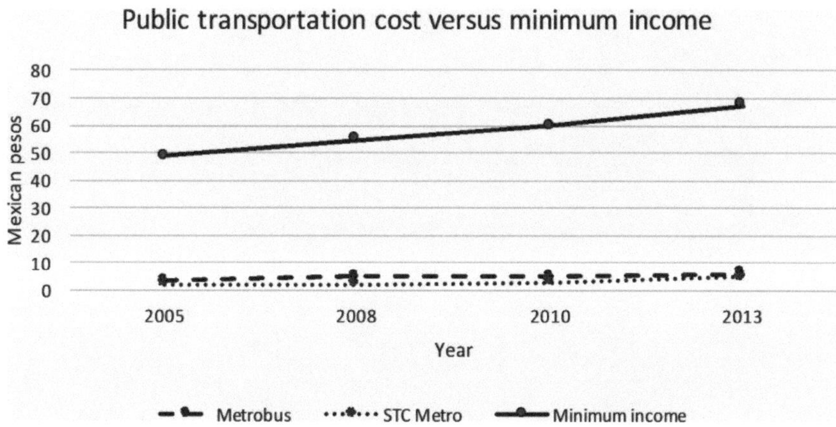

Figure 5. Evolution of public transportation cost versus minimum income in Mexico City.

in their socioeconomic production as suggested by [9]. Additionally, in 2013, the cost of public transport represented almost the 9% of the minimum income in the case of Metrobus. For the same year, the cost of Metro represented almost the 8% (see **Figure 6**).

It is important to note that citizens in Mexico City prefer to use private vehicles for mobility purposes instead of the public transport networks, such as Metro and Metrobus, because these networks offers a poor service coverage, a lack of modal transfer centers and a high vulnerability in terms of connectivity and accessibility [2]. The big problem using private vehicles is the cost of fuel associate (see **Figure 7**). In general, the sheer distance to everyday destinations means some families spend 25% of their income on transport so Mexico City's greenhouses gas (GHG) emissions by up to 70% and cost USD 2.5 billion (33 billion Mexican pesos) each year in lost economic productivity [28].

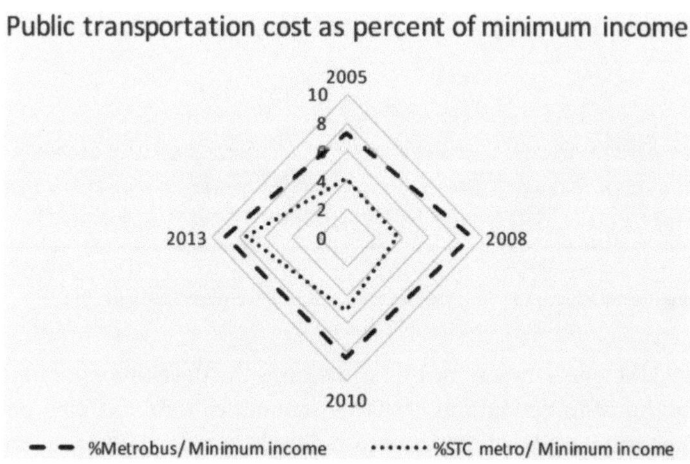

Figure 6. Public transportation cost as percent of minimum income in Mexico City.

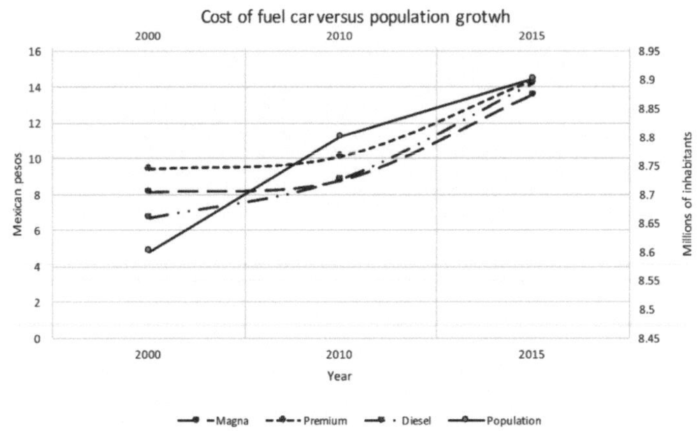

Figure 7. Evolution of cost fuel car versus population growth in Mexico City.

Patents	Year
166	2000
215	2001
206	2002
167	2003
179	2004
212	2005
181	2006
219	2007
219	2008
233	2009
321	2010
308	2011
427	2012
390	2013
337	2014
367	2015

Table 4. Patents requested by Mexico City as residence of the inventor between 2000 and 2014.

As suggested by [28], one solution for this situation is the development of mixed-use areas, were we can find housing, restaurants, services, schools, cultural facilities, parks, and more, so under this scenario if connectivity in Mexico City is increased, the need for private vehicles will be reduced, thus increasing the viability of public transport, walking, and bicycling.

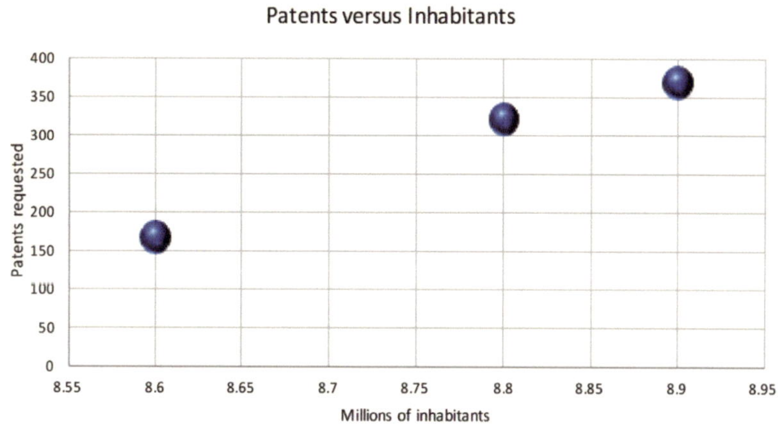

Figure 8. Patents requested versus population growth in Mexico City.

Following [28], other benefits of mixed-use areas are the shared community spaces that foster interactions among community members: interactions that would not be possible under a car-centric model, and the saving individuals money on transportation by reducing the length and number of every day trips and eliminating the need for car ownership. In short, mixed-use areas support local business.

Other agglomeration effect in Mexico City is manifested as increase in per capita temporal rates of socioeconomic activities, such as innovation [22]. One dimension of technological aspects in Mexico City is the number of patents [29]. **Table 4** shows the evolution in the number of patents requested by Mexico City as residence of the inventor from 2000 to 2014. **Figure 8** shows the relationship between the patents requested by Mexico City as residence of the inventor and the population growth for three different years, 2000, 2010, and 2015. We observe that as increase in in population size corresponds to an increase in the number of patents requested.

4. The energy necessary to maintain the Mexico City connected

In this section, we present the estimation of energy necessary to maintain the city connected as well as the impact of social connections on the socio-economic-environmental indicators. We use the data from UN-HABITAT Global Urban Observatory [30]. For the period 1999–2004, transport was the highest consumer of energy in Mexico City followed by industry and residential and commercial buildings (see **Figure 9**).

Residential buildings include all energy used for activities by households, except for transport. Industry includes a combination of all industrial sub-sectors, such as mining and quarrying, iron and steel, and construction. Energy used for transport by industry is not included, but is reported under transportation. Transport includes all fuels used in road vehicles and non-road transport. In the first case, military, as well as agricultural and

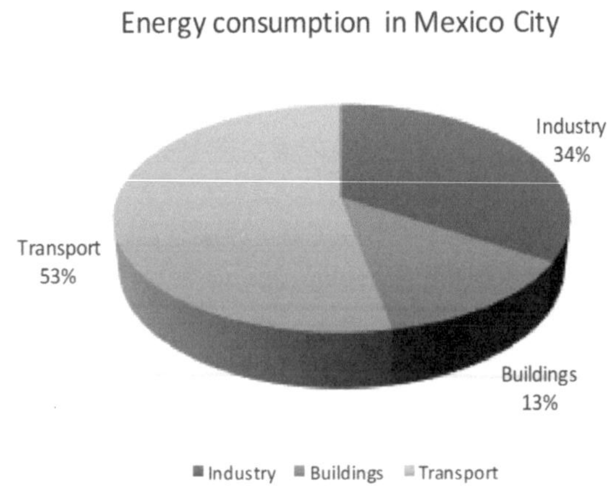

Figure 9. Energy consumption in Mexico City, data from various sources 1999–2004 [30].

industrial highway use are included. The second case includes transport in the industry sector and covers railways, air, internal navigation, fuel used for transport of minerals by pipeline and non-specified transport. In 2008, the annual per capita energy consumption tendency in Mexico City was major than 1900 kilowatt-hours (KWh) and included transport, public lighting, and water pumping (see **Figure 10**). Demand for water has doubled every 20 years, at a rate exceeding that of population growth suggesting that losses, increased connections to the water supply network and increased per capita consumption all play a part [32]. According to Valdez [33], to meet demand for water, Mexico City has to withdraw water from ever more distant sources, often over 100 km away. This situation imposes a considerable energy cost on the system: at 4.5 kWh/m^3, water transported from a distance uses almost 20 times the amount of energy as the withdrawn from aquifers below the city [32].

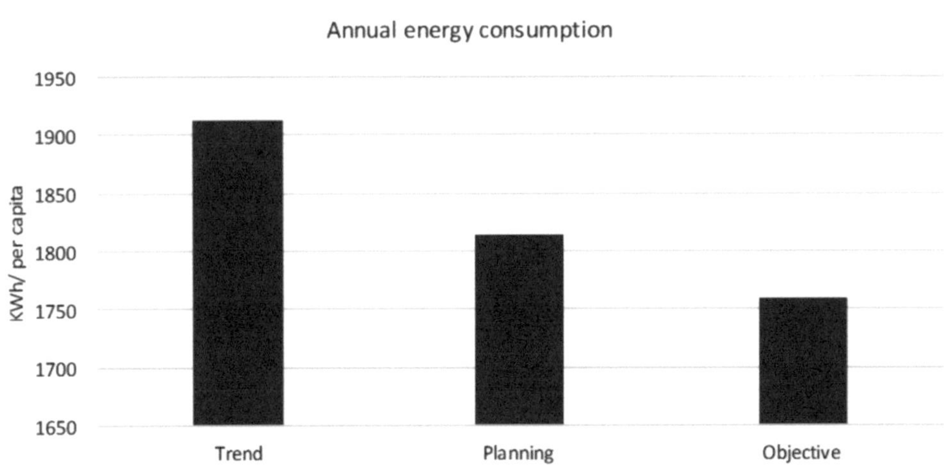

Figure 10. Energy consumption per capita in Mexico City, transport, public lighting, and water pumping [31].

5. Concluding remarks

The principal purpose of this study was to analyze the metabolic scaling of socio-cultural and technological aspects in the context of Mexico City in order to predict the energy necessary to maintain the city socially connected and to estimate the impact of such social connections on the socio-economic-environmental indicators. Agglomeration effects in Mexico City are manifested, as an increase in population, on overall travel times when compared to a free flow situation, in the cost of public transportation, in the number of patents, in the demand for water, and as a decrease in social cohesion. We observed that the exponential population growth in Mexico City started in 1950 and continues in 1970. Based on the social cohesion values estimated by CONEVAL for three years: 1990, 2000, and 2010, we observed that the social cohesion in Mexico City decreased from the last 27 years, with the consequences associated. In general, the sheer distance to everyday destinations means some families spend 25% of their income on transport. Demand for water has doubled every 20 years, at a rate exceeding that of population growth. All these socio-economic outputs of Mexico City are proportional to the number of social interactions realized per unit time. Mexico City to realize the full socioeconomic potential need to expand connectivity per person and of social inclusions but the big problem is that such connectivity is supported in many cases by transport causing high congestion levels. We consider that an initiative to expand the connectivity in Mexico City need to be based in the use of the Information and Communication Technologies (ICT's) and updating the use areas from non-mixed-use to the mixed-use approach in order to favor the social interactions.

Acknowledgements

The author appreciates the partial support by National Council for Science and Technology of Mexico (CONACyT) (SNI number 65477).

Author details

Aida Huerta-Barrientos[1,2*]

*Address all correspondence to: aida.huerta@comunidad.unam.mx

1 Faculty of Engineering, National Autonomous University of Mexico, Mexico City, Mexico

2 The Complexity Sciences Center, National Autonomous University of Mexico, Mexico City, Mexico

References

[1] PROMEXICO. 2016. Available from: http://mim.promexico.gob.mx/work/models/mim/Documentos/PDF/mim/FE_CDMX_vfi.pdf [Accessed: August 01, 2017]

[2] De la Mota I, Huerta-Barrientos A. Simulation-optimization of the Mexico City public transportation network: A complex network analysis framework. In: Mujica Mota M, Flores De La Mota I, editors. Applied Simulation and Optimization 2. 1st ed. Switzerland: Springer; 2017. p. 43-80. DOI: 10.1007/978-3-319-55810-3_2

[3] Lara-Rosano F. Teoría, métodos y modelos de la complejidad social I. In: Seminario de Investigación; CCADET; 2012

[4] Caballero L. La Búsqueda del Comienzo: El Pensamiento Complejo en Biología. Ciudad de México: CopIt ArXives; 2008

[5] Boccara N. Modeling Complex Systems. 1st ed. New York: Springer-Verlag; 2004

[6] Byrne D. Complexity Theory and the Social Sciences. An Introduction. 1st ed. London: Routledge; 1998

[7] Padgham M. Are Cities Complex? 2017. Available from: http://huckg.is/gisruk2017/GISRUK_2017_paper_79.pdf [Accessed: September 01, 2017]

[8] Bettencourt L. Cities as Complex Objects. USA; Santa Fe Institute; 2012

[9] Betterncourt LMA. The kind of problem a city is. SFI Working Paper: 2013-03-008; 2013

[10] Rapoport E. Interdisciplinary perspectives on urban metabolism. A review of the literature. UCL Environmental Institute Working Paper; 2011

[11] Fisher-Kowalski M. Social and ecological systems interacting: Historical and future transitions. In: Proceedings of the HENVI Science Days, 2011; Helsinki, Finland; 2011

[12] Geddes P, Ferreira J, Jha S. The Outlook Tower: Essays on Urbanization in Memory of Patrick Geddes. Popular Prakashan: Bombay; 1976

[13] Geddes P. Cities in Evolution: An Introduction to the Town Planning Movement and to the Study of Civics. London: Williams & Norgate; 1915. Available from: https://archive.org/stream/citiesinevolutio00gedduoft#page/n9/mode/2up Accessed: August 15, 2017

[14] Gonzalez de Molina M. A guide to studying the socio-ecological transition in European agriculture. DT-SEHA No. 10-06; 2010

[15] Krausman F. The socio-metabolic transition. Long term historical trends and patterns in global material and energy use. Social Ecology Working Paper No. 131; Institute of Social Ecology; 2011

[16] Bringezu S, van der Sand I, Schütz H, Bleischwitz R, Moll S. Analysing global resource use of national and regional economies across various levels. In: Bringezu S, Bleischwitz R, editors. Sustainable Resource Management. Global Trends, Visions and Policies. Sheffield: Greenleaf; 2009. p. 10-52

[17] Kennedy C, et al. The study of urban metabolism and its applications to urban planning and design. Environmental Pollution; 2010. DOI: 10.1016/j.envpol.2010.10.022

[18] Minx JC, Creutzig F, Medinger V, Ziegler T, Owen A, Baiocchi G. Developing a Prgamatic approach to assess urban metabolism in Europe. A Report to the Environment Agency

Prepared by Technische Universität Berlin and Stockholm Environment Institute; Climatecon Working Paper No. 01/2011; Technische Universität Berlin; 2011

[19] Guibrunet L, Castán-Broto V. The sustainability of the informal city: An urban metabolism approach. In: Proceedings of the 11th International Conference of the European Society for Ecological Economics (ESEE); 30 June–3 July, 2015; Leeds, UK; 2015. Available from: http://www.esee2015.org/wp-content/uploads/2015/10/0220.pdf

[20] Hyman K. Urban infrastructure and natural resource flows: Evidence from Cape Town. The Science of the Total Environment. 2013;**461-462**:839-845

[21] Castán Broto V, Allen A, Rapoport E. Interdisciplinary perspectives on urban metabolism. Journal of Industrial Ecology. 2012;**16**(6):851-861

[22] Bettencourt LMA, Lobo J, Strumsky D, West GB. Urban scaling and its deviations: revealing the structure of wealth, innovation and crime across cities. PLoS One. 2010;**5**(11):e13541

[23] INEGI. 2017. Available from: http://cuentame.inegi.org.mx/monografias/informacion/df/poblacion/dinamica.aspx?tema=me [Accessed: September 01, 2017]

[24] Secretaría de Cultura de la Ciudad de México. Programa de Fomento y Desarrollo Cultural 2014-2018. México; 2014

[25] Consejo Nacional para la Cultura y las Artes (Conaculta). Atlas de Infraestructura y Patrimonio Cultural de México 2010. In: Sistema de Información Cultural (SIC). 1st ed. Ciudad de México: Conaculta; 2010

[26] CONEVAL. 2017. Available from: http://www.coneval.org.mx/ [Accessed: September 03, 2017]

[27] TOMTOM. 2017. https://www.tomtom.com/en_gb/trafficindex/city/MEX [Accessed: September 10, 2017]

[28] WRI. 2017. Available from: http://www.wri.org/blog/2014/07/people-oriented-cities-mixed-use-development-creates-social-and-economic-benefits [Accessed: September 01, 2017]

[29] INEGI. 2017. http://www3.inegi.org.mx/sistemas/sisept/default.aspx?t=etec11&s=est&c=19173 [Accessed: September 01, 2017]

[30] UN-HABITAT. State of the World's Cities 2008/2009. Harmonious Cities. Earthscan, London, UK; 2008

[31] Barnés F. Taking action on energy efficient cities: The Mexico case. In: International Conference on Energy Efficiency in Cities; Centro Mario Molina; 2016

[32] International Energy Agency - IEA. Mexico Energy Outlook. OECD/IEA, Paris, France; 2016

[33] Valdez M, et al. The water-energy-carbon nexus: Optimising rainwater harvesting in Mexico City. Environmental Processes. 2016;**3**:307-323

Section 2

New Urban Trends

Chapter 7

Promotion of Smart Community Strategy in Vietnam's Binh Duong Province

Tetsuro Saisho

Additional information is available at the end of the chapter

http://dx.doi.org/10.5772/intechopen.73102

Abstract

Vietnam's Binh Duong Province is located in southeastern Vietnam, immediately to the north of Ho Chi Minh City (HCMC). The province constitutes the southern economic zone of Vietnam, and many Japanese companies—mainly manufacturing enterprises, including small and medium enterprises—are investing there. Becamex IDC Corp (a Vietnamese state-owned developer) and Tokyu Corporation (a Japanese company) established Becamex Tokyu, a joint venture company, in March 2012. As a smart community strategy, Becamex Tokyu is promoting Tokyu Binh Duong Garden City, which is a development plan integrating new urban development. The Becamex Tokyu Bus, which is a wholly owned subsidiary of Becamex Tokyu, operates the new transportation system KAZE Shuttle. In addition, Binh Duong Province has developed a smart community strategy centered on energy establishment in urban development, information and communications technology infrastructure development and utilization, the elimination of physical waste in infrastructure construction, and smart traffic. In this chapter, by using his field work as a basis, the author examines the current status of, and issues faced by, development of the smart community strategy in Binh Duong Province in Vietnam.

Keywords: digital transformation society, new city, regional development, smart community, Vietnam

1. Introduction

A smart community is a social system that introduces information and communication technology (ICT) such as the promotion of energy conservation, the diversification of energy sources, the reduction of CO_2 (carbon dioxide) emissions and the use of electric vehicles, and combines advanced and comprehensive cutting-edge technologies.

A smart community actively utilizes new technologies such as renewable energy and energy management systems (EMSs) to achieve more efficient use of energy. A smart community implements new urban development, investigates urban environmental problems, introduces local transportation systems, optimizes residents' lifestyles, and conducts regional development in the form of housing, buildings, factories, and new urban areas [10].

In recent years, the emerging country of the Socialist Republic of Vietnam has concentrated its efforts on regional and urban development, based on the concept of smart communities, in various parts of the country [6].

There have been a very few studies of smart communities (including Smart Cities) in Asia (including Vietnam), although this is a valuable research area. This type of development is also a reference case for other efforts such as regional development and urban development in other Asian countries.

Therefore, in this chapter[1], the author considers the current status of new urban development in Vietnam's Binh Duong Province from the perspective of smart communities.

2. Outline of Binh Duong Province and urban development

2.1. Outline of Binh Duong Province

Vietnam's Binh Duong Province is located 17 km north of Ho Chi Minh City (HCMC) in southeastern Vietnam and consists of one city, four towns, and four prefectures (**Figure 1**) [2].

The provincial capital of Binh Duong Province is the city of Thu Dau Mot, with an area of 2694 km^2. In 2015, the population of the province was about 2 million (of which 60% work) and the gross domestic product growth rate is 14.5% [2].

As of December 31, 2016, in Binh Duong Province, there were a total of 3050 approvals for foreign direct investment (FDI)[2] (composition ratio 13.5%), with a cumulative total investment of 26,600 million USD (composition ratio 9.1%) [12, 13].

The current (2017) monthly minimum wage[3] of workers in Binh Duong Province is 7.15 million VND (about 17,300 yen) in the same area as HCMC and Hanoi. This figure was 3.5 million VND (about 16,100 yen) in the previous year and has thus increased by about 50% in a year. Binh Duong Province, along with Dong Nai Province (mainly HCMC) and Ba Ria-Vung Tau Province, constitutes the Southern Focal Economic Zone of Vietnam.

In Binh Duong province, a number of Japanese companies (large enterprises and small and medium enterprises) are entering the market, mainly in the manufacturing industry (**Table 1**) In March 2012, Tokyu Corporation (a Japanese company) and the Investment and Industrial

[1]This chapter has been drawn up and modified from a large part of the author's paper, A Study of the Smart Community in Vietnam: Case of Binh Duong Province in the Southern Regional City; "Information Management · 73rd National Conferences Proceedings" [Autumn Edition] [11].

[2]Foreign Direct Investment (FDI) means the acquisition of permanent interests (predominantly through management) in foreign companies.

[3]The minimum wage system in Vietnam was adopted in 1997. A minimum wage was formulated for each region (1–4); since October 2011, the wages of domestic companies and foreign-funded enterprises have been unified [4].

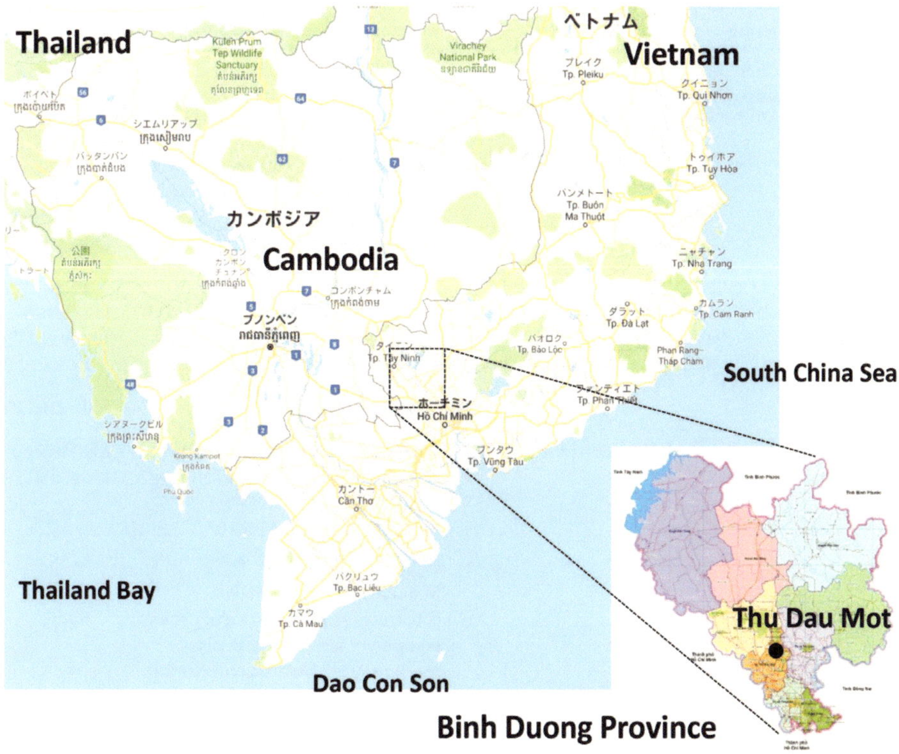

Figure 1. Location of Binh Duong Province. Source. Created from the HP material of Binh Duong Province.

Development Corporation (abbreviated Becamex IDC Corp)[4] (a Vietnamese state-owned developer) established Becamex Tokyu Co., Ltd.

2.2. Binh Duong Province and urban development

With a capital of 8600 billion VND (about 32.7 billion yen), Becamex Tokyu is promoting a unified development plan for new city development and new region development[5] in Binh Duong Province. The core facility for the development is Tokyu Binh Duong Garden City in Binh Duong New City.[6]

The development plan for Binh Duong New City includes a huge project to set up SORA gardens, which include low-, middle-, and high-rise condominiums and will cover about 71 ha of a 110-ha site owned by Becamex Tokyu. In addition, SORA gardens will have a commercial building with a Hikari shopping precinct along with residential buildings and business and commercial facilities.

[4]Established in 1976, Becamex IDC Corp, as a developer, is investing in, developing and managing 100% Vietnamese government capital in industrial parks, residential areas, urban areas, and development of traffic infrastructures [17].

[5]Tokyu Group utilizes the railway line development know-how that the company has accumulated over many years to conduct overseas town development projects such as the New City in Binh Duong Province, Vietnam, a rental housing business for Sriracha Japanese in Thailand, and residential land development in the Yanchep District of Western Australia [16].

[6]Binh Duong New City is located in the Binh Dun Industrial, Urban, and Service Multiple District and is scheduled to be upgraded to a central government shipping city in 2020.

	Company name	Industry	Business contents	Industrial park
1	Sakata Inx Vietnam Co., Ltd. Head office	Other products	Production of printing ink	Vietnam Singapore Industrial Park 2
2	Asti Electronics Corporation	Electronic products	Two-wheel wire harness, four-wheel wire harness, electric control board for home appliances	Dong An St., Tan Dong Hiep Ward, Di An Town, Binh Duong Province
3	Vietnam LSI Cooler Co., Ltd.	Electric machine	Production of heat sink	Vietnam Singapore Industrial Park 2
4	Astee Horie VN Co., Ltd.	Service industry	Plastics painting/vapor deposition/printing/ASSY	Dong An Industrial Park
5	Iwai Plant Tech Vietnam Co., Ltd.	Machinery products	Designing, manufacturing, selling and repairing food manufacturing machines, pharmaceutical manufacturing machines and chemical manufacturing machines Manufacture, procurement, construction, sale, service of general equipment mainly comprising machinery for food manufacturing, pharmaceutical manufacturing machinery, and chemical manufacturing machinery	My Phuoc 3 Industrial Park
6	Inter Art Saigon Co., Ltd.	Construction industry	Interior design, construction	Viet Hung 1 Industrial Park
7	Uchihashi Vietnam Co., Ltd.	Electronic products	Manufacture of electronic equipment	Vietnam Singapore Industrial Park 1
8	MHI Engine System Vietnam Co., Ltd.	Machinery products	Manufacture and sale of general purpose machines and special vehicles	Vietnam Singapore Industrial Park 1
9	Ohmori Vietnam Co., Ltd.	Electric appliances	Parts processing for semiconductor devices, part machining for machine tools	Vietnam Singapore Industrial Park 2
10	Oda Vietnam Co., Ltd.	Other products	Production and sales of ready-made curtains, product exports by Vietnamese fabric	Dong An Industrial Park
11	Kubota Vietnam Co., Ltd.	Machinery products	Manufacture and sale of tractor/combine and its implementation	My Phuoc 3 Industrial Park
12	Saigon Stec Co., Ltd.	Packaging	COF assembling/inspection for camera module/liquid crystal driver	Vietnam Singapore Industrial Park 2
13	Sakai Chemical (Vietnam) Co., Ltd.	Plastic products	Production of scientific materials for plastic products	My Phuoc 3 Industrial Park
14	Taisei Bijutsu Printing Vietnam Co., Ltd.	Pulp and paper	Print services to Japanese companies operating in Vietnam	My Phuoc 3 Industrial Park

http://dx.doi.org/10.5772/intechopen.73102

	Company name	Industry	Business contents	Industrial park
15	Daido Vietnam Co., Ltd.	Clothing industry	Button production	145/46 Tan Hoa Blk, Dong Hoa Ward, Di An Township, Binh Duong Province
16	GF Vietnam Co., Ltd.	Clothing production	Clothing production	Song Than 2 Industrial Park
17	Hayabusa (Vietnam) Co., Ltd.	Other products	Manufacture of fishing tackle	My Phuoc 2 Industrial Park
18	Vina Kraft Paper Co., Ltd.	Pulp and paper	Manufacture and sale of cardboard	My Phuoc 3 Industrial Park
19	V-Eikou Co., Ltd.	Electronic products	Manufacture of parts and finished products such as mobile phone parts, desktop holder, PDA charger, and so on.	My Phuoc 2 Industrial Park
20	Fujiya Manufacturing Vietnam Co., Ltd.	Metal products	Manufacture of pliers, nippers	My Phuoc 3 Industrial Park
21	Vietnam Success Co., Ltd.	Machine parts production	Manufacture of mechanical parts	Dong An Industrial Park
22	Honda Logicom (Vietnam) Co., LTD.	Warehouse/transport related business	Warehouse and commodity storage service, transportation agency service	Vietnam Singapore Industrial Park 2
23	Yamaken Apparel Vietnam Ltd.	Clothing production	Clothing production	Dong An Industrial Park
24	Yamaken Concrete Vietnam Co., Ltd.	Other products	Production of concrete	Song Than 3 Industrial Park
25	Makita Vietnam Co., Ltd.	Electric appliances	Sales of electric tools, air tools, engine tools and parts and after-sales service	Vietnam Singapore Industrial Park 1
26	Maruichi Sun Steel Corp	Steel	Manufacture and sale of steel pipes	DT743Rd., Dong Tac Quarter, Tan Dong Hiep Ward, Di An County, Binh Duong Province
27	NSK Precision Co., Ltd.	Precision mechanical equipment	Manufacture of rotating equipment for dental use	Vietnam Singapore Industrial Park 2
28	Yamaichi Special Steel Vietnam Co., Ltd.	Metal products	Sales, processing, heat treatment of special steel	Dong An 2 Industrial Park
29	Eon Mall Vietnam Co., Ltd.	Service industry	Retail, sales	No. 01 Dai Lo, Khu Pho Binh Giao, Thuan Giao Ward, Thuan An District, Binh Duong Province

	Company name	Industry	Business contents	Industrial park
30	Becamex Tokyu Co., Ltd.	Urban development	Real estate development/real estate sales/real estate leasing	Lot C18, Hung Vuong Boulevard, Hoa Phu ward, Thu Dau Mot city, Binh Duong Province

Table 1. Major Japanese companies expanding into Binh Duong Province. Source. Created from materials such as [18–48].

The total cost of this project (the whole development project) is about 100 billion yen. Development is proceeding in three steps, namely SORA gardens I (Gate City, which covers 15 ha), SORA gardens II (Core City, covering about 15 ha), and SORA gardens III (Garden City, of about 40 ha). In March 2015, SORA gardens I, which is a high-rise apartment building and is the first project, was completed, and in January 2015, a commercial *Hikari* containing food courts and family marts began operating.

SORA gardens I has a total land area of 9082 m^2, a total floor area of 31,287 m^2, 24 floors, 406 houses, and a total of about 1500 apartments. The permanent facilities and facilities of SORA gardens include lounges, gyms, rooftop pools, rooftop gardens, parking lots, and motorbike storage. The final development will have business parks, financial centers, international conference halls, commercial facilities, and universities on the premises. By about 2020, the complex will house 120,000 people and, including the surrounding industrial park, there will be about 400,000 people as part of the plan to develop a new city.

In December 2014, Becamex Tokyu Bus Co., Ltd., a wholly owned subsidiary of Becamex Tokyu, began operating the KAZE shuttle as a route connecting the provincial capital city of Thu Dau Mot and Binh Duong New City. In the KAZE shuttle, we have brought Japanese bus operation know-how to Vietnam. As part of this novel transportation system, we provide thoroughly safe operation, scheduled operations based on timetables posted at bus stops, polite customer service, clean bus interiors, comprehensive safety and accident prevention, a user-friendly riding method[7], and detailed user guidance.

On March 5, 2016, Becamex Tokyu Bus opened five bus routes connecting the major parts of the region and the surrounding areas from the center of Binh Duong New City. Along with this expansion of bus lines, we have newly established a bus terminal in front of *Hikari*, a commercial facility in Binh Duong New City. With the bus terminal as a transport hub, we have connected major sites in the provincial capital Thu Dau Moy and Binh Duong New City via a bus route to improve the convenience of travelers in the area.[8]

The bus business in Binh Duong Province is a new form of public transport that is easy for local residents to use and is an indispensable function of advanced town planning. The bus route has become indispensible as a means of travel for the administrative staff of Binh Duong

[7]In Binh Duong New City, (1) check the time of the bus in the timetable; (2) when the bus arrives, enter via the front door (the rear door is exclusively for alighting); (3) after getting on the bus, pay the fare at the fare box; and (4) when you reach your stop, press the nearby buzzer to inform the driver.

[8]Becamex Tokyu Bus aims to create new value in Binh Duong New City.

Figure 2. Facilities of Binh Duong Province. Source. Created from the HP material of Binh Duong Province.

new government center and administrative center, visiting officials at the administrative center, and commuting local residents.

2.3. Urban development in Binh Duong New City and the surrounding area

In the vicinity of Binh Duong New City, Korea's Lotte Vietnam Shopping Center (Lotte Mart) opened its fifth Vietnam shop on November 23, 2013 (**Figure 2**) [2]. Lotte Mart plans to increase the number of its domestic stores in Vietnam to 60 by 2020.

On November 1, 2014, AEON Vietnam Co., Ltd., a Vietnamese corporation of the Japanese company AEON Group, implemented its dominant store strategy[9] in Binh Duong Province. Following the opening of AEON Mall Tan Phu Celadon on January 11, 2014, the corporation opened a suburban-type AEON Mall Binh Duong Canary, its second Vietnamese store [7].

AEON Mall Binh Duong Canary has shopping facilities, amusement facilities, and a Japan Zone area, providing a style of shopping mall that encourages customers to stay and emphasizes Japanese quality and enhanced services.

In addition, Vietnam Singapore Industrial Park 1 (VSIP 1) was founded in 1996 in the vicinity of the Binh Duong new city. VSIP 1 is an industrial cluster of the largest Japanese companies in southern Vietnam; it also includes small and medium Japanese enterprises.

[9] "Dominant store" opening is the strategy of intensively opening stores in areas where retailers are located.

3. Binh Duong Province and smart community

3.1. What is a smart community?

The Japan Smart Community Alliance (JSCA)[10] defines a smart community as "a community where various next-generation technologies and advanced social systems are effectively integrated and utilized, including the efficient use of energy, utilization of heat and unused energy sources, improvement of local transportation systems, and transformation of the everyday lives of citizens"[5].

The Agency for Natural Resources and Energy[11] and the Ministry of Public Management, Home Affairs, Posts and Telecommunications define the smart community as "an energy-efficient system that utilizes IT technologies such as energy management systems and storage batteries, as well as electricity and heat and traffic" [14].

In Vietnam's Binh Duong Province, "a community that realizes integrated management and optimal control of all infrastructure, including electricity, water, transportation/logistics, medical care, information, etc.," is regarded as "a community of the next generation" [18].

At the moment, the definition of a smart community has a variety of interpretations by industry, academia, and government, and an academic definition has not been established.

Promoting energy conservation by improving the efficiency of energy use[12], diversifying energy sources by introducing renewable energy, reducing CO_2 emissions, and introducing electric vehicles) by using an energy management system (EMS), which is a social system that combines energy in a comprehensive and comprehensive manner throughout the region.

EMS[13] utilizes ICT. This type of management system includes, for example, visualization of electricity consumption, equipment control for energy saving and CO_2 emissions reduction, and control of renewable energy and electric storage (**Figure 3**) [9]. EMS encompasses fields such as HEMS (housing), BEMS (building), FEMS (factories), and CEMS (communities), depending on the objects to be managed (**Table 2**) [1].

In the residential field, as part of the home energy management system (HEMS), household electrical products are connected to a home area network (HAN) to manage energy in the home. The HEMS may include the concepts of the net zero energy house (ZEH) and mansion energy management system (MEMS), depending on the home management target.

In the building field, a building energy management system (BEMS) includes responses to the monitoring of electricity demand. The energy management system of the entire building

[10]The Smart Community Alliance (JSCA), established in April 2010, is the secretariat of the New Energy and Industrial Technology Development Organization (NEDO).

[11]The Agency for Natural Resources and Energy, an external agency of the Ministry of Economy, Trade, and Industry established in 1973, is an administrative agency of the national government as stipulated by the National Administrative Organization Act and the Ministry of Economy, Trade and Industry Establishment Law.

[12]Energy is classified into electrical energy, atomic nuclear energy, thermal energy, light energy, stationary energy, sound energy, dark energy, and so on according to its forms of movement and preservation.

[13]The author considers the energy management system (EMS) in his 2017 paper, A Study of the Smart Community in Vietnam: A Case of Smart City Community Strategy in Da Nang City, "Kokushikan Business Review" [8, 11].

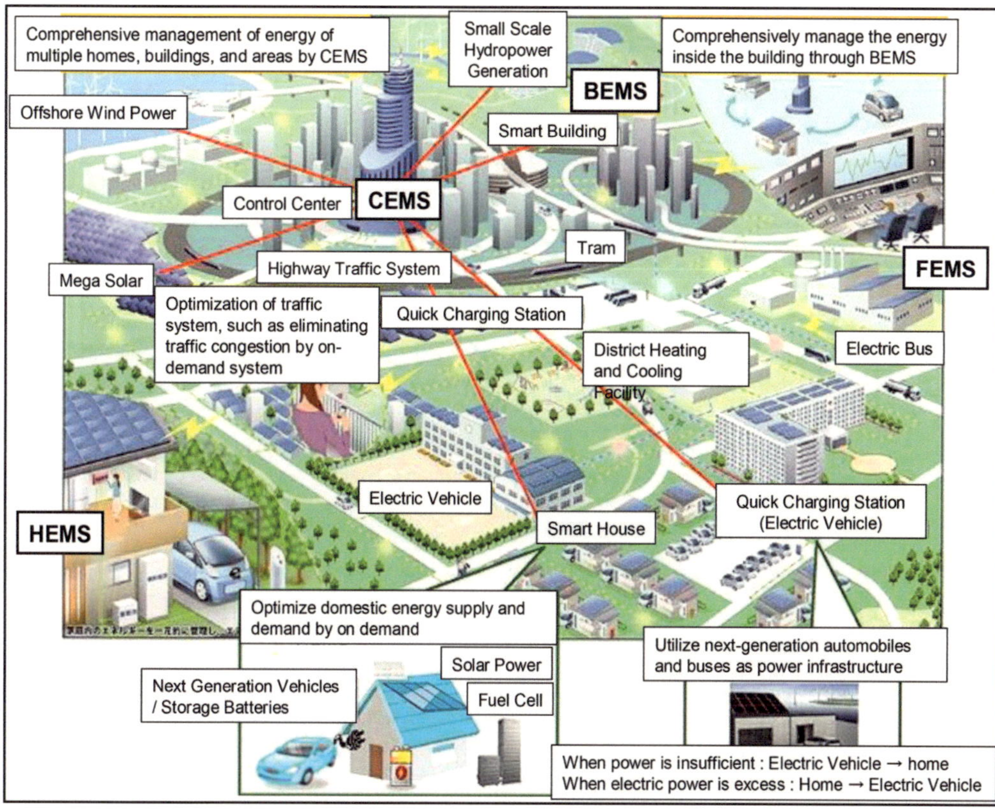

Figure 3. Smart community overview. Source. Ministry of Economy, Trade and Industry, "Smart Grid/Smart Community."

monitors and controls the amount of power used by building distribution equipment, air-conditioning equipment, lighting equipment, ventilation equipment, and office automation (OA) equipment. The BEMS may include the concept of the zero energy building (ZEB), depending on the building management target.

In the factory field, the factory energy management system (FEMS) includes the response to the monitoring of electricity demand. FEMS is an energy management system for the entire factory and controls the amount of power used by, for example, power distribution and distribution facilities in the factory, air-conditioning and sanitation equipment, lighting equipment, utilities, and manufacturing and production facilities. The FEMS may also include a manufacturing management system (MMS), depending on the plant management target.

The regional community energy management system (CEMS) involves the mechanism of the smart grid, which is a power grid with added communication and control functions. The CEMS is used to manage the energy of an entire region and monitors and controls the power supply to the area, including the power managed by the HEMS, BEMS, and FEMS. CEMS may also include a regional energy management system (REMS), depending on the regional management target.

In an information society, the EMS is a mechanism for building new energy infrastructure utilizing ICT. Essentially, the EMS combines various elements such as people, machinery,

Field	System name	Overview
House	Home Energy Management System (HEMS)	Connect various home electric appliances and household OA equipment existing in the household to the network and control energy used in the home in conjunction with HAN.
	Net Zero Energy House (ZEH)	It is a house that generates energy such as photovoltaic power generation and consumption of fuel energy such as petroleum and coal used at home is almost zero (net zero) annually
	Mansion Energy Management System (MEMS)	Measure and accumulate the electricity consumption of the condominium, make it visible, control the connected equipment such as air conditioning and lighting equipment, and suppress and control the demand peak.
Building	Building Energy Management System (BEMS)	Monitor and control the amount of electricity used in buildings such as power distribution equipment, air conditioning equipment, lighting equipment, ventilation equipment, OA equipment, and so on, and control the amount of energy in the entire building.
	Zero Energy Building (ZEB)	Reduce energy consumption at building operation stage through energy conservation and use of renewable energy and limit energy to zero as much as possible.
Factory	Factory Energy Management System (FEMS)	Monitor and control the amount of power usage such as power distribution and distribution facilities in the factory, air conditioning/sanitation equipment, lighting equipment, and so on, and control the energy used in the whole factory.
	Manufacturing Management System (MMS)	Respond to the monitoring and control of production facilities in various fields within the factory, realize improvement in production efficiency in the factory, and control the reduction of energy consumption.
Region	Community Energy Management System (CEMS)	Monitor and control the amount of power supply at power plants including renewable energy within the region, and control the energy used throughout the region.
	Regional Energy Management System (REMS)	Optimize and improve energy use of target area, control energy consumption and CO_2 emissions, and control energy cost reduction.

Table 2. Classification of energy management system (EMS). Source. Created by the author.

facilities, organizations, and functions in a modern information society, and it manages the coordination of energy among these elements. Through the cooperation of these various elements, the EMS can be used to achieve excellent overall functionality, effects, and reactions in homes, buildings, factories, and regions.

EMS supports a wide variety of forms of power, such as hydroelectric power generation, wind power generation, solar power generation, geothermal power generation, thermal power generation, nuclear power generation, wave power generation, and pumped storage power generation.

EMS utilizes ICT to combine all elements (generation, conversion, accumulation, transmission, consumption) organically, from power generation to power transmission, electric power conversion, and distribution, and ultimately to electric power consumers. Through collaboration with a variety of various organizations, it realizes effective and efficient use of energy.

3.2. Binh Duong Province and power energy

Ahead of other cities in Vietnam, Binh Duong Province is promoting a smart community strategy that is an innovative approach to energy, local transportation, industrial waste, and water treatment.

Electricity demand has continued to increase in Vietnam, despite the 2008 global economic crisis. In Binh Duong Province, priority is being placed on the installation of power transmission lines and distribution lines from newly constructed power plants in order to cope with the electric power demand within the area. As a consequence, there has been no progress in the expansion of the existing electric power system, which integrates power generation, electric power conversion, power transmission, and power distribution to customers. In other words, because of this situation, the load factor of transmission and transformation/distribution equipment is rising in Vietnam.

Binh Duong Province therefore has the same problems with electrical energy as in other regions of Vietnam. To address these problems, the province is making a number of efforts. The system average interruption duration index (SAIDI) for individuals and organizations is about 70 h a year. In addition, the system average interruption frequency index (SAIFI) is as high as 17 times a year. As described earlier, power supply reliability has been poor for a long time because of frequent occurrence of power failures.

About 90% of Binh Duong's electric power SAIDI and SAIFI have human impacts due to accidental blackouts and repair blackouts. SAIDI and SAIFI are used to evaluate the reliability of supply of Binh Duong's electric power; efforts are being made to improve the reliability of the electricity supply by decreasing the numerical values of these indexes.

In Binh Duong Province, in buildings in which importance is placed on the reliability of power supply (such as factories, hospitals, provincial government buildings, government offices, and schools) measures such as the installation of in-house power generation facilities are being taken. In the province, there are many foreign companies entering industrial clusters such as industrial parks, and the demand for electricity is remarkable.

For this reason, Binh Duong Province is taking measures to ensure that the infrastructures, such as the new establishment, expansion, or renewal of transmission lines, substations, and substations, are adequate.

In addition, Binh Duong Province is working on energy conservation measures based on the "Law on Economical and Efficient Use of Energy and Measures for its Implementation,"[14] which came into force in 2011. For example, a reduction target of 2% of electricity consumption compared with the previous year has been set, and each month, the Ministry of Commerce and Industry of Binh Duong Province announces the degree of achievement of the target. In

[14]Government Ordinance No. 21 gives the details and enforcement method of the Energy Conservation Law of Vietnam (21/2011 / ND-CP), as applied to energy-intensive facilities, factories, agriculture, and transportation agencies (over 1000 petroleum-equivalent tons). Large-scale customers with energy consumption corresponding to that of offices (500 oil-equivalent tons or more) are required to announce energy-saving targets, formulate and report plans, and so on [3].

addition, Binh Duong Province has implemented measures to support the companies that promote energy conservation.

The following support is provided as part of the Binh Duong Province company support policy:

1. Provide subsidies to small- or medium-sized enterprises that conduct energy audits (the upper limit is 50 million VND.)
2. Provide subsidies for introducing energy-saving equipment [30% of the renewal cost in the case of renewal of the facility (the upper limit is 70 million VND) and up to 100 million VND for new installation)].
3. Support companies that implement energy conservation and projects.

Regional issues such as electrical energy requirements, local transportation, industrial waste, or water treatment do not require individual solutions. We need the concept of a smart community to deal with regional problems by tackling them in a complex and comprehensive way. In response to the smart community concept, we need to improve the reliability of power supply in buildings such as buildings, factories, hospitals, provincial government buildings, government offices, and schools and in neighboring areas.

For example, as part of the promotion of a smart community in Binh Duong Province, the Bus Rapid Transit (BRT)[15] project is being developed. Other examples in Binh Duong New City are the establishment of high-quality electricity in urban development, the development and utilization of ICT infrastructure, the elimination of physical waste in constructing infrastructure, and the development of a smart community strategy by introducing smart traffic.

3.3. Smartization in Binh Duong New City

Tokyu Corporation has used an abundance of know-how in its urban development at Tama Denen on the Tokyu Denen Toshi Line. Tokyu Corporation takes advantage of its own know-how, exploring its strategies for building towns and new cities' and making its town planning packages smart [15].

In smartization by using the Tokyu Corporation city development package, we aim to contribute to town development by forming convenient infrastructure and comfortable communities.

We are exploring the following applied packages [7]:

1. City development that coexists with nature is surrounded by nature. In our urban development, we provide a rich urban space surrounded by water and greenery, and an environment that can interact with nature.
2. In city planning centered on public transportation, the station (or bus terminal) is the center. In this type of urban development, we concentrate urban functions, create comfortable environments, and introduce low-carbon type traffic.

[15]Bus Rapid Transit (BRT) is a high-speed bus transportation system that paves the orbit of a railroad, creates a bus exclusive road, and enables fast operation on a regular basis.

3. As part of resource recycling in town development, water and wastes are recycled. In this type of city planning, we provide safety and security, towns that are resistant to flooding and rising waters, and cities that are strong against disasters and have little crime.

4. Utilization of state-of-the-art technology in energy-saving town planning. In town development, we explore the relevant technologies based on the concept of regional traditions and climate, utilization of renewable energy, and optimum use of energy.

To establish energy and high-quality electricity supply and to improve the reliability of supply in Binh Duong Province, we need to explore state-of-the-art technology.

To establish a reliable energy supply, it is necessary to establish a company that provides an EMS service, construct a system with less power failure, construct a power distribution system with less power loss, improve energy efficiency, implement energy-saving measures for the entire region, and use renewable energy.

As part of the core technology of a smart community, on-site combination of EMS, smart meters,[16] power storage devices, and renewable energy generation (such as solar, wind power, hydropower, geothermal, solar heating, atmospheric heating, natural heating, and biomass[17] power generation) is needed. Both power supply and demand must be controlled.

In the development and utilization of ICT infrastructure in Binh Duong New City, smartization has been applied to the information network by using a new Wi-Fi system.[18] In addition, there is a need to build a Wi-Fi system that can disseminate the Internet and improve the processing speed of computers. It must also be able to handle the huge and complicated big data[19] generated by an information society.

Smartization can also be applied by using artificial intelligence (AI)[20] functions. In addition, by improving the efficiency of data collection and introducing surveillance cameras, a system can be constructed that eliminates waste by using personal identification functions.

There is a need to eliminate waste in the physical (real world) as well as in cyberspace.[21] For example, we need to avoid wasting fuel and time by preventing traffic congestion on the roads, install wide roads that will not need to be expanded in the future, and improve infrastructure by eliminating future waste from a variety of perspectives.

3.4. Smartization target of Binh Duong New City

Completion of the smart community in Binh Duong New City is planned for 2022 and sets the following objectives:

[16]Unlike the conventional analog inductive watt-hour meter, the smart meter visualizes electricity by digital measurement. The smart meter is a next-generation watt-hour meter managed by a communications function in the meter.

[17]Biomass is a concept representing the amount of biological resources; it is a renewable organic resource that excludes fossil resources.

[18]Wi-Fi is a network based on the standard of the industry group Wi-Fi Alliance in the USA.

[19]"Big data" refers to large capacity digital data generated as the Internet spreads and computer processing speeds improve.

[20]Artificial Intelligence (AI) is a technology that allows computers to perform intellectual behaviors such as understanding and reasoning of languages or problem-solving on behalf of human beings.

[21]Cyberspace is a virtual data space that can be used by many people such as computer software and computer networks.

1. Introduction of supervisory control and data acquisition (SCADA)/distribution management systems (DMSs) to distribution companies in Binh Duong New City, and installation of smart meters (with automatic meter reading) to all major customers.

   With smart meters, electricity customers can digitally measure their power consumption at 30-min intervals and can thus understand their electricity usage.

2. Achievement of a 10% reduction in SAIFI in Binh Duong New City.

   By understanding the SAIFI of Binh Duong New City, we can respond by comparing, analyzing, and evaluating the figures for Vietnam as a whole.

3. Achievement of a 20% reduction over 5 years in SAIDI in Binh Duong New City.

   By understanding the SAIDI of Binh Duong New City, we can respond by comparing, analyzing, and evaluating the numerical values for Vietnam as a whole.

4. Reduction in the number of workers at substations in Binh Duong New City.

   The minimum wage in Binh Duong Province is classified as the "first type[22]" (highest) among domestic workers. By creating a smart electric power service, we can reduce the labor costs of electric power workers at the domestic wage level in Vietnam.

5. Improvement of the power supply/demand prediction in Binh Duong New City to realize a 2% peak cut in electric power supply and demand.

   To do this, we will create a demand monitoring system and demand control system that uses advanced metering infrastructure (AMI).

   By introducing the demand monitoring system, electricity consumption is measured in real time. Power demand is realized and monitored such that it does not exceed a set amount of electricity.

   Also, when demand control system is introduced, an attempt will be made to reduce electricity charges by lowering the maximum value of demand electricity in the region.

6. Reduction of transmission and distribution losses in Binh Duong New City.

In electric power transmission and distribution to customers via electric wires, loss of power occurs because of electrical resistance and is accompanied by transmission/distribution loss.

Also, when power is transmitted and distributed to distant places, the voltage drops; power is converted to heat and is lost.

As ways to reduce transmission and distribution losses, the following have been implemented: a method of generating electricity locally, in the vicinity of the area where it was consumed; a method of reducing the electrical resistance by thickening the electric power cables; and a method of high-voltage transmission to reduce the electrical resistance by increasing the electric power.

[22]The first type of area includes Hanoi City, Hai Phong City, Ho Chi Minh City, Binh Duong Province, Ba Ria-Vung Tau Province, and Dong Nai Province.

In addition to these goals, it is necessary to utilize a smart grid,[23] to promote the domestic production of smart-grid-related products in Vietnam, and to visualize electricity usage by using smart meters in association with renewable energy systems.

4. Conclusion

In the Binh Duong new city developed by Tokyu Corporation (Japanese Company), we aim to form a smart community that utilizes the know-how that has been realized in regional development in Japan, exporting smartization by city planning package.

Generally, with the promotion of the smart community, development using existing buildings and facilities is difficult. In other words, it is not necessary to consider existing buildings and facilities in developing new urban areas from the ground, so it is possible to promote an optimal and ideal smart community.

Unlike smartization in Da Nang City, which transforms existing old facilities and facilities, the Vietnam's Binh Duong Province smart approach facilitates the development of new cities. Also, unlike HCMC and Hanoi's major cities, in the Binh Duong Province, where population is low, development of new cities is easy to promote.

Furthermore, in the control of energy centered on the EMS by the Binh Duong New City, the neighborhood area is also a great merit to utilize its convenience. Therefore, the Binh Duong Province has attempted new attempts to develop industrial cities and new urban areas including neighboring areas.

The smart community strategy in the Binh Duong province has only just begun. At the present time, it is a situation in which it is impossible to make a judgment on the success or failure of the smart community strategy. Future tasks include strict capital planning accompanying the smart community strategy, and the inability of SORA gardens to sell real estate.

Meanwhile, in real estate in the high-class residential area of PHU MY HUNG in HCMC 7 ward, the sale did not proceed at first.

However, PHU MY HUNG attracted attention after 5 years; there are cases where urban development advanced. Also, in Hoa Lac Hi-Tech Park in Hanoi, there are cases where the urban development advanced more than 10 years ago. For the future development of new urban areas, it is necessary to promote a smart community strategy that cannot be confirmed by immediate earnings and to evaluate future potential.

In Vietnam, dynamic economic activities and corporate actions not seen in industrialized countries have shown that the activities of companies and related organizations showed active movements and showed high economic growth. One of the sources of economic growth in Vietnam is the effort to form a smart community (Smart Community Strategy).

[23]The smart grid is a next-generation electric grid that has a mechanism to constantly optimize the demand and supply of electric power by utilizing ICT. In the smart grid, we aim to build a power supply system with high efficiency, quality, and reliability by controlling existing power-generation facilities such as hydropower and thermal power plants and distributed energy sources using new forms of energy such as wind power and solar power generation. Progress is being made in smart grid in each country as one of the countermeasures against global warming.

The smart community strategy in Vietnam's Binh Duong Province is urban development utilizing "know-how" realized in regional development in Japan. In addition, the smart community strategy of Vietnam's Binh Duong Province has just been developed, and it is a situation in which it is impossible to make a judgment of the right or wrong at the present moment.

However, the development of smart community strategy in the future is a very useful reference not only in Vietnam but also in urban development of developed countries and emerging countries, and it can be a useful urban accumulation model. Vietnamese smart community research is a very valuable research content, including not only Vietnamese literature but also English literature from other countries. In addition, smart community research in emerging countries has few approaches so far and is a very valuable field as a research area.

In recent years, the development of large cities of ASEAN countries, emerging countries such as Jakarta (Indonesia), Manila (Philippines), Bangkok (Thailand), Yangon (Myanmar), Kuala Lumpur (Malaysia), and so on is remarkable. The scenery of major cities in emerging countries is comparable to that of Tokyo, Beijing, Shanghai, Seoul, Taipei, and other big cities.

In the large cities of emerging economies, problems of environments and energy due to enormous urban areas are getting worse, and both have similar problems. The promotion of the smart community strategy in Vietnam's Binh Duong province, along with Hanoi and Ho Chi Minh City in Vietnam, is one example of a solution to the big city problem of emerging economies.

Research on smart communities, so far, mainly focuses on technical contents such as energy, environment, and electric vehicle (EV). In this chapter, we examined the smart community of Vietnam's Binh Duong Province from the viewpoint of social science. Studies of smart communities in Asian countries are hardly approached and have great academic contributions.

Acknowledgements

This work was supported by Japan Society for the Promotion of Science (JSPS) Grants-in-Aid for Scientific Research (KAKENHI)(B) Grant Numbers JP16H03664, and JSPS KAKENHI (B) Grant Numbers JP17H03327.

Author details

Tetsuro Saisho

Address all correspondence to: tsaisho@kokushikan.ac.jp

Kokushikan University, Tokyo, Japan

References

[1] Becamex Tokyu. Location, "Sora Gardens". Becamex Tokyu; 2015. <http://sora-gardens.com/location.html>

[2] Binh Duong Province <http://www.binhduong.gov.vn/>

[3] CỘNG HÒA XÃ HỘI CHỦ NGHĨA VIỆT NAM, CHÍNH PHỦ. QUY ĐỊNH CHI TIẾT VÀ BIỆN PHÁP THI HÀNH LUẬT SỬ DỤNG NĂNG LƯỢNG TIẾT KIỆM VÀ HIỆU QUẢ. 21/2011/NĐ-CP, CỘNG HÒA XÃ HỘI CHỦ NGHĨA VIỆT NAM; 2011 <https://thuvienphapluat.vn/van-ban/Tai-nguyen-Moi-truong/Decree-No-21-2011-ND-CP-detailing-the-law-on-economical-and-efficient-use-126217.aspx>

[4] CỘNG HÒA XÃ HỘI CHỦ NGHĨA VIỆT NAM, CHÍNH PHỦ. QUY ĐỊNH MỨC LƯƠNG TỐI THIỂU VÙNG ĐỐI VỚI NGƯỜI LAO ĐỘNG LÀM VIỆC THEO HỢP ĐỒNG LAO ĐỘNG. 153/2016/NĐ-CP, CỘNG HÒA XÃ HỘI CHỦ NGHĨA VIỆT NAM; 2016. <https://thuvienphapluat.vn/van-ban/Doanh-nghiep/Nghi-dinh-153-2016-ND-CP-muc-luong-toi-thieu-vung-doi-voi-nguoi-lao-dong-lam-viec-theo-hop-dong-lao-dong-319987.aspx>

[5] Japan Smart Community Alliance (JSCA). Toward Realization of Smart Community. <https://www.smart -japan.org/index.html>

[6] Tetsuro Saisho. Possibility of forming industrial clusters in Vietnam—Case study of FPT City Da Nang in Da Nang. Journal of Business and Economics. 2015;**6**(11):1813-1823

[7] Aeon Investor Relations Office. Supplementary Statement of Financial Statements for the Fiscal Year Ending February 2017 (92th Stage). Settlement Material. Aeon; 2017. <http://v4.eir-parts.net/v4Contents/View.aspx?cat=tdnet&sid=1457783>

[8] Ishii K. Smart City That Creates Sustainable Cities and Overseas. Ministry of Land, Infrastructure, Transport and Tourism; 2014. <https://www.mlit.go.jp/common/001029873.pdf>

[9] Ministry of Economy. Trade and Industry, Smart Grid·Smart Community. Ministry of Economy, Trade and Industry. <http://www.meti.go.jp/policy/energy_environment/smart_community/>

[10] Saisho T. A study of the smart community in Vietnam: A case of Binh Duong Province in the southern regional city. Information Management 73th National Conferences Proceedings [Autumn Edition]. Japan Society for Information and Management; 2016. pp. 183-186

[11] Saisho T. A study of smart communities in Vietnam: A case of smart community strategy in Da Nang City. Kokushikan Business Review. 2017;**6**(2):1-34. The Society of Business, Kokushika University

[12] Jetro H. Vietnam General Overview: Vietnam economy viewed by numbers. Basic Information of Vietnam. JETRO. 2017. <https://www.jetro.go.jp/ext_images/world/asia/vn/data/vn_overview201704.pdf>

[13] Jetro H. Vietnam Power Survey 2016 (March 2017). "Survey Report". JETRO. 2017. <https://www.jetro.go.jp/ext_images/_Reports/02/2017/401bfd604d232c7d/rpvn-el2016-201703.pdf>

[14] Ministry of Economy, Trade and Industry, Energy Saving Division. New Energy Division New Industry and Social System Promotion Office, Ministry of Internal Affairs and Communication International Communication Standards Division, Communication Standards Division.

Efforts towards Building a Smart Community. Cabinet Office, Government of Japan; 2014. <http://www8.cao.go.jp/cstp/tyousakai/juyoukadai/energy/4kai/siryo3-3.pdf>

[15] Tokyu Corporation. International Business (Town Planning Abroad). Business Contents. Tokyu Corporation; 2017. <http://www.tokyu.co.jp/company/business/urban_development/abroad/>

[16] Nihon Keizai Shimbun. Tokyu, Investing Large-Scale Housing Land of 88 billion yen in Vietnam, 9,000 units. Nihon Keizai Shimbun electronic version. June 2, 2016, Nihon Keizai Shimbun; 2016. <http://www.nikkei.com/article/DGXLZO03105430S6A600C1MM8000/>

[17] Becamex Tokyu. About Binh Duong Province. Becamex Tokyu. <http://www.becamex-tokyu.com/ja/introduction_of_binh_duong_province/>

[18] Becamex Idc Corp. Smart Community Research Project in Binh Duong Province. Becamex; 2013

[19] Aeon Co., Ltd. <https://www.aeon.info/>

[20] ASTI Corporation. <http://www.asti.co.jp/>

[21] Astee Horie Co., Ltd. <http://www.astee-horie.co.jp/>

[22] Daido Vietnam Co., Ltd. <http://vtown.vn/company/daido-vietnam.html>

[23] Eikou Co., Ltd. <http://www.j-eikou.jp/index.html>

[24] Fujiya Co., Ltd. <http://www.fujiya-kk.com/ja/>

[25] GF Co., Ltd. <http://www.g-f.co.jp/>

[26] Hayabusa Fishing Hooks Co., Ltd. <http://www.hayabusa.co.jp/>

[27] Hondalogicom Co., Ltd. <http://honda-logi.com/>

[28] Inter Art Saigon Co., Ltd. <http://www.interart-saigon.com/en/index.php>

[29] Iwai Kikai Kogyo Co., Ltd. <http://www.iwai.co.jp/>

[30] Kubota Corporation. <https://www.kubota.co.jp/>

[31] LSI Cooler Co., Ltd. <http://www.lsi-cooler.co.jp/>

[32] Makita Corporation <https://www.makita.co.jp/>

[33] Maruichi Steel Tube Ltd. <http://www.maruichikokan.co.jp/>

[34] Mitsubishi Heavy Industries, Ltd. <http://www.mhi.co.jp/>

[35] Nakanishi Inc. <http://www.japan.nsk-dental.com/>

[36] Suncorona Oda Group <http://www.sunoda.co.jp/>

[37] Ohmori Co., Ltd. <http://www.ohhmori.co.jp/>

[38] Rengo Co.,Ltd. <http://www.rengo.co.jp/>

[39] Sakai Chemical Industry Co., Ltd. <http://www.sakai-chem.co.jp/jp/>

[40] Sakata Inx Corporation. <http://www.inx.co.jp/>

[41] Sharp Takaya Electronic Industry Co., Ltd. <http://www.s-takaya.co.jp/>

[42] Taisei Bijutsu Printing Co., Ltd. <http://www.taiseibijutsu.jp/>

[43] Tokyu Corporation <http://www.tokyu.co.jp/>

[44] Uchihashi Estec Co., Ltd. <http://www.uchihashi.co.jp/>

[45] Yamaguchi Seiko <http://www.yamaguchi-sk.com/>

[46] Yamaichi Special Steel Co., Ltd. <http://www.yamaichi-hagane.co.jp/>

[47] Yamaken Apparel Ltd. <http://yamaken-ap.jp/>

[48] Yamaken Company Limited <https://www.yamaken-con.jp/>

Chapter 8

Wicked Water Systems: A Review of Challenges and Opportunities

Tamara Anna Streefland and Yoram Krozer

Additional information is available at the end of the chapter

http://dx.doi.org/10.5772/intechopen.71914

Abstract

The contemporary urban water system is under extreme pressure due to growing demand, climate change, and social inequality. Conventional methods to mitigating extreme water events have proven to be insufficient to safeguard our growing urban centers. Unless the competing demands and pressures of the urban water system are addressed in holistic manners, we will soon lack of access to good-quality water, and extreme water events will increasingly affect our metropoles, with most severe consequences for communities already living in marginal conditions. This chapter takes as point of departure that the future urban waterscape is a wicked problem in which actions taken to mitigate the problems are often inadequate and temporary, even when they are the result of the public debate and shared concern. The current urban water system has reached a critical threshold, but how can innovative urban water design or planning solutions be implemented when there is so much at stake? The chapter will address the urban waterscape, its contemporary challenges and what we can expect in future climate conditions. Furthermore, we will discuss contemporary solutions as well as highlight the sociopolitical, economic, and ecological barriers to their implementation. To illustrate the challenges as well as the range of solutions, we will present the algal blooms in Lake Erie, USA, as a case study. We will end with an elaboration on how to innovate in the case of wicked problems.

Keywords: wicked problems, water, urban design, innovation, algal blooms

1. Introduction

On 25th of August, 2017, Hurricane Harvey made the first landfall in Texas. Although Houston, a city of approximately 6.5 million inhabitants, lays above sea level, it experienced massive floods over the following days due to the extreme amounts of rain and wind and storm surge [1]. Simultaneously, Nepal, India, and Bangladesh are struck by heavy monsoon rains, causing Mumbai's streets to turn into rivers and more than 1200 deaths throughout the entire region [2]. Houston's extreme floods were caused by the combination of heavy rains causing interior waterways to overflow, a storm surge that prevented them from emptying out in the ocean, and a general lack of pervious surface in the city [3]. Mumbai vulnerability lays in its rapid expansion that is accompanied by construction and development on floodplains as well as clogged waterways due to an excess of plastic waste [2].

The contemporary urban water system is under extreme pressure of rapid urbanization, growing demand, climate change, and social inequality. It has become apparent that water is no longer a free endless resource and that waste cannot just be transported to other places, because water systems are globally connected [4]. Unless the competing demands are addressed in holistic manners, we will soon lack access to good-quality water, and extreme flooding, droughts, and tsunamis will increasingly affect the planet and its people, with most severe consequences for those already living in marginal conditions. Disempowered communities, minorities, and the poor are at the frontlines [5].

The urgency of this has been recognized at the recent Climate Summit in Paris. Commitments have been made by governments and other stakeholders, but the tensions underlying our failure to address these issues in the past have not been resolved, and the implementation of radical policies and integrated innovations remains to be seen. Interventions and implementations are challenged by the uncertainty of climate models, cycles of capital investment and governmental risk aversion, fragmentation of governmental responsibility and inclusion of knowledge, the definition of innovation, and scale.

We can think of transforming the urban water ecologies to meet the needs of our future cities as a wicked problem, a term first introduced by Horst Rittel, a German scholar, in the 1960s [6]. Wicked problems are not easy to formulate or to reach a consensus on a solution. Actions taken to mitigate the problems are often inadequate and temporary, even when they are the result of the public debate and shared concerns [7]. Rittel defines some characteristics of wicked problems. First, one cannot understand the problem without understanding its context. There is no way of first understanding and then solving since the existing pieces might shift in a solution.

Second, there is no stopping rule. In the case of a wicked problem, it is not clear when a solution is found. The intervention does not stop when the problem is solved but is limited by external factors such as time or money.

Third, there are no true or false answers since the criteria to determine that do not exist. Any judgment by actors is determined by the values and objectives of the group that they are part of, the place they take within the realm of the existing problem.

Finally, there is no test (except for real-life implementation) to a wicked problem. Every attempt will have consequences that are either desirable or not, which will then trigger another set of consequences. Every implemented "solution" will leave its mark in the system [6].

For example, the Dutch and the storm surge barrier they decided to build in the 1960s after a devastating storm with 5-m waves broke the existing dikes and had affected many lives in the South of Holland in 1953. Holland had lost the battle with water in a storm that was beyond their expectation. This called for an intervention outside of what they already knew. During the process of implementing the barrier, all types of issues arose such as a changing hydrology causing strong tides as the project proceeded, complicating further intervention. Although, until this day, that levee has protected the low-lying south of the country, it is a continuous train of intervention that has been implemented to secure the Dutch with dry feet [8].

The current urban water system has reached a critical threshold, but how can innovative urban water design or planning solutions be implemented when there is so much at stake? This chapter will first briefly discuss the value of water. Second, it will address the urban waterscape and its contemporary challenges. What are contemporary urban centers facing in terms of water pressures, and what can we expect in future climate conditions. Furthermore, we will discuss contemporary design and planning of solutions as well as highlight the socio-political, economic, and ecological barriers to their implementation. To illustrate the challenges as well as the range of solutions, we will present the algal blooms in Lake Erie, USA, as a case study. The chapter will end with an elaboration on how to innovate in the case of wicked problems.

2. The value of water and what is at stake

There is a certain dualism at stake when thinking about water and cities. Water needs to get into our urban systems and flow through them, for otherwise urban populations cannot exist; think about drinking water or water needed for urban industries. On the other side, we also need to be protected against it. Too much water, in the form a natural hazard or a polluted body like sewage, is a constant threat to society resulting in inadequate attempts of controlling the flows.

Water has multiple physical attributes that affect its relation to the human body and environment and that shape its use [9]. Its fluidity and plurality are reflected in the ways we refer to it. Water is experienced (tasteless, odorless, cold, salty), and it is volatile in its form (rain, ice, and gas). It moves at different speed (current, wave, mirror), has a carrying capacity (of nutrients and pollutants), and is an active agent (flows, shapes, erodes, moisturizes), which can also be contained (in soil pores, pathways, and rivers).

The fluidity and plural attributes of water are at the core of distribution of quality and quantity of water that is of key importance to maintain the integrity of ecosystems and that will preserve their ability to provide services valuable to humans [10]. Natural water supplies vary over times as these flows cause scarcity and in other places floods. The fact that water has a certain carrying capacity allows it to carry nutrients, but it also allows it to erode our coasts resulting in turbid currents. Large technical infrastructures, such as dams, reservoirs, and artificial storages, shield us from the variability of the resource and protect us against floods [10].

Historically, all large urban centers were built in the proximity of water for it to survive. When ancient Rome around 300 BC grew beyond its capacity to provide fresh water to its citizens locally, the Romans engineered an extensive system, including large aqueducts, underground pipes, as well as measures to control the velocity of the water to import water into the city [11]. This infrastructure had a combined length of over 400 km and brought water from places as far away as 80 km [12]. There was one major pitfall to their ingenious water system; many of the pipes were made from lead, a strong neurotoxin, which contaminated the water and poisoned Rome's inhabitants [13]. Some scholars advocate that lead poisoning was at the base of the fall of the Roman empire [14]. Lead poisoning is not only a problem of the past as was seen in Flint Michigan in 2015. The local government switched their water resource to a more affordable but corrosive option, the Flint River, which caused the pipes to leak lead into the drinking water [15].

3. Contemporary urban water systems and future threats

All water is connected and so are the issues surrounding our water systems. What are the pressures threatening our urban centers exactly? As mentioned before urban areas face issues regarding the quantity and quality of water manifesting itself in scarcity, pollution, and flood events.

First, flooding events are caused by a multitude of factors, including water body flooding and heavy rains, but also factors that are related to the urban fabric itself such as clogging of drainage systems and land use change. Although there is no exact prediction of what will happen in the light of climate change, most scientists agree that extreme weather events such as hurricanes will occur more frequently and will be of higher intensity. Additionally, more extremes in precipitation will occur (both droughts and excess) and a rise of sea level which will most rapidly affect coastal cities and small island communities [16]. At the same time, rapid urbanization leads to expansion of the urban surface which often results in a lack of stormwater retention space due to imperviousness of the surface that simultaneously accelerates stormwater flow, keeping the water in the streets. Excess waste is at the base of clogging the existing man-made drainage channels and natural systems such as mangroves or wetlands [17].

Jongman and colleagues [18] address the rise of costs (infrastructure, housing, businesses, etc.) as a result of flooding throughout Europe. They suggest that by 2050, those annual costs will rise from 4.5 Billion to 23 Billion Euros, under the anticipated climate change and socio-economic development [18].

An abundance of water can be problematic but so is scarcity. In some cases, cities experience both. Cities often rely on sources in their hinterlands to supply their population, industries, and agriculture with fresh water. However, our fresh water resources are decreasing. We have consumed water without limits under the assumption that it is a renewable resource. Moreover, growing cities along with a growing population increase the need for fresh water. In response, cities have created complex large-scale infrastructures trying to reach new sources of underground water or surface water [19]. When surface water is insufficient for the demand, cities often turn to groundwater; however, when more water is consumed than what is refilling the source, this will not only result in depletion and water stress in the future but also to saltwater intrusion and soil subsidence of lower lying areas as well as to negative effects on natural streamflows and depended ecosystems [20, 21]. In a capitalist society, consuming less water is hard to achieve, since it is in direct opposition to the principles on which such society is based [22].

Furthermore, human activities such as irrigation and industrial use threaten the global water systems. The benefits of our water use to economic growth go hand in hand with degradation of our (aquatic) ecosystems that will result in major costs in the future [23]. Think about micro plastic that enters the water system through our waste systems. These fragmented particles are almost impossible to filter out of the systems, and we have yet to understand the largely unknown long-term consequences [24].

Apart from synthetic pollution, pollution can also occur by a disbalance of nutrients in the water. An example is Toledo, a city of a little over 500,000 inhabitants in the US. The drinking water source for the city is the adjacent Lake Erie, but in 2014 the inhabitants had to close their taps because of the presence of microcystin, a toxin that emerged from algae blooms occurring in the lake [25]. These blooms are caused by nutrient runoff, phosphorus, in particular, that flows from urban sewage overflow (point source) and predominantly agricultural runoff (nonpoint source) through the watershed into the lake and are intensified by heavy rain. Climate change, extreme weather events, and hot summers will cause harmful algae blooms to occur more often and more intense within the following years [26]. We will review this more in-depth in the following sections.

Summarizing, floods, pollution, and scarcity are all interconnected, e.g., in the case of floods, it can pollute water bodies when hitting a chemical factory. Pollution in its turn leads to scarcity of freshwater.

How can we secure healthy water system on our planet in a changing world?

4. Mitigating water stress

Stable socio-environmental ecosystems do not exist; these systems are always on the move. Society wants to influence the ecological system by implementing measures to alter the hydrological response and to keep it at its highest stage of productivity, preventing it from collapse [27]. The implementation measures are often incentivized from an economic efficiency agenda and are predominantly technocratic as they often fail to fully integrate social dimensions of water. Furthermore, they are often narrow solutions that are unsynchronized with the dynamics of the physical landscape.

Mollinga distinguishes three dimensions at the heart of water control: technical/physical, organizational/managerial, and a socioeconomic and regulatory [28].

Historically, water infrastructure and projects to alter the hydrological response have been created by scientific and technical experts, emphasizing rational science and economic efficiency. With the increase of tension between social, environmental, and economic elements in urban areas, this is no longer viable. Implemented infrastructures have more often redirected the system than stopped it from developing. Again and again, the effect of one intervention has asked for another one as systems are successional and dynamic [29].

There are four dominant ways of altering the hydrological response through physical implementations:

1. inverting the course of the water flow to get water supply (e.g., irrigation);
2. altering the stream network (wetlands);
3. altering the drainage basin (dams); and
4. changing the global climate (overusing water) [28].

Furthermore, innovation in urban water systems comes in three ways: (1) new technologies (desalination or new waste water systems), (2) new management approaches (business models, policy implementations), and (3) techniques to increase the functioning of the current system (monitoring, etc.) [30].

The large-scale infrastructure-engineering approach was overtaken by more recent notions of conservation or restoration. In both, system ecology remains entrenched within the same modern paradigm that many argue is the structural cause of environmental and social decline, as Corner puts it "whereas conservation utilizes ecology to facilitate further control over the human environment, restoration uses it to provide rhetorical force to emotional feelings about the primacy of nature and the errors of the Anthropocene" [31]. Thus, conservation aims at keeping the same ecosystem services of a system and tries to keep a system from moving away from its current state. It has faith that adaptive and transformative measures will save the day [32].

In both the large infrastructure/engineering and conservation/restoration trains of thought, the symptoms of ecological distress are dealt with, while causal and cultural foundations—the social structures that underlie at these problems—are often not addressed.

The problematic at stake with the drinking water systems of Toledo, Ohio, provides a good case study of proposed interventions and barriers while working with unprecedented issues and wicked problems. In 2016, as part of Enaegon, a collaborative urban design company with a focus on water, we proposed an implementation that would contribute to diverting the algae blooms in Lake Erie. We took as a point of departure that wicked problems can be challenged throughout the journey toward a design intervention that can change the interactions in the systems. The outcome is not necessarily a solution but can help better understand the wickedness of the problem and form the basis of new experiments.

4.1. Case study: algae blooms in Lake Erie

In the first weekend of August 2014, the residents of Toledo, Ohio, were advised not to drink the water from their faucets due to high levels of microcystin, a toxin that is caused by algae blooms in the source of their drinking water, Lake Erie. People prepared for days without water, standing in lines at grocery stores. Some even crossing the border to Canada, packed with as many empty containers as they could find [33].

When the local authorities announced the drinking water ban, no indication of its duration was given—local government simply did not know. They had to wait for the US Environmental Protection Agency (EPA) to run tests and approve the drinking water. This was problematic since there was no standard for what was a safe amount of microcystin in the water. The only standard that did exist was the outdated standard instituted by the World Health Organization [34].

This process that causes harmful algal blooms in Lake Erie is called eutrophication, which can be defined as an increase of organic carbon production (this is basically everything that is alive from plants to animals to humans) in an ecosystem [35]. Although higher production is often seen as a good thing (as it is sometimes referred to as nutrient enrichment), in this case, its side effect is algal blooms that in the end lead to hypoxia, the death of a water body.

Eutrophication is caused by nutrient runoff from farming operations that involve intensive fertilization with both manure and chemical fertilizer, as well as sewer overflow from adjacent urban areas. The International Joint Commission (IJC), a binational advisory body on the great lakes water quality agreement, estimates that 61–84% of the nutrient runoff is caused by agricultural runoff whereas only 16% flows from combined sewers of the surrounding cities [36].

Most of the runoff happens in spring due to extreme rain events, flushing the phosphorus into the agricultural fields and drainage ditches due to surface and subsurface flow. In the case of Lake Erie, it is the Maumee watershed that is the main tributary due to extensive farming operations (80% of the land is under cultivation) in the region. All the ditches in one watershed flow toward the same point, the Maumee River, and then into Lake Erie at the Toledo waterfront [36].

Although 2014 was the first time that drinking water system got severely affected, algal blooms have been happening for decades. Due to ongoing urbanization and the dumping of untreated sewage around the lake's shoreline, more extreme algal blooms started to emerge in the 1950s and 1960s. Lake Erie's oxygen-depleted waters and its changing ecosystems even caused it to be declared dead by the beginning of the 1970s [37].

It is not only drinking water that is jeopardized. Algal blooms will trigger economic losses through a decrease in tourism (who wants to swim in a toxic lake?) and a declining fishing sector since these algal blooms pretty much affect the whole ecosystem by depleting oxygen, killing other species and ultimately "killing" the lake [38].

Thus, the main problem lays with the farming practices and farming policy, but the consequences lay in the drinking water supply for the communities that are reliant on Lake Erie as their drinking water source.

So what is the plan? The Ohio Sea Grant on Social Network Analysis of Lake Erie HAB's Stakeholder Groups has identified more than 150 stakeholders that are more or less involved (and yes we are only talking about the Maumee Watershed) [39].

The International Joint Commission (IJC), a collaborative advisory body between Canada and the USA, existing of a variety of experts ranging from scholars to politicians to experts from the private sector, has stressed the urgency of a 41% decrease in nutrient runoff to alleviate future algal blooms and safeguard drinking water quality. The majority of agricultural runoff is generated by farmers in the Maumee Watershed which borders the states of Indiana, Michigan, and Ohio; the IJC has designated this watershed the number one area of concern [36].

However, this is a concern without a solid action plan or timeframe for its resolution.

The federal government claims to have acted on the urgency with a response through bilateral agreements and complex funding mechanisms. In 2012, the US EPA and Environment Canada signed an updated version of the in Great Lakes Water Quality Act (GLWQA); the agreement bound them to develop detailed commitments in 2016 [40].

Strangely enough, these new commitments have been made to achieve the 41% reductions in nutrients but fail to put a deadline to the proposed action and therefore also lack sanctions if the deadline is not met. The agreement does specify that "domestic action plans" are being developed. However, it is the question of what kind of measures they will propose.

4.1.1. *Strategies to mitigate the algal blooms*

Historically, many different control measures to divert the algal blooms have been mechanical, varying from bio-manipulation, aeration of the lake, liming (a process where limestones alter the pH of the lake, creating a more hostile environment for algal), and dredging to improve water flows. These techniques are incredibly expensive and therefore not resilient, as they do not solve the problem—they are merely attempts at mopping with the tap open [41].

The following strategies are currently existing in the Maumee watershed:

4.1.1.1. *Policy*

Over the last 2 years, there has been some movement in terms of legislation. The Governor of Ohio (John Kasich) has passed a Farm Bill in 2015 that regulates the application of fertilizer and manure by farmers. The law limits application on snow-covered soil, saturated soil, and when the weather forecast predicts rain within the next 12 h [42].

Another bill that was adopted is Farm Bill 150; it requires all farmers that apply fertilizer to undergo a certification process that educates them in fertilizer application, encourages the adoption of nutrient management plan, and allows the Ohio Department of Agriculture to better track fertilizer distribution [42].

Although the impact of this legislation is hard to measure, it does seem a step in the right direction.

4.1.1.2. Changing behavior

The 4R strategy was put forward by the global fertilizer industry in 2009. The Ohio Department of Agriculture has adopted this strategy and now enforces the training under farm bill 150 [43].

Farmers earn the certification by earning 5 h of education credits. This training is organized by a provider that needs to have at least the level of certified crop advisor. This provider then needs to keep track of the amount of participants and document the progress of these farms in their use of fertilizer [43]. As expected, the strategy involves 4Rs regarding applying manure and fertilizer in the right way (right source, right rate, right time, right place). This is the main strategy that is adopted by the US Department of Agriculture (USDA), the National Resources and Conservation Council (NRCS), and the Ohio Department of Agriculture (ODA) through funding by the Great Lakes Restoration Initiative, a funding mechanism instituted by the federal government, which is the main source of funding for algal bloom mitigation projects instituted by governmental agencies.

4.1.1.3. Green infrastructure and farming practices

There is funding available for farmers who want to take land out of production and use it for strategies such as crop rotations or buffers, two-stage ditches, or other water management installations that will catch the phosphorus before it runs off. Farmers can apply for farm adaptations that will lead to a reduction in nutrient runoff. This is the largest source of funding available and is based on the number of farm acres; it is a set amount of funding each year. It works through a 3–5-year contract that is more likely a mortgage and is competitive. When a farmer applies, they get a technical report from the Soil and Water Conservation District. The farmer then gets a 50–100% match to their own funding proposed. This is a fund that comes from congress and exists until the money is finished [44].

4.1.1.4. Monitoring

Currently, Toledo's drinking water is monitored every 10 min, and people have the option of getting an alert from the water company in case of a threat.

Monitoring also exists in the water surface water bodies going into the lake. The Great Lakes Restoration Initiative differentiates between sensors on different scales in the larger Lake system, the tributaries of the Maumee as well as Edge of field monitoring, which is in collaboration with local farmers. Since there is a lot of skepticism about where the nutrients actually come from, which farmer is contributing and how much, this is a key component. Monitoring is a big part of assessing which strategies could work, as it provides the grounds for a solid impact analysis [45].

4.1.1.5. Technological retrofitting

In the 1980s the municipal governments of the urban centers surrounding Lake Erie pushed back the phosphorus loading levels by implementing retrofits to the wastewater treatment plants. An investment of eight billion dollars was made [46]. Currently, the city is a new phase of retrofitting the drinking water plant [47]. Although there is funding available both governmental and nongovernmental and there is a definite need for systemic change, an integrated response to the algal bloom has yet to be seen. In the summer of 2017, more algal blooms occurred in Lake Erie although the drinking water was safe this time [48].

4.1.1.6. Enaegon's proposal

Enaegon proposed to build of the existing incentive of creating edge-of-field buffers for farmers. Through a productive ditch filter of efficient crops with high phosphorus uptake, the nutrient runoff can be decreased intensively. Participating farmers will profit from the sale of the phosphorus-rich crop. The ditch can be implemented through existing mechanisms such as funding and distribution, enabling a farmer to shift the problem of overfertilization into an economic opportunity. Simultaneously, the relationships between rural and urban populations will be strengthened through documentation. The project entailed an interwoven system design of an environmental strategy, economic strategy, and a social strategy and built of existing funding, monitoring, research, and relationship.

4.1.2. Challenges to systemic change in the Maumee watershed

The complex process of trying to implement this project provided us with many insights into the Lake Erie algal bloom and the barriers to implementing any integral innovation.

4.1.2.1. Skepticism

Farmers are skeptical of their share of the problem. Who is the actual source of the nutrient runoff causing the blooms, is it the Concentrated Animal Feeding Operations or the regular farming operation, and if so, which farmers are to blame? The percentage of farming contribution to the issue of nutrient pollution depends on to which stakeholder you talk too. In terms of advice, farmers take it rather from industry actors such as fertilizer companies and their consultants, instead of the government, e.g., their local soil and water district. Also, Toledo residents are skeptical of the water that comes out of their taps. Since microcystin toxin does color the water. Public cases like Flint, Michigan, further induce governmental mistrust.

4.1.2.2. Monitoring and modeling

When there is a perceived source, you may identify the polluter and hold him responsible, i.e. the polluter pays structure. Monitoring is a timely and expensive manner; only watching is and not acting is a problem in the public perception. Although Heidelberg University and other individual researchers are installing edge-of-field monitoring, this practice can be used for impact analysis, but wide implementation for finding out the exact source at this point seems unattainable. Furthermore, using models to predict the flow of nutrients or predict future weather is covered in uncertainty. How do we plan the measures that will also be a fit for future conditions?

4.1.2.3. Funding

Whereas we have identified four main funding sources (the Ohio Sea Grant, EQIP, CRP, and GLRI), there are barriers to these mechanisms. First is the uncertainty of continuation of this funding source; Ohio's department of higher education might stop funding the Ohio Sea Grant, which would decrease research done on this subject. Also, with the change of office in 2017, the Great Lakes Restoration Initiative could very well dry up, taking away funding for projects for local governments. Then there is the question of yield. Farmers are reluctant to adopt because it might affect the yield of their land. We must not forget that farming is a for-profit business.

4.1.2.4. Collaboration and implementation

There are more than 150 stakeholders involved in only the Maumee Watershed. One of the main problems is the fact that commitments have been made on federal and state level, but the actual action, besides policy, needs to be undertaken on micro level, at the level of county or municipal government. Furthermore, conflicting and overlapping administrative boundaries further trouble implementation of meaningful policy. Lack of communication leads to an abundance of reports, covering the same topics, just slightly changing the scope. It is important to notice that an important part of this entanglement lays at the many different types of actors that are trying to work together in different ways and their diverse sets of values and objectives. Finally, it seems that the core issue is the lack of overview of existing incentive and organization.

4.1.2.5. Ownership

Land ownership is a crucial component of this system. Not only is the land fragmented by different types of land use or administrative boundaries but the quantity of owners mostly fragments it. The water that carries the nutrients travels through these different sheds of ownership. One important takeaway is that both ditch and buffer strip are owned and maintained by the farmer but have to meet EPA regulations. However, there are minor parts where county governments have jurisdiction over the ditches and streams.

4.1.2.6. Communication and awareness

There seems to be a serious lack of media coverage. Farmers are not aware of all the progress that has been made by the municipality, and the media is not willing to cover it. In many conversations I had, the fact that farmers will only implement if they know that something will have an impact, seeing is believing. Incentives such as demonstration farms, no-tillage breakfasts, and research demonstration, as well as word of mouth between farmers, cater toward adaptation.

4.1.2.7. Land use

The Maumee Watershed and surrounding areas are overwhelmed by the amount of industrial corn that has been planted, induced by the era of intensification and industrialization. Federal subsidies on mono-crop corn production are maintaining this system, and farmers will take action if the benefits of taking action outweigh the benefits of not taking action. In the long run, changing the system will probably only happen when the intensity of production lowers in the area.

These barriers are specific to the Maumee Watershed, Toledo, and the Lake Erie algal blooms, but thinking of this as a wicked problem, can we place them in a wider context? Which high-level challenges can we define that apply to innovating for systemic change while dealing with a wicked problem?

5. Societal challenges to implementation

Extreme weather events are not disasters; they become social disasters when they affect people [5]. Often the extremity of the consequences of a natural event is caused by deep-rooted societal problematic. There is a lack of integrated responses that link to geo-physical understandings of water systems with political, social, and cultural analyses and vice versa. We try to control the entire earth system to ensure the ecosystem services we need for growth, but society is still dependent on the geophysical processes that are in place. In our attempts to optimize our benefits and resource availability, we alter the cycles that inform our system hydrology (sediment, carbon, nutrient, and water) to our disadvantage. These cycles are what inform the hydrological movements within the geophysical fabric of time and space.

What challenges are at the base of societies' inability to plan for future conditions?

First, hydrological and climate models have increasingly become agents that inform that policy landscape, by putting the spatiotemporal dynamics of the geophysical system into a data model. However, most climate models do not accurately represent risk since they are often based upscaling of scenarios or historical events [49, 50]. Edwards (2003) identifies a scale of force that runs through the human body into the geo-physical sphere. The force of water can only be controlled within a certain range of natural variability. In other words, we cannot plan or predict for forces of strength or character that we have yet to experience and do not understand. As a political issue, climate change represents the dawning awareness that geophysical scales of force must be included in any complete analysis of infrastructure. However, we must realize and account for our limitations in predicting future ecosystem states [51].

Second, every implementation has an investment associated with it. A challenge in implanting large-scale water innovations is that investment cycles of capital are often too short, e.g., a government that is in place for 4 years [52]. Furthermore, governments are often risk averse; they are likely to invest in proven technologies that have guidelines and projections. After all experimentation with new measures can result in catastrophic social implications and loss of economic resources [50]. However, the solution for a wicked problem is not one that has been tested before.

Additionally, which institution is responsible, how are decisions being made, and based on what knowledge are they made? Edwards (2003) points out that multilevel governance slows down the implementation of measures, as different institutions get entangled with each other in decision-making and responsibility. Furthermore, administrative boundaries and the fact that water flows require collaboration between different regimes complicate the response. Watershed borders are different borders than administrative borders [51]. Then, legal control of water resources can be a constraint to innovation (e.g., water supply, permits) [50]. Finally, the political waterscape is largely influenced by the knowledge that is available. There are multiple, sometimes conflicting, disciplines through which water is studied. What scientific practices, infrastructures, and organizations shape water in different places? Several socio-political infrastructures have been built to address water issues such as the Environmental Protection Agency in the USA or the Waterboards in the Netherlands; it is important to consider which knowledge disciplines and actors are included and how these models shape the system (e.g., local knowledge). What historical, social, and cultural factors shape the development and use of scientific research and governance styles? After all, what counts as science and how are these disciplines framed changes through the decades?

Fourth, what are actual innovations and what is re-branding? Green infrastructure is a good example. One of the issues with the urban fabric is the fact that it covers what is underneath, limiting the infiltration of water. In some cities, hard surfaces can account for as much as 84% of the total surface [53]. Andersson et al. ([54], p. 156) define green infrastructure as "an interconnected network of multifunctional green-spaces that are strategically planned and managed to provide a range of ecological, social, and economic benefits." Examples include green roofs, public parks, urban wetlands, green streets, and bioswales [54]. The high-level benefits of green infrastructure are twofold; they regulate stormwater runoff as well as decrease the heat island effect but also allow for recreational benefits [55]. A key challenge regarding the implementation of green infrastructure lays in its perception. How can urban planners utilize green infrastructure to mitigate pressures on the urban ecosystem instead of merely using them to re-brand existing incentive as being "green" [56]?

Finally, although we attempt to shape the geophysical influences in our human time frame, these efforts will in the end not shape the geophysical indefinitely, as the system over millennia will evolve to different states. Current infrastructures on geophysical scale are fragile structures. Although we try to control the geophysical system within an era, we call the Anthropocene, in order to get water to our taps, we still rely on the cycles that make up the geophysical systems. What are scalable solutions, and do these even exist [50]?

So how do we move forward? How can we overcome these wicked problems and create resilient urban centers that will be able to deal with the pressure of climate change and rapid urbanization stake? How can we accomplish real innovation with visible results?

6. Innovating in wicked problems

Wicked problems pose serious challenge to the conventional view on innovation. The convention comprehension of an innovation process starts with the definition of clear ideas for solutions and go-no go decision-making in five subsequent main steps, sometimes subdivided into more. In this view, researchers elaborate on specified issue, and engineers define workable concepts. This research and development phase should deliver inventions. Then, the proof of concept is demonstrated, pilot production starts, and product is commercialized. This business development phase should deliver an innovation. The decisions are made in the selection process, referred to as a funnel. All costs of this process can be covered only if each step is successfully fulfilled; it is if the researchers elaborated on right questions, designers attained a novelty, customer are satisfied with it, pilot delivered sufficient quality, and sales generated profit. This demand-pull model is vested in the public administration and management handbooks [57]. It can reflect processes in large public institutions and corporations but rarely applies to small- and medium-scale businesses and individual experts who usually develop by trial and error using their knowledge and picking ideas available for free. Nevertheless, they are main sources of innovations, particularly when environmental issues are addressed [58]. Since the wicked problems are contextual, they can only be mitigated case by case without claim of applications to many other cases; the demand-pull model is rarely practical in such cases.

A different train of thought is embraced in the entrepreneurial model of innovations. Uncertainty, herewith, is considered a source of entrepreneurial operations and necessary for discoveries. The essential entrepreneurial skill is scanning and finding opportunities for discoveries due to differences in information and understanding between interests. Errors of some decision-makers are resources of innovation for others [59]. The innovating can be comprehended as an individual capability in using knowledge and skills that are available in society for creating new practical solutions. This use of knowledge and skills refers to knowledge spillovers being a metaphor for valuable interactions about the knowledge issues and solutions between people, which can vary from exchange about a cooking receipt to space travels. Although the value of interactions is rarely predictable, the conditions for knowledge spillovers can be fostered through education, cultural diversity, creativity, freedom of expressions, and suchlike, and the entrepreneurial skills in discovering opportunities for innovations can be enhanced through engagement of interests in networks, awarding of outstanding ideas and suchlike social relations [60]. In this view, creating conditions for the knowledge spillover and risk bearing is advocated rather than creating bureaucracy with the aim to reduce uncertainties through more specific selection of innovators, which is close to gambling and deplores innovation. Taleb wrote "Innovation is precisely something that gains from uncertainty: and some people sit around and wait for uncertainty and using it as raw material, just as our ancestral hunters" [61]. This notion of unpredictable innovation processes, meanwhile, entered into the mainstream of management theory, referred to as the effectuation theory [62]. Herewith, the entrepreneur is not striving toward a clearly defined goal based on the probabilities of success deliberated with regard to all relevant factors but has a multiple options to choose at a time depending on entrepreneurial personal preferences and given external conditions. Wicked problems, herewith, are tackled based on concepts of possible solutions designed for the specific situations and tested in cooperation with relevant interests. The successful solutions can be adapted and applied in other contexts. This trial-and-error process evolves when broad knowledge basis can be used.

Author details

Tamara Anna Streefland[1]* and Yoram Krozer[2,3,4]

*Address all correspondence to: tamarastreefland@gmail.com

1 Flatlands, Amsterdam, Netherlands

2 CSTM, University of Twente, Enschede, Netherlands

3 Jyothy Institute of Technology, Bangalore, India

4 Sustainable Innovations Academy, Amsterdam, Netherlands

References

[1] Ny Times. How Hurricane Harvey Became so Destructive. 28th August. Available from: https://www.nytimes.com/2017/08/28/climate/how-hurricane-harvey-became-so-destructive.h] [Accessed: 30th August]

[2] The Guardian. South East Asia Floods Kill 1,200 and Shut 8 Million Out of Schools [Internet]. 30th August 2017. Available from: https://www.theguardian.com/world/2017/aug/30/mumbai-paralysed-by-floods-as-india-and-region-hit-by-worst-monsoon-rains-in-years. [Accessed: 30th August]

[3] The Washington Post. Hurricane Harvey Shows. 28th August. Available from: https://www.washingtonpost.com/news/energy-environment/wp/2017/08/29/hurricane-harvey-shows-how-we-underestimate-flooding-risks-in-coastal-cities-scientists-say/?utm_term=.4affc0fc88e3 [Accessed: 30th August]

[4] Kaika M, Swyngedouw E. Fetishizing the modern city: The phantasmagoria of urban technological networks. International Journal of Urban and Regional Research. 2000 Mar 1; **24**(1):120-138

[5] Smith N. There's no such thing as a natural disaster. Understanding Katrina: Perspectives from the Social Sciences. 2006 Jun 11;**11**

[6] Rittel HW, Webber MM. Dilemmas in a general theory of planning. Policy Sciences. 1973 Jun 1;**4**(2):155-169

[7] Walker WE, Loucks DP, Carr G. Social responses to water management decisions. Environmental Processes. 2015 Sep 1;**2**(3):485-509

[8] Gerritsen H. What happened in 1953? The big flood in the Netherlands in retrospect. Philosophical Transactions of the Royal Society of London A: Mathematical, Physical and Engineering Sciences. 2005 Jun 15;**363**(1831):1271-1291

[9] Orlove B, Caton SC. Water sustainability: Anthropological approaches and prospects. Annual Review of Anthropology. 2010 Oct 21;**39**:401-415

[10] Dyson M, Bergkamp G, Scanlon J. Flow: The Essentials of Environmental Flows. Gland, Switzerland and Cambridge, UK: IUCN; 2003. pp. 20-87

[11] Sedlak D. Water 4.0: The Past, Present, and Future of the World's Most Vital Resource. New Haven & London: Yale University Press; 2014 Jan 28

[12] Chanson H. Hydraulics of roman aqueducts: Steep chutes, cascades, and dropshafts. American Journal of Archaeology. 2000 Jan;**1**:47-72

[13] Hodge AT. Vitruvius, lead pipes and lead poisoning. American Journal of Archaeology. 1981 Oct 1;**85**(4):486-491

[14] Nriagu JO. Saturnine gout among Roman aristocrats: did lead poisoning contribute to the fall of the Empire? New England Journal of Medicine. 1983;**308**(11)

[15] Wines M, Schwartz J. Unsafe lead levels in tap water not limited to Flint. The New York Times. 2016 Feb 8

[16] IPPC. In: Field CB, editor. Managing the Risks of Extreme Events and Disasters to Advance Climate Change Adaptation: Special report of the Intergovernmental Panel on Climate Change. New York: Cambridge University Press; 2012 May 28

[17] Tingsanchali T. Urban flood disaster management. Procedia Engineering. 2012 Jan 1;**32**: 25-37

[18] Jongman B, Hochrainer-Stigler S, Feyen L, Aerts JC, Mechler R, Botzen WW, Bouwer LM, Pflug G, Rojas R, Ward PJ. Increasing stress on disaster-risk finance due to large floods. Nature Climate Change. 2014 Apr 1;**4**(4):264

[19] McDonald RI, Weber K, Padowski J, Flörke M, Schneider C, Green PA, Gleeson T, Eckman S, Lehner B, Balk D, Boucher T. Water on an urban planet: Urbanization and the reach of urban water infrastructure. Global Environmental Change. 2014 Jul 31;**27**:96-105

[20] Gleeson T, VanderSteen J, Sophocleous MA, Taniguchi M, Alley WM, Allen DM, Zhou Y. Groundwater sustainability strategies. Nature Geoscience. 2010 Jun 1;**3**(6):378

[21] Wada Y, van Beek LPH, van Kempen CM, Reckman JWTM, Vasak S, Bierkens MFP. Global depletion of groundwater resources. Geophysical Research Letters. 2010 Oct 1;**37**(20)

[22] Klein N. This Changes Everything: Capitalism vs. the Climate. New York: Simon and Schuster; 2015 Aug 4

[23] Vörösmarty CJ, McIntyre PB, Gessner MO, Dudgeon D, Prusevich A, Green P, Glidden S, Bunn SE, Sullivan CA, Reidy Liermann CE, Davies PM. Global threats to human water security and river biodiversity. Nature. 2010;**467**(7315):555

[24] Eerkes-Medrano D, Thompson RC, Aldridge DC. Microplastics in freshwater systems: A review of the emerging threats, identification of knowledge gaps and prioritisation of research needs. Water Research. 2015 May 15;**75**:63-82

[25] Fitzsimmons EG. Tap water ban for Toledo residents. New York Times. 2014 Aug 3;**3**:A12

[26] Lake Erie Algae [Internet]. Available from: www.lakeeriealgae.com [Accessed: 30th August 2017]

[27] Holling CS, Goldberg MA. Ecology and planning. Journal of the American Institute of Planners. 1971 Jul 1;**37**(4):221-230

[28] Mollinga PP, Bhat A, Cleaver F, Meinzen-Dick R, Molle F, Neef A, Subramanian S, Wester P. Water, politics and development: Framing a political sociology of water resources management. Water Alternatives. 2008;**1**(1):7-23

[29] Wolff J. Cultural landscapes and dynamic ecologies: Lessons from New Orleans. In: Reed C, Lister NM, editors. Projective Ecologies. New York: Harvard Graduate School of Design and ACTAR; 2014. pp. 184-203

[30] Kiparsky M, Sedlak DL, Thompson Jr BH, Truffer B. The innovation deficit in urban water: The need for an integrated perspective on institutions, organizations, and technology. Environmental Engineering Science. 2013 Aug 1;**30**(8):395-408

[31] Corner J. Ecology and landscape as agents of creativity. In: Thompson GF, Steiner FR. editors. Ecological Design and Planning. New York: Wiley; 1997. pp. 81-108

[32] Fletcher D. Flood control freakology: Los Angeles river watershed. In: Kazys V editor. Barcelona: Actar Distribution: The Infrastructural City: Networked Ecologies in Los Angeles. 2008:36-51

[33] Fitzsimmons EG. Tap water ban for Toledo residents. New York Times. 2014 Aug 3;**3**:A12

[34] NRCD. https://www.nrdc.org/onearth/algae-almost-ate-toledo. [Accessed: 30th August]

[35] Nixon SW. Coastal marine eutrophication: A definition, social causes, and future concerns. Ophelia. 1995 Feb 1;**41**(1):199-219

[36] International Joint Commission. A Balanced Diet for Lake Erie: Reducing Phosphorus Loadings and Harmful Algal Blooms. International Joint Commission; 2014

[37] Force OL. Ohio Lake Erie Phosphorus Task Force Final Report. Ohio Environmental Protection Agency, Division of Surface Water; 2010

[38] Dodds WK, Bouska WW, Eitzmann JL, Pilger TJ, Pitts KL, Riley AJ, Schloesser JT, Thornbrugh DJ. Eutrophication of US freshwaters: Analysis of potential economic damages

[39] Turner K. Social Network Analysis of Lake Erie HABS Stakeholder Groups, Work in Progress. Kent State University; 2016

[40] Available from: www.binational.net

[41] Schindler DW. Recent advances in the understanding and management of eutrophication. Limnology and Oceanography. 2006 Jan 1;**51**(1part2):356-363

[42] Ohio EPA. Existing Nutrient Management Strategies. 2014. Retrieved from: http://www.epa.ohio.gov/Portals/35/wqs/NutrientMangementInitiaitives.pdf [April 16, 2016]

[43] Lecture Tom Bruulsema. Why the Fertilizer Industry Promotes 4R Nutrient Stewardship. International Plant Nutrition Institute, University of Guelph; November 25, 2013. http://phosphorus.ipni.net/ipniweb/region/nane.nsf/0/9FF16696EE81BD8B85257C2E004433BB/$FILE/Bruulsema%204R%20economics%202013.pdf

[44] NRCS. Available from: http://www.nrcs.usda.gov/wps/portal/nrcs/main/national/programs/financial/eqip/ [Accessed: 15th July 2016]

[45] Great Lakes Restoration Initiative. Great Lakes Restoration Initiative Action Plan. 2010. Retrieved from: https://www.glri.us/actionplan/pdfs/glri-action-plan-2.pdf

[46] EPA Great Lakes National Program Office. United States Great Lakes Program Report on the Great Lakes Water Quality Agreement. US Environmental Protection Agency; Dec 1997. Retrieved from: http://www.epa.gov/greatlakes/glwqa/usreport/part6.html [October 16, 2014]

[47] http://www.toledoblade.com/local/2017/07/22/Upgrades-to-Toledo-s-water-treatment-plant-over-budget.html

[48] Toledo Blade. Lake Erie Algal Bloom Likely to 1 of the Largest. July 13, 2017. Available from: http://www.toledoblade.com/local/2017/07/13/Scientists-predict-Lake-Erie-algae-will-not-cause-drinking-water-problems-this-summer.html [Accessed: 26th August 2017]

[49] Jongman B, Hochrainer-Stigler S, Feyen L, Aerts JC, Mechler R, Botzen WW, Bouwer LM, Pflug G, Rojas R, Ward PJ. Increasing stress on disaster-risk finance due to large floods. Nature Climate Change. 2014 Apr 1;**4**(4):264

[50] Kiparsky M, Sedlak DL, Thompson Jr BH, Truffer B. The innovation deficit in urban water: The need for an integrated perspective on institutions, organizations, and technology. Environmental Engineering Science. 2013 Aug 1;**30**(8):395-408

[51] Edwards PN. Infrastructure and modernity: Force, time, and social organization in the history of sociotechnical systems. Modernity and Technology. 2003;**1**:185-225

[52] de Graaf R, van der Brugge R. Transforming water infrastructure by linking water management and urban renewal in Rotterdam. Technological Forecasting and Social Change. 2010 Oct 31;**77**(8):1282-1291

[53] Gartland L. Heat Islands. London: Earthscan; 2011

[54] Andersson E, Barthel S, Borgström S, Colding J, Elmqvist T, Folke C, Gren Å. Reconnecting cities to the biosphere: Stewardship of green infrastructure and urban ecosystem services. Ambio. 2014 May 1;**43**(4):445-453, p. 156

[55] Pataki DE, Carreiro MM, Cherrier J, Grulke NE, Jennings V, Pincetl S, Pouyat RV, Whitlow TH, Zipperer WC. Coupling biogeochemical cycles in urban environments: Ecosystem services, green solutions, and misconceptions. Frontiers in Ecology and the Environment. 2011 Feb 1;**9**(1):27-36

[56] Heazle M, Tangney P, Burton P, Howes M, Grant-Smith D, Reis K, Bosomworth K. Mainstreaming climate change adaptation: An incremental approach to disaster risk management in Australia. Environmental Science & Policy. 2013 Nov 30;**33**:162-170

[57] Godin B. The linear model of innovation: The historical construction of an analytical framework. Science, Technology, & Human Values. 2006 Nov;**31**(6):639-667

[58] Freeman C. The greening of technology and models of innovation. Technological Forecasting and Social Change. 1996 Sep 1;**53**(1):27-39

[59] Kirzner IM. Entrepreneurial discovery and the competitive market process: An Austrian approach. Journal of Economic Literature. 1997 Mar 1;**35**(1):60-85

[60] Shane S. Prior knowledge and the discovery of entrepreneurial opportunities. Organization Science. 2000 Aug;**11**(4):448-469

[61] Taleb NN. Antifragile: Things That Gain from Disorder. New York: Random House; 2012 Nov 27. p. 405

[62] Sarasvathy SD. Causation and effectuation: Toward a theoretical shift from economic inevitability to entrepreneurial contingency. Academy of Management Review. 2001 Apr 1;**26**(2):243-263

Chapter 9

Waste in the City: Challenges and Opportunities for Urban Agglomerations

Jutta Gutberlet

Additional information is available at the end of the chapter

http://dx.doi.org/10.5772/intechopen.72047

Abstract

Worldwide cities are rapidly expanding, creating visible environmental and social challenges. The generation of waste is one of the central concerns in urban agglomerations, particularly in the global South, where inadequacies, absences and weaknesses shape the local waste management system. Uneven geographic development has created obvious spaces of exclusion and neglect. In response, informal and organized waste pickers engage in selective waste collection and recycling, serving their community and the environment. These contributions are still mostly unrecognized and unaccounted for. This chapter begins with emphasizing the challenges of urban growth, consumption, poverty and waste. In the global South, every day millions of informal waste pickers reclaim recyclables from household waste to earn their living. In doing so they make an important contribution to reducing the carbon footprint of cities, recovering resources, improving environmental conditions and health creating jobs and income among the poor, particularly in low-income residential areas. This chapter discusses the organization of these initiatives into networks and examines the challenges and benefits of such practices that promote grassroots resilience and contribute to reducing both the adverse impacts of cities on climate and environmental change (UN sustainable development target # 11.6) as well as urban poverty (Goal # 8).

Keywords: urban growth, global South, urban development, waste management, sustainability

1. Introduction: rapid urban growth, waste and inequality

Since 1950 the world's urban population has grown from 746 million to 3.9 billion in 2014 [1]. In the global South, most cities, particularly the metropolitan areas are rapidly expanding into large urban and suburban agglomerations, with so called "in-between cities," where some of the rural characteristics are still mixed into the urban fabric. Cities attract people for many reasons, and most often unemployment and the prospect of a better life with improved and safer living conditions is the key driver to urban growth. Migration, particularly from rural to urban, but also different forms of population movement from other cities, regions and even from other countries are responsible for rapidly changing the urban population [2]. In some parts of the world natural population increase is still on the rise and coupled with higher average life expectancies, population sizes are still becoming bigger. This dynamic urban growth can generate significant stress on city administrations who need to provide the necessary basic infrastructure and public services to expanding neighborhoods and new settlements. As a result of the incapacity to provide these, part of the population lives in extreme poverty and under critically neglected living conditions, often causing sever health challenges to their families and surrounding community [3].

This chapter draws on many years of research and outreach experience with informal and organized waste pickers in different cities of the world. I have learned through participatory action research lenses and in community based research approaches focusing on everyday praxis in the city in the global South, with a particular eye on waste. This reseach practice uncovers post-colonial contexts of waste and value, including gender, class and race perspectives, urban transformation and infrastructure impacts or related challenges in the global South. I am informed by feminist theory, which uncovers power relations and embraces the concepts of equality and equity as crucial in the outcomes of urban development. The research seeks to empower vulnerable populations and value their knowledge grounded in everyday experiences and takes into consideration masculinist power and representation. I acknowledge that the local expertise and understanding cannot be fully realized from the outside [4]. Political Ecology is relevant to urban analysis, because it is inclusive of these multiple layers and actors that shape urban landscapes over time.

Section 1 of the chapter introduces the concept global South and the Urban Political Ecology (UPE) framework. Then, in Section 2 I provide a brief contextualization of waste management in urban agglomerations in the global South, in terms of characteristics of waste and prevailing forms of dealing with waste. I present some of the current social and environmental challenges linked to waste. Section 3 presents the idea of social grassroots innovations, coming from waste pickers. The UPE lens situated in the global South context looks at household waste and some of the grassroots actors, the processes and transformative practices they bring to waste management. There are concrete livelihood opportunities attached to collecting, separating, trading, adding value, and in performing environmental education and technical training in waste management. The final Section 4, highlights some of the insights gained from waste pickers and their organizations that contribute to a place-based understanding of working with waste, grounded in their concrete experiences. The key recommendation in this chapter underlines the important role of public policies in stimulating grassroots development and to address the serious challenges waste and disposal pose in urban agglomerations.

1.1. Contextualizing the global South

The global South is a spatial and historical concept used to facilitate the understanding of commonalities and differences between countries. However, as a category of places, there is the risk of presenting a rather dualist perspective on development, opposing the South with the North, or even interpreting the term as geographic location, which of course is not the objective. The term recognizes the shared characteristics related to the historical processes experienced under colonialism and imperialism, which have strongly shaped their economies and cultures. The term recognizes situated differences in the multi-scalar processes and transformative practices observed among countries, regions and places. Global South is a term that provides a telling difference from countries we call the global North. Yet, the lived experiences in these locations (both in the North and in the South) are multiple, temporal and place specific. Cities differ immensely from each other and cannot be put together under the same banner. Therefore, a dichotomous division between two worlds would not be tenable empirically and also not desirable politically. It is a contested term, but yet it helps us grasp common causes and consequences of unequal power relations, manifested in everyday urban politics with high levels of inequality and persistent poverty.

Conceptualizing the global South brings to life the specific historical social, economic and political processes unfolding, that find their epicenter in urban experiences in the global South. The bulk of urban growth is now happening in that part of the world and we see urban imaginations, based on processes that are primarily taking place in the global North shifting to patterns that evolve from the global South, as becoming more relevant [5, 6].

With urban growth consumption rates are also on the rise globally. Worldwide cities generate over 720 billion tons of wastes every year [3]. In cities people mostly rely on industrialized and heavily packaged food, significantly adding to the quantity of household waste generated every day. Waste is not yet perceived as a critical challenge, as a socio-ecological issue of highest priority to city administrators nor to the community, and waste is treated mostly with "end of pipe" measures, rather than pro-actively curbing generation and discard of waste, thus reducing the use of virgin resources and stimulating circular resources flows. Yet, in many cities waste is an obvious and visible problem, with uncollected waste amounting in public space, affecting the water quality and environmental health in the city. Waste collection services are often unequally provided within cities, with observable patterns of social and environmental injustices related to waste accumulation and availability of waste infrastructure and services. Those services that are provided usually focus primarily on collection and disposal [3].

1.2. Taking a Situated Urban Political Ecology lens

Situated Urban Political Ecology (UPE) is a detailed framework to better understand the ways in which society, politics and cultural behavior influence the city's metabolic processes, such as flows of resources and outputs as well as the related actors in the city [7]. It recognizes the fact that cities are built with resources from nature and that in turn cities also shape nature into urban landscapes. Waste flows and those actors that work with waste and in waste management, for example, bring to life the combination of natural and societal forces involved in urban metabolic processes that create urban landscapes and geographies of inequality in terms of social and environmental justice. In such a framework, the challenges and opportunities that come with the recovery of recyclable materials from waste become obvious. The situated UPE lens helps us understand how the legacies of colonialism have shaped who benefits and in which ways they benefit from the urban environment. Power structures manifest in urban development and urban policies equally define who has access to resources and services. Waste management is increasingly becoming one of the greatest challenges, particularly for a highly urbanizing world. It is closely connected to physical *and* social processes taking place in the city, but also in the expanding suburbia and in the "in-between city" [8].

UPE sees urbanization as a political process of socio-ecological change, which can also be studied as a process of socio-metabolic transformations [9]. The metaphor of *urban metabolism* sees the city as a living organism with flows of resources going in and out of the city, actors intervening in the transformation and consumption of these recourses, and with related services and product outputs. This is a systems perspective of the city, where social processes, spatial form, and the material and energy metabolism are equally connected and interwoven [10]. The analogy helps us identify and map waste flows and key actors involved in the collection, transformation and final destination of the resources embedded in waste. We can, for example, recognize social relationships and power relations that underpin activities related to waste under different forms of "waste regimes" and waste management systems. The concept of "waste regime" seeks to understand the economic, political, and material dynamic through which waste is produced, conceptualized, and politicized [11]. Waste regimes are bound to specific historical, cultural and geographic contexts. Waste management systems involve different forms of technology, automatisation and practices in waste collection, transformation and final destination.

How is it, that certain values prevail, whereas others are undermined, and, how do these "value regimes" [12] operate in different ontological, cultural, material, and political settings? Urban metabolism analysis studies the entry, transformation and storage of materials and energy and the discharge of any kind of waste and unwanted products. Here, infrastructures and services play crucial roles in maintaining cities and providing for the residents. Cities surely are complex systems. With a dynamic and cyclical perspective applied to planning and development, this approach shows where cities are not livable, are unhealthy and unsustainable or are unjust and inequitable [13].

The UPE focus directs attention to social power relationships and how these produce historically specific social and physical natures. Related to waste management different actors, with more or less levels of inclusion and power can be mapped. The scope of those dealing with waste is wide, ranging from small to large and even multinational contractors, government officials, recycling businesses, middlemen (scrap dealers), organized recycling cooperatives and associations to informal waste pickers. In addition, there are the everyday experiences with waste of ordinary people, governmental and non-governmental actors, contractors, developers, and so on. What are the values embedded in the roles played by the diverse institutions and actors? Where do they locate and where do they position themselves, in the local and global processes of treating, sorting, trading, and recycling waste? There are apparent and hidden social justice issues related to control, ownership, and appropriation of waste management resources and technologies. As already hinted, there are uneven geographical processes at play, inherent to the production of urban environments. In the formal part of the city waste is regularly collected, while in the informal neighborhoods these services are neglected. Sometimes the infrastructure and service gap is filled by grassroots initiatives. The following section will describe some of the key challenges city dwellers in the global South are currently facing.

2. Waste challenges in the global South

Waste constitutes a key developmental and environmental issue. It is an almost unavoidable consequence of human activity. Today humans generate more waste than ever before, not only because of dramatic population increase over the past centuries, but also because of the changed nature of consumption and the different composition of solid waste. A shift toward waste minimization and away from depositing it at landfills is important. Per capita consumption of packaged goods and consumer products has skyrocketed after World War II, with the rapidly expanding adoption of growth and consumption oriented economic development. This is when material consumption gained momentum on a global scale [14]. Waste in the city is a transversal theme; it affects water quality, causes flooding (e.g., urban storm waterlogging due to trapped waste in water drainages), generates public health issues by hosting disease vectors, affects the perception of public space (e.g., as a space of neglect and lack of citizenship) and furthers the sense of exclusion. But waste also has other social, economic and environmental facets, which will be discussed further on.

Post-consumer waste generation has more than doubled worldwide, between 1971 and 2002. In the global South, growth in municipal solid waste generation has become exponential from the 1980s onwards, and it continues to steadily grow in most of the global North, except for Central and Eastern European countries and the Former Soviet Union [15]. While Western Europe and North America on average already experienced municipal solid waste (MSW) rates between 1.4 and 1.8 kg/capita/day over the past decade, the population in many large cities in the global South is now also reaching values between 1 and 1.4 kg/capita/day [16]. The urban lifestyle contributes to higher waste generation not only in people's homes but also outside. Particularly the food service industry thrives on disposables. Today, people consume more in the streets and their consumption leaves more disposable waste in public waste bins. In 2012, urban residents globally generated about 1.2 kg/capita/day of MSW, compared to 0.64 kg in 2002 [17]. In Brazil, the average daily quantity of MSW generated per person is currently about 1.1 kg. For major cities in Africa MSW generation is estimated to range from 0.3 to 1.4 kg/capita/day [18]. Differences in waste generation can be large, as demonstrated by data for Bamenda and Yaounde (the capital) in Cameroon, which generate 0.5 and 0.8 kg/capita/day, respectively [18]. Population size and growth rates are important factors that influence municipal solid waste management. There is a positive correlation between population size and both, the rate of waste produced and the percentage of households enjoying regular waste collection. Yet, it is clear that rapidly growing cities have a hard time in providing consistent waste collection services.

Under the current era, industrial production of consumer goods is characterized by a reduction in product life spans, growing product variety, material component diversity, and increased packaging. All these characteristics are drivers for increased use of natural resources and are responsible for generating waste and producing water, soil and air contaminants. The rise in solid waste is linked to increased levels of urbanization and wealth. Between 1997 and 2007, the Gross Domestic Product (GDP) in India has increased by 7%, while estimates indicate a rise in municipal solid waste over these 10 years by 45%, from a total of 48 million to 70 million tons [19]. The figures for Brazil demonstrate a similar correlation between wealth and solid waste generation. From 2009 to 2010, GDP rose by 7.5%, while MSW increased by 6.8%. In the following year, GDP slowed down with an increase of 2.7%, and MSW generation increased only by 1.8% [20].

Population growth comes with an increase in consumption and waste. More affluent segments of the population consume more and generally their consumption also produces a larger environmental impact. China, India and Brazil alone have added another 509 million new consumers between 1990 and 2000, with an average purchasing power of 839 billion US$ [21]. These "new consumers" are defined as *"people within typically four member households with purchasing power of at least PPP $10,000 per year, i.e., at least PPP $2,500 per person … (PPP dollars are between 1.3 and 5.3 times higher than conventional dollars in 20 countries* - 17 developing and 3 transitional countries)" [21], p. 4963. Increased income enables consumers to purchase household appliances, electronics, cars, and other items that mark affluent lifestyles, including the consumption of more packaged food items and meat. *Our current waste regime is characterized by an exponential increase in volume and material diversification of discarded objects and substances, as a consequence of increases in packaging, shorter product durability, programmed obsolescence, economic growth logic, consumerism, and mass consumption* [22], p. 58. All these factors are responsible for driving solid waste generation. Consequently, cities have to cope with large quantities of solid waste. This includes household waste, construction waste, industrial waste organic waste (e.g. from public parks and other green spaces), and often also diverse forms of toxic waste produced in the city. Of particular concern is the waste that is not collected and that accumulates in illegal dumps, in streets, riverbeds or unoccupied spaces, often following certain patterns delimiting distinguished formal and informal spaces in the city.x

2.1. Composition of household waste

Waste composition reflects cultural and technological trends and varies greatly between different continents and regions over time. There are many technical aspects involved in creating more sustainable and equitable waste management services. While ashes from heating and cooking, e.g., were reported as large components of household waste in North America until the middle of the last century, plastic appears only since the 1970s as a separately recorded substance [23]. Urban waste in the global North currently contains more recyclable goods and electronics, while municipal waste in the global South still has a larger biodegradable fraction and less recyclable material content. Often these valuable materials have already been reclaimed by the household or by informal recyclers for reuse or trading.

In African cities, the organic content of household waste is still much higher and tops 70% [18]. The household waste composition in Brazil is still typical for the global South, with large

http://dx.doi.org/10.5772/intechopen.72047

fractions of organic (51.4%) and recyclable (31.9%) materials (metals, paper and cardboard, plastics, and glass), and a small proportion classified as other materials (16.7%) [24]. Yet, here the amount of electronic waste is quickly growing, increasing the demand for E-waste recycling.

2.2. Characteristics of waste management

Most municipal solid waste generated worldwide is still deposited at landfills and waste dumps (70%), while 19% is officially recycled or treated by mechanical or biological treatments and a small proportion is incinerated (11%) [25]. Landfill technologies differ from open dumping to sanitary landfills, with methane capturing. The burning of waste is common, particularly in and around informal settlements and in rural areas. Although worldwide many countries are upgrading their landfills to sanitary landfills, as has happened, for example, in South Africa, Uganda, Ghana and Egypt a decade ago, at the time raised the concern that most landfills in Africa are *"owned and operated by the very body that is supposed to enforce standards. The philosophy of getting waste out of sight and consequently out of mind seems to be the overriding consideration of these authorities"* [18], p. 17. As a consequence, most resources which are limited are spent only on the removal of waste, particularly in formal neighborhoods, and little investment is done in the infrastructure for more sustainable waste management.

Some cities in the global South also adopt expensive waste management models, e.g., mechanized separation systems for recycling or high tech *Waste to Energy* incineration. These waste management options generate very little employment and are not financially sustainable; often locking governments into long-term waste management contracts, preventing the use of more appropriate technologies. Yet, policy makers are increasingly interested in the social aspects of waste. They have learned through experience that not considering the social aspect of waste compromises the implementation of their policy goals and often results in detrimental and costly social effects, particularly for vulnerable social groups.

Informal collection of recyclable and reusable materials is widespread in the global South and significant amounts are recovered. At the same time formal recycling programs are still rare and are most often insignificant in terms of the percentage of recovered materials. There are environmental (and health) impacts as well as benefits of various degrees involved in the act of informally collecting, separating, redirecting and recycling materials contained in waste. Organized door-to-door selective collection of recyclable materials, in particular, embodies opportunities for environmental education in the community; helping shift attitudes and values away from current wasteful consumption patterns and habits, toward reuse and informed, educated consumption and disposal.

In the case of Brazil, 80% of the country's household waste is regularly collected, and the primary final destination for it is sanitary landfills (58.1%) and controlled landfills (24.2%). The rest gets deposited at unprotected waste dumps (17.7%) [24]. In 2016, only 927 municipalities (17%) in Brazil had some sort of official selective waste collection in place [26]. As in most countries in the global South, selective waste collection happens primarily through informal waste collectors. They have historically been stigmatized and denied epistemic agency. It is crucial that research interrogates how shifts in the waste and recycling systems can change how society perceives waste pickers and also how waste pickers construct themselves and their praxis, in order to build up an efficient and inclusive waste management system.

2.3. Impacts from waste infrastructure

Landfills are still necessary, but when uncontrolled they are a source for environmental impacts on soil, water and air. They are located close to urban agglomerations, sometimes competing with environmentally protected areas. Landfills and dumps generate significant greenhouse gases (GHGs), primarily methane (5–10% of global methane is emitted by landfills) and carbon dioxide, as microbial communities decompose the organic matter contained in the waste [27]. Converting open dumping and burning to sanitary landfills implies *"control of waste placement, compaction, the use of cover materials, implementation of surface water diversion and drainage, and management of leachate and gas"* [15], p. 595, thus improving the carbon footprint of waste disposal. Ironically, the landfill upgrading process now creates a shift from mostly CO_2 emissions from aerobic decomposition and burning to CH_4 emissions, which continue for several decades after waste disposal. However, methane emissions from landfills can be stabilized with gas recovery technology, as is already widely implemented in the global North and beginning to be adopted in many countries in the global South. Landfills further impact the soil and groundwater with leachate produced as water percolates intermittently through the refuse pile. Leachate can contain high levels of nutrients (nitrogen, phosphorous, potassium) heavy metals, toxins (such as cyanide) and dissolved organic compounds. One of the big challenges in most cities is to ensure that all operating landfills are designed properly and are monitored once they are closed. For that, local governments need to access funding programs that seek to improve the condition of landfills.

Mismanaged and uncollected waste is a public health hazard. Abandoned waste attracts disease vectors (including rats, mosquitoes) and if carried into waterways leads to storm waterlogging, causing inundations [28] and consequent public health hazards. When burned, a number of toxic substances are emitted, impacting local neighborhoods.

Waste incineration (including *Waste to Energy*) and other thermal processes are local sources of air pollution, constituting additional health risk factors to city dwellers, who often already have to cope with serious air contamination issues. These installations produce CO_2 from fossil carbon sources and generate other contaminants such as dioxins, furans, polycyclic aromatic hydrocarbons (PAHs), volatile organic compounds (VOCs), mercury (Hg) and many other GHGs. Particularly, the fly ashes and slags become a hazardous output [16, 29]. *Waste to Energy* is attractive as a quick solution addressing the growing solid waste generation and increasing energy requirements in the city. It is an expensive technology often only viable through public private partnerships (PPPs). Waste management is a multibillion dollar industry. One of the leading waste technology corporations worldwide is *Veolia*. This company's main new activity is *Waste to Energy*, generating many billion dollars of revenue every year. According to their forecast the global market in this sector will be worth about 30 billion euros in 2020. They already have contracts in 33 countries and are recently expanding in the global South [30]. In many countries waste incineration has become a threat to waste pickers, for whom household waste is a resource. The city's commons make their livelihoods from collecting recyclable materials and feeding them into the circular economy. Yet, even in cities where organized waste pickers perform this service, waste incineration is under consideration, funded through PPPs.

http://dx.doi.org/10.5772/intechopen.72047

Morris [31] argues that recycling mixed solid waste saves more energy than generated by *Waste to Energy* facilities. His findings underline that recycling conserves energy that would be used to extracting natural resources and transforming them to produce goods that can also be manufactured from recycled materials. Mining, extraction, transportation and transformation of natural resources, generates environmental impacts (often in pristine environments) and also emits greenhouse gases (GHG), affecting the global climate. Recycling values the resources and energy incorporated in the making of these products, often allowing for new goods to be manufactured from the recovered materials. It is global consensus that *"the climate benefits of waste avoidance [reuse] and recycling far outweigh the benefits from any waste treatment technology, even where energy is recovered during the process"* [32], p. 1. Recycling values labour and generates income to the many different actors along the value chain. With well-designed and functioning recycling operations, all resources can be recovered from the waste stream for re-use and often for up-cycling in the generation of new products.

Another urban environmental issue relates to the fact that waste and recyclable materials often travel long distances. De-regulation and globalization re-shape the movements of these materials. Transportation uses energy and adds to air pollution, traffic and noise in large urban agglomerations. Worldwide, half of all plastics, paper and scrap metals are exported to South East Asia. China is leading dealing with recyclable material, with importing over 7.4 million tons of plastic waste, 28 million tons of waste paper and 5.8 million tons of steel scrap; mostly treated in backyard shops or small-scale industries [25]. More recently, particularly the transcontinental shipping of electrical and electronic equipment waste (WEEE) has become a serious challenge, especially as it is shipped to global South cities. 70% of the global WEEE ends up in Chinese cities [33]. While the rough dismantling of E-waste (recovering plastics, copper and other metals, etc.) happens in the global South, reclaiming the high value components (rare earths) happens in the global North, who is in possession of the specific recycling technology. Waste trafficking is often illegal and *"has become institutionalized practice among certain corporations that pollute, dump toxic waste and make environmental crime victims of various global minorities"* [34], p. 103. This section has outlined a serious of challenges that can be avoided or addressed in good waste governance, as will be discussed further in this chapter.

3. Resource recovery and social grassroots innovation

The bulk of material recovery in the global South is informal, grassroots and involves a wide spectrum of domestic reuse of bottles, cans, plastics, paper, cardboard and many other discarded materials. Yet, its role is largely unrecognized in waste management and by city authorities. In Delhi, India 15–20% of the MSW (daily 1,275 to 1,700 tons) is collected by informal recyclers. The waste pickers also redirect 200 tons per day of separated organic material to a large-scale composting plant. They collect organic waste from households in the affluent neighborhoods, where they compost it in a series of community composting pits [35]. Often, the lack of local markets for recyclables is still a prevailing limitation for the recycling activity to further flourish [18].

A well-known example for informal grassroots recycling is the work of the *Zabaleens* in Cairo, recovering approximately 6,000 tons of MSW per day (up to 80% of the waste generated in the city) of the material entering the solid waste system, compared to 11% formal recycling [36]. In many cities around the world a large number of workers recover significant quantities of recyclable material from the waste stream, generating savings of around 20% or more to the municipal waste management budget, which in large cities can represent many millions of dollars per year [37]. The city of Lima, Peru has no formal recycling program and relies entirely on the local *recicladores*, including informal service providers, street collectors and dump pickers, to divert 20% of the MSW into recycling, as a report by GIZ/CWG found [14]. These examples illustrate a common situation in waste management in urban agglomerations in the South.

The study by GIZ/CWG has translated the environmental benefits associated with informal material recovery as reduced negative externality costs, expressed in Euros. According to their studies the informal recyclers generate 97.6% of these externality costs in the case of Lima, Peru and 83.4% in Cairo, Egypt [38], p. 21. There is evidence in most big cities that informal workers perform a service that saves city expenditures.

Innovations in waste management from the grassroots level bring many social and environmental benefits that tackle the UN sustainable development target # 11.6, *making cities and human settlements inclusive, safe, resilient and sustainable*, as well UN sustainable development target # 8, *promoting sustained, inclusive and sustainable economic growth, full and productive employment and decent work for all* [39].

Several questions remain prominent for a paradigm shift in waste management. One of these questions is how we can get the true recognition for the creation of jobs and improvement of livelihoods from informal and organized recycling. Particularly organized waste pickers are a grassroots source of innovation.

Another question addresses how we can stimulate behavioral change toward prevention, reuse and recycling. Informal sector recycler are those individuals or enterprises that are involved in private sector recycling and waste management activities which are not sponsored, financed, recognized, supported, organized or acknowledged by the formal solid waste authorities, or which operate in violation of or in competition with formal authorities [40]. Waste pickers are carriers of grassroots innovations and have many lessons to share that can help improve municipal waste management systems. In many countries waste pickers have organized in cooperatives, associations, networks or social movements. *"Grassroots innovation movements seek innovation processes that are socially inclusive towards local communities in terms of the knowledge, processes and outcomes involved"* [41], p. 114.

3.1. Place based informal and organized recycling

Amid the pressures of climate change, population growth, industrialization and urbanization, one of the major challenges faced in global communities is the sustainable and equitable access to infrastructures, services and resources. There is usually a complex network of actors in waste governance, including residents, waste pickers, waste managers, engineers, bureaucrats, consultants, businesses, but also activists, journalists and scientists. These actors often do not agree on how waste related problems are defined or get solved, nor do all of these actors unanimously recognize that different sources of knowledge are needed to solve these problems. There might even be divergence on what type of knowledge to use, how it is produced and communicated across different societal sectors and actors.

People's relationships to waste and the meanings attributed to waste reveal about culture and society. In order to achieve a fundamental shift in how we see, generate and manage material waste we need to involve other stakeholders and their knowledge. Waste pickers contribute to developing, understanding and solving waste management problems. Innovative governance models can potentially emerge from a dialog with organized waste pickers creating collaborative relationships in providing waste services. Transdisciplinary understanding of waste encompasses this collective approach, bringing together the formal and non-formal actors for creation, communication and use of waste-related knowledge.

In this chapter, I have provided diverse examples for informal recycling activities, highlighted within different situated contexts. *"Waste picking is a key occupation in solid waste management in most cities in the global South. Waste pickers can range from poor people rummaging through garbage in search of food, clothing and other basic, daily needs to informal private collectors of recyclables for sale to middlemen or businesses, as well as organized collectors/sorters of recyclables linked to unions, cooperatives or associations"* [42], p. 6. Waste picker contribute environmentally, by reclaiming resources and channeling these into the circular economy. In their work, they are constantly challenged by everyday life decision-making and the daily challenges shape them as grassroots innovators. They work locally in the community, with high levels of participation in decision-making and flat hierarchies [43]. Despite, often being dismissed, this population hosts key actors in waste management.

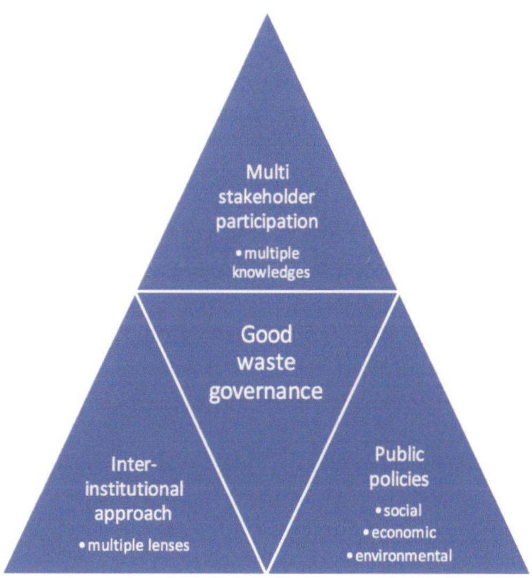

Figure 1. Major components of good waste governance.

Social aspects of waste management, or the socio-economic advantages of recycling, as highlighted by [44–46], are not yet widely recognized and comprehensive social indicators demonstrating the social contributions of organizing waste pickers are yet to be developed, in order to be able to clearly measure the benefits deriving from that work to society. From practice, we know that inclusive waste management generates positive contributions to democracy. During the negotiation process between recycling cooperative and local government for waste management service contacts, e.g., waste pickers as citizens affirm their rights to have a voice and to participate in these decisions, thus strengthening democracy. Waste governance decisions can also undermine democratic relations between citizens and the state and even further deepen inequality and poverty. In contrast, good waste governance embraces the following building blocks, as shown in **Figure 1**.

4. Final considerations

Waste constitutes a major challenge to city administrators and urban populations at large. However, waste is not perceived as an "issue" yet. Waste is treated through the engineering lens rather than from an interdisciplinary perspective. We need to move beyond seeing waste as a merely technical issue and move towards a complex socio-environmental-technical understanding of waste. Learning from the praxis of a wider range of stakeholders (including waste pickers, elected officials, waste managers, private companies and middlemen or scrap dealers) is critical to either facilitating or hindering transformations in the waste and recycling systems.

Urban communities have a say in what happens to their waste and who has access to waste. They must have a say in the decision-making whether to invest in expensive waste management technology, without prioritizing job creation or whether to support labor intensive, inclusive forms of waste management and resource reclamation. Cities can promote a shift towards waste minimization and resource recovery. Waste governance decisions need to also be based on "good governance" principles, including democracy and consensus orientation, participation, accountability, transparency, responsiveness, equity and inclusiveness, be effective and efficient and following the rule of law [1]. These guiding principles should also be applied to waste governance and specifically applied in waste management.

When it comes to deciding over which waste management process and technology to favor and the design of specific policies, the following questions are relevant for local governments.

(1) Who should be involved in policy and decision-making (key stakeholder, e.g., waste picker organizations, local business associations, educational sector, NGOs, experts)?

Participation is not without challenge and stakeholders have to ask what is their mandate? What are the local political realities? What is the available budget? What are the priorities within the city? and so on.

(2) What technology is most appropriate in terms of:

- environmental concerns (air pollution, water and soil contamination)

- poverty reduction and employment generation
- economic sustainability (cost – benefit, short to long term)
- environmental sustainability (resource savings and reclamation, reduction in GHG emissions, etc.)

According to the Intergovernmental Panel on Climate Change (IPCC), solid waste and its management are considered key contributors to climate change. Greenhouse gases are emitted or avoided in the upstream and downstream stages in the life cycle of municipal solid waste management systems [47]. Upstream emissions can be avoided when recycled resources replace virgin resources in the fabrication of metal, glass, plastic and paper products. In addition, landfill gas (CH_4) and deforestation represent other upstream impacts that are reduced with recycling [48–51]. Fossil fuel greenhouse gas emissions are of course also associated with recycling operations, as energy and some virgin resources are consumed during the collection and transportation of materials, processing, and re-manufacturing [52]. With recycling, however, both methane (CH_4) and carbon dioxide (CO_2) emissions are avoided through the diversion of resources from landfills, through resource recovery and recycling of paper, cardboards and other biodegradable material [47, 53], and through reducing the amount of waste to be deposited at landfills.

Research underlines the need to redefine clean development mechanisms (CDMs) to allow for the recognition of resource recovery for reuse and recycling as measures to reduce GHG emissions, save natural resources and energy [54]. Recycling has not yet been considered a CDM, while *Waste to Energy* and Methane to Energy projects, associated to landfills, are already funded under this mechanism [55]. These shortcomings need to be addressed by including a social development agenda within CDM policy frameworks. Taking a social perspective on thermal treatment of solid waste the outcome is aggravated by the fact that this form of waste management destroys the resources in waste and thus, the source of income for waste pickers and recyclers.

There are challenges and limitations related to recycling (down-cycling, up-cycling) which governments should discuss and act on. There are often not enough down-cycling alternatives for many waste materials and waste flows. Here too, cities can become drivers for innovative forms of reuse and recycling. Not to forget is the fact that collection, transportation and processing of waste and recyclables also generate fossil-derived carbon dioxide and other pollutants from the fuel used in transportation, and therefore also needs to enter the equation.

Millions of informal waste pickers collect household waste daily in cities around the globe to earn a living. In doing so they contribute to reducing the carbon footprint of cities, recover resources, improve the environmental conditions and health in the city. The research discussed in this chapter points towards a radical economic and social shift away from growth centered urban development and *end of pipe* waste management; towards steady state development, embracing de-acceleration, stability, sufficiency and sustainability. Recovering materials for reuse or recycling is a step towards that direction and helps avoid natural resource extraction. Inclusive recycling addresses poverty and unemployment issues. The work of informal recyclers, as resource reclaimers—for recycling and reuse—and the role they play as environmental educators can be extended to gradually recover more of the materials that are still wasted in landfills or incinerators, progressing towards more resilient and healthy communities; this implies a different, participatory waste governance. Waste management is an important field of urban governance. The success or failure of governments is linked to how they deal with waste and with the responses society is already producing. Waste governance is more than just having the right laws and policies in place and having institutions enforce the policies. It is also about levels of democratic participation, recognizing other forms of knowledge, and understanding the links between waste, value and society in order to tackle broader social, political, cultural and economic issues that affect the urban agglomerations. Inclusive waste management has the potential of bridging a social gap. Finally, weak urban governance (government that lacks participation, democracy, transparency, equity and inclusiveness) is probably the major impediment to proper solid waste management.

Author details

Jutta Gutberlet

Address all correspondence to: gutber@uvic.ca

University of Victoria, Victoria, B.C., Canada

References

[1] United Nations-Economic and Social Commission for Asia and the Pacific. What is Good Governance. Bangkok: UNESCAP; 2014. http://www.unescap.org/search/node/good%20governance [Accessed: 10-10-2017]

[2] WBGU (German Advisory Council on Global Change). Humanity on the Move: Unlocking the Transformative Power of Cities. Berlin: WBGU; 2016

[3] UN-HABITAT (United Nations Human Settlements Programme). State of the Art of the World's Cities 2012/2013 - Prosperity of Cities. Nairobi: UN-Habitat & New York: Routledge; 2013

[4] Hesse-Biber S, Yaiser M. Feminist Perspectives on Social Research. New York: Oxford University Press; 2004

[5] Arabindoo P. Rhetoric of the 'slum': Rethinking urban poverty. City. 2011;**15**(6):636-646. DOI: 10.1080/13604813.2011.609002

[6] Parnell S, Robinson J. (Re)theorizing cities from the global South: Looking beyond neoliberalism. Urban Geography. 2012;**33**:4

[7] Heynen N, Kaika M, Swyngedouw E, editors. In the Nature of Cities: Urban Political Ecology and the Politics of Urban Metabolism. London: Routledge; 2006

[8] Keil R, Young D, Wood PB, editors. In-Between Infrastructure: Urban Connectivity in an Age of Vulnerability. Kelowna, British Columbia: Praxis (e) Press; 2011

[9] Rapoport E. Interdisciplinary Perspectives on Urban Metabolism. A review of the literature, UCL Environmental Institute Working Paper. Development Planning Unit, UCL; 2011

[10] Swyngedouw E, Heynen NC. Urban political ecology, justice and the politics of scale. Antipode. 2003;**35**(5):898-918

[11] Gille Z. From the Cult of Waste to the Trash Heap of History: The Politics of Waste in Socialist and Postsocialist Hungary. Bloomington: Indiana University Press; 2007

[12] Appadurai A. Introduction: Commodities and the politics of value. Appadurai, A. (Ed.). In: The Social Life of Things: Commodities in Cultural Perspectives. Cambridge: Cambridge University Press; 11th Edition; 2013

[13] Newman P, Jennings I. Cities as Sustainable Ecosystems: Principles and Practices. Washington, DC: Island Press; 2008

[14] United Nations Human Settlement Program (UN-HABITAT). Solid Waste Management in the World's Cities. London, UK: Gutenberg Press; 2010

[15] Bogner J, Abdelrafie AM, Diaz C, Faaij A, Gao Q, Hashimoto S, Mareckova K, Pipatti R, Zhang T. Waste management. In: Metz B, Davidson OR, Bosch PR, Dave R, Meyer LA, editors. Climate Change 2007: Mitigation. Contribution of Working Group III to the Fourth Assessment Report of the Intergovernmental Panel on Climate Change. Cambridge: Cambridge University Press; 2007

[16] Vergara SE, Tchobanoglous G. Municipal solid waste and the environment: A global perspective. Annual Review of Environment and Resources. 2013;**37**:277-309

[17] Hoornweg D, Bhada-Tata P. What a Waste: A Global Review of Solid Waste Management. Urban Development Series Knowledge Papers. Nr. 15. Urban Development & Local Government Unit. Washington DC: World Bank; 2012. https://siteresources.worldbank.org/INTURBANDEVELOPMENT/Resources/336387-1334852610766/What_a_Waste2012_Final.pdf [Accessed: 10-10-2017]

[18] Achankeng E. Globalization, urbanization and municipal solid waste management in Africa. In: African Studies Association of Australasia and the Pacific 2003 Conference Proceedings - African on a Global Stage; 2003. http://wiego.org/sites/wiego.org/files/publications/files/Achankeng_Globalization_Urbanization_MSWMgmt_Africa.pdf [Accessed: 10-10-2017]

[19] United Nations Environment Programme (UNEP). Waste Investing in energy and resource efficiency. Green Economy. Online report; 2011. http://sawic.environment.gov.za/documents/3636.PDF [Accessed: 10-10-2017]

[20] BNDS (Banco Nacional de Desenvolvimento Social). Relatóriofinal do perfil institucional, quadro legal e políticas públicas relacionados a residuos sólidos urbanos no exterior e no Brasil. Fundação de Apoio ao Desenvolvimento da Universidade Federal de Pernambuco – FADE. Pesquisa Científica BNDS FEP No 02/2010 (report); 2012

[21] Myers N, Kent J. New consumers: The influence of affluence on the environment. Proceedings of the National Academy of Sciences, USA. 2003;**100**:4963-4968. DOI: 10.1073/pnas.0438061100

[22] Gutberlet J. Urban Recycling Cooperatives: Building Resilient Communities. London, New York: Routledge Taylor & Francis Group; 2016. 183 pp

[23] Walsh DC. Urban residential refuse composition and generation rates for the 20th century. Environmental Science & Technology. 2002;**36**(22):4936-4942

[24] ABRELPE (Associação Brasileira de Empresas de Limpeza Pública e Resíduos Especiais). Panorama dos Resíduos Sólidos no Brasil 2011. São Paulo (report): ABRELPE; 2011

[25] International Solid Waste Association (ISWA). Globalization and Waste Management. Phase 1: Concepts and Facts. Vienna: ISWA; 2012. www.iswa.org/index.php?eID=tx_iswatfg_download&fileUid=36 [PDF Accessed: 10-10-2017]

[26] CEMPRE. CICLOSOFT; CEMPRE. CICLOSOFT; 2014. http://cempre.org.br/ciclosoft/id/2 [Accessed: 10-10-2017]

[27] Matthews E, Themelis NJ. Potential for reducing global methane emissions from landfills, 2000-2030. In: Proceedings Sardinia 2007, 11. International Waste Management and Landfill Symposium S. Margherita di Pula, Cagliari, Italy, 1-5 Oct. 2007; 2007. http://www.necec.org/files/Matthews_Themelis_Sardinia2007.pdf [Accessed: 10-10-2017]

[28] Akter A, Mohit SA, Chowdhury MAHC. Predicting urban storm water-logging for Chittagong city in Bangladesh. International Journal of Sustainable Built Environment. 2016;**6**:238-249

[29] Porta D, Milani S, Lazzarino A, Perucci CA, Forastiere F. Systematic review of epidemiological studies on health effects associated with management of solid waste. Environmental Health. 2009;**8**(60):1-14

[30] Reuters, FEBRUARY 10, 2015. https://www.reuters.com/article/veolia-strategy/update-1-veolia-to-get-half-of-revenue-from-industrial-clients-by-2020-idUSL5N0VK4RG20150210 [Accessed: 24-10-2017]

[31] Morris J. Recycling versus incineration: An energy conservation analysis. Journal of Hazardous Materials. 1996;**47**:277-293

[32] United Nations Environment Program (UNEP). Waste and Climate Change: Global Trends and Strategy Framework. United Nations Environmental Programme Division of Technology, Industry and Economics International Environmental Technology Centre Osaka/Shiga; 2010. http://wedocs.unep.org/handle/20.500.11822/8648 [Accessed: 10-10-2017]

[33] Bartl A. Ways and entanglements of the waste hierarchy. Waste Management. 2014;**34**(1):1-2

[34] Simon DR. Corporate environmental crimes and social inequality: New directions for environmental justice research. In: White R, editor. Environmental Crime: A Reader. UK: Willan Publishing; 2000, 2009. pp. 103-116

[35] CHINTAN. Cooling Agents: An Analysis of Climate Change Mitigation by the Informal Recycling Sector in India. Report prepared in association with The Advocacy Project, Washington DC; 2009

[36] Fahmi W, Sutton K. Cairo's contested garbage: Sustainable solid waste management and the Zabaleen's right to the city. Sustainability. 2010;**2**:1765-1783. DOI: 10.3390/su2061765

[37] Wilson DC, Velis C, Cheeseman C. Role of informal sector recycling in waste management in developing countries. Habitat International. 2006;**30**:797-808

[38] Scheinberg A, Simpson M, Gupt YE, et al. Economic Aspects of the Informal Sector in Solid Waste. Eschborn, Germany: GTZ; 2010

[39] UN Sustainable Development Goals. http://www.un.org/sustainabledevelopment/sustainable-development-goals/ [Accessed: 20-10-2017]

[40] Scheinberg A, Wilson DC, Rodic-Wiersma L. Solid Waste Management in the World's Cities. London; Washington, DC: Earthscan for UN-Habitat; 2010

[41] Smith A, Fressoli M, Thomas H. Grassroots innovation movements: Challenges and contributions. Journal of Cleaner Production. 2014;**63**:114-124

[42] Dias S, Samson M. Informal Economy Monitoring Study Sector Report: Waste Pickers. Cambridge, MA, USA: WIEGO; 2016

[43] Grabs J, Langen N, Maschkowski G, Schäpke N. Understanding role models for change: A multilevel analysis of success factors of grassroots initiatives for sustainable consumption. Journal of Cleaner Production. 2016;**134**:98-111

[44] Baumann H, Tillman AM. The Hitch Hiker's Guide to LCA. An Orientation in Life Cycle Assessment Methodology and Application. Lund: Studentliteratur; 2004

[45] Reddy BS, Assenza GB. Climate change - A developing country perspective. Current Science. 2009;**97**(1):50-62

[46] Rogger C, Beaurain F, Schmidt TS. Composting projects under the Clean Development Mechanism: Sustainable contribution to mitigate climate change. Waste Management. 2011;**31**(1):138-146

[47] United States Environmental Protection Agency (US EPA). Solid Waste Management and Greenhouse Gases. A Life-Cycle Assessment of Emissions and Sinks; 2006. http://purl.access.gpo.gov/GPO/LPS76916 [Accessed: 10-10-2017]

[48] Bogner J, Pipatti R, Hashimoto S, Diaz C, Mareckova K, Diaz L, Kjeldsen P, Monni S, Faaij A, Gao Q, Zhang T, Ahmed MA, Sutamihardja RT, Gregory R. Mitigation of global greenhouse

gas emissions from waste: Conclusions and strategies from the Intergovernmental Panel on Climate Change (IPCC) Fourth Assessment Report. Working Group III (Mitigation). Waste Management Resources. 2008;**26**(1):11-32

[49] Chester M, Martin E, Sathaye N. Energy, greenhouse gas, and cost reductions for municipal recycling systems. Environmental Science & Technology. 2008;**42**:2142-2149

[50] Pimenteira CAP, Pereira AS, Oliveira LB, Rosa LP, Reis MM, Henriques RM. Energy conservation and CO_2 emission reductions due to recycling in Brazil. Waste Management. 2004;**24**:889-897

[51] Thorneloe SA, Weitz KA, Nishtala SR, Yarkosky S, Zannes M. The impact of municipal solid waste management on greenhouse gas emissions in the United States. Air & Waste Management Association. 2002;**52**:1000-1011

[52] ICF Consulting. Determination of the Impact of Waste Management Activities on Greenhouse Gas Emissions: 2005 Update, Final Report. Submitted to: Environment Canada and Natural Resources Canada. Toronto: ICF Consulting; 2005. https://www.rcbc.ca/files/u3/ICF-final-report.pdf [Accessed: 20-10-2017]

[53] Morris J. Comparative LCAs for curbside recycling versus either landfilling or incineration with energy recovery. International Journal of LCA. 2005;**10**(4):273-284

[54] King M, Gutberlet J. Contribution of cooperative sector recycling to greenhouse gas emissions reduction: A case study of Ribeirão Pires, Brazil. Waste Management. 2013;**33**(12):2771-2780

[55] Forsyth T. Building deliberative public – Private partnerships for waste management in Asia. Science. 2005;**36**(4):429-439

Section 3

Urban Landscapes

The Corbusier Dream and Frank Lloyd Wright Vision: Cliff Detritus Vs. Urban Savanna

Yuemin Jin, Jon Burley, Patrica Machemer,
Pat Crawford, Haoxuan Xu, Zhen Wu and
Luis Loures

Additional information is available at the end of the chapter

http://dx.doi.org/10.5772/intechopen.71908

Abstract

Investigators are seeking methods to assess the visual and environmental quality of the landscape across urban areas. In addition investigators are interested in applying these predictors to study landscape transformation and change. In our study we employed an environmental quality prediction equation, which assesses environmental quality to create a visual quality map of southern Michigan and then evaluated the map's ability to determine the map's reliability. Through the Kendall's coefficient of concordance statistical test, we determined that the map is significantly reliable ($p \leq 0.01$) and conclude that constructing such a map of a large area is possible. We then applied this approach to quantify environmental quality change to southeast Michigan (Detroit metropolitan area) from land-use maps in the 1800s, and from a map constructed in 2008. Only areas with cliff detritus had statistically significant changes. Many of these cliff detritus areas are now being transformed back to pastoral urban savanna environments, a vision that had been embraced by Frank Lloyd Wright. Wright's approach compares differently with the grand vision Le Corbusier had for urban areas, a series of multiple-use towers spaced across an urban forest. The sprawling towers of Shanghai, P.R. of China exemplify this model in a modern manner.

Keywords: environmental psychology, landscape architecture, land-use planning landscape planning, urban sprawl

1. Introduction

For the last 50 years, investigators have been seeking quantitative methods to predict and assess the visual and environmental quality of the landscape. The literature on this subject is vast. Mo *et al.*, reviews much of this pertinent literature and discusses the perceptions of respondents in North America, France, Portugal, and P.R. of China. One of her and her colleagues' findings suggests that Europeans and North Americans may have broad similar perceptions about landscape [1]. Asians may have a different sensibility concerning environmental preference. Concurrently, Partin *et al.*, studied the response of participants to computer generated images and reported that the perception of computer images was similar to the perception of photographs of landscapes [2]. In other words it was possible for investigators to present computer generated images to respondents and obtain a similar response as if the respondents were examining photographs. He also demonstrated how a small study could be folded into a larger and widely studied set of images to obtain stable and reliable results. In addition, Burley *et al.*, presented a model that predicted a high proportion of explained respondent preference [3]. The model included esthetic, economic, ecological, functional, and cultural requirements. Their work provides the current background and theory concerning the types of variables presently used in developing predictive models. In their study, they specifically addressed the reclamation of Detroit's "gray-fields". The study team employed an investigative approach to evaluate Frank Lloyd Wright's vision for cities, called "Broadacre City" to current visions for Detroit, and with the past environmental conditions of Detroit. They concluded that the new visions for Detroit and Frank Lloyd Wright's vision was indeed significantly preferred over past built environments. Liu and Burley and Yilmaz and Burley provide additional background concerning the variables that Mo *et al.* and Partin *et al.*, have employed [4, 5]. Liu and Burley noted that respondents can have widely dispersed expectations concerning values and criteria to assess landscapes; yet Yilmaz and Burley note that collectively and statistically, strong and reliable predictive equations evaluating landscapes can be generated from respondents, explaining up to 90% of the variance in respondent preferences [5].

During this timeframe, Lu *et al.*, examined the Lower Muskegon watershed, located on the west side of the lower peninsula in Michigan [6]. He and his colleagues studied images of urban areas, farmland, wetlands, and forests and attempted to construct an environmental quality map of his study area. The result he obtained through statistical analysis revealed that the relationship between his predictions in the map and the real photographs are in concordance and at a reasonable (95%) confidence level. He concluded that visual/environmental quality could be mapped and reliably predicted in the Lower Muskegon Watershed. These recent studies provide a setting for our investigation.

As an extension of Lu's *et al.*, research, we were interested in applying this approach to all of the southern part of Michigan. We wanted to make a validated map of southern Michigan. In addition we were interested in examining the transformation of environmental quality in the Detroit metropolitan area, comparing pre-settlement environments from the 1800s to current conditions. In addition, we desired to compare the environments of greater Detroit with the vision held by Le Corbusier [7]. Le Corbusier envisioned a grand sea of urban forest penetrated by multiple-function buildings spreading endlessly across the urban landscape. Two structures were build in France applying his principles, one in Marseilles and the other in Rezé-les-Nantes. The experiment ended there. However, the broad expanse of towers forming the city of Shanghai, with a population of 25 million people, is a modern version, expressing this vision. We were interested in assessing the perceived environmental quality of Detroit and Shanghai with the evaluation equation.

2. Study area and methodology

Michigan is located in the Great Lakes Region of the United States of America. Michigan is the only state to consist of two peninsulas. These two peninsulas are linked by the Mackinac Bridge. The Upper Peninsula is separated from the Lower Peninsula by the Straits of Mackinac. The Lower Peninsula whose shape looks like a mitten was chosen as the study area. The area of this research is known as Southern Michigan, which is no further north than N. 44.2 in latitude (**Figure 1**). The majority of people live in this southern portion of Michigan containing many more people than the other part of Michigan. Cities in the study area include: Grand

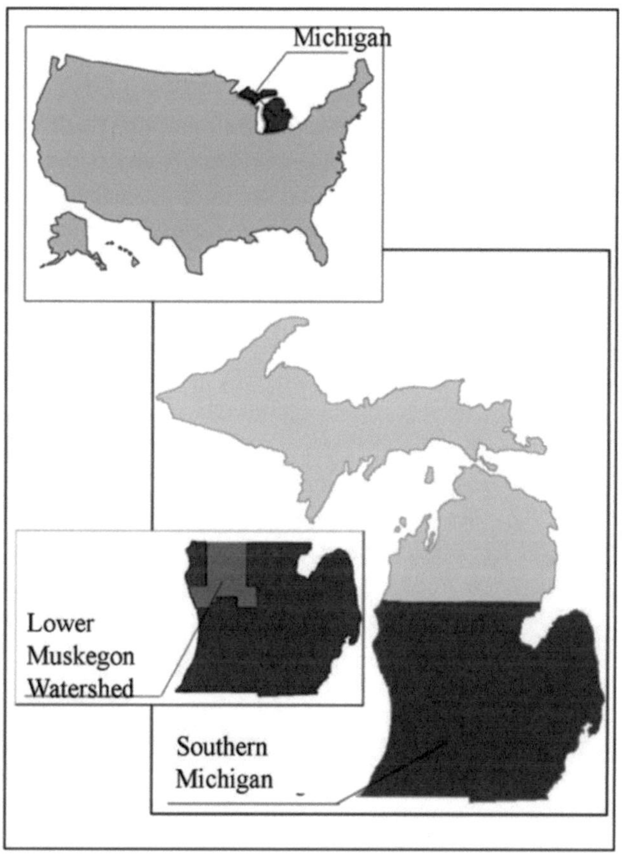

Figure 1. Location of the study area and related study areas (north is at the top of the figure).

Rapids, Battle Creek, Benton Harbor, Jackson, Kalamazoo, Flint, Pontiac, Bay City, Midland, Holland, Saginaw, Muskegon, East Lansing, Ann Arbor, and Metropolitan Detroit.

Le Corbusier's Unité d'Habitation of Nantes-Rezé, France was built in 1955 [7]. Nantes is a French port along the Loire River, in western France. The structure contains a school at the top level. From this basic idea emerged in Shanghai a vast collection of towers, many multi-function. The architectural towers emerged over 100 years after the a French Concession trading

Figure 2. A model of the central urban core of shanghai in 2007 copyright © 2007, Jon Bryan burley, all rights reserved, used by permission.

port south of the Chang Jiang (Yangtze River), along the sea coast was established. Since the 1990s, Shanghai has experienced extensive redevelopment with towers extending sometimes as far as the eyes can see (**Figure 2**).

The methodology to evaluate large areas of urban and rural landscape was similar to Lu *et al.* [6]. Lu *et al.* explains in great detail the methodology. In summary, images of various landscapes across the Michigan study area were collected and randomly sorted into two groups: one group to assist in making a prediction and another group to validate or refute the prediction. From the first group, scores for the images were generated by employing an Eq. (1) developed by Burley [8]. This paper by Burley, published in 1997 is the formative paper in this line of work and investigators interested in understanding the fundamentals of this line of research are urged to examine this paper. Once one has obtained the scores, then the scores are applied to similar land-uses to form a map predicting environmental quality. Next the second group of images are compared to predictions made by the map through the use of Kendall's Concordance, a statistical technique that examines and tests for significant agreement/similarity [9, 10]. If the scores agree, then it is possible to make a reliable visual quality prediction map. Jin also provides more detail, step-by-step concerning this methodology [11]. Investigators seriously interested in applying this methodology should obtain copies of Lu *et al.*, and Jin for a full and complete explanation [6, 11]. Single images can also be scored and compared with other images with this equation. Scores below 50 are very biospheric, meaning the image contains predominantly plants, water, sky, and flowers. Images in the 60s contain some roads or buildings. Images that are 70 and above are rather noospheric with some images scoring in ranges over 110 (highly un-preferred images).

$$Y = 68.30 - (1.878*HEALTH) - (0.131*X1) - (0.064*X6) + (0.020*X9) + (0.036*X10) + (0.129*X15) - (0.129*X19) - (0.006*X32) + (0.00003*X34) + (0.032*X52) + (0.0008*X1*X1) + (0.00006*X6*X6) - (0.0003*X15*X15) + (0.0002*X19*X19) - (0.0009*X2*X14) - (0.00003*X52*X52) - (0.0000001*X52*X34) \tag{1}$$

where (See **Table 1**):

The test statistics are provided in Eqs. 2 and 3. The statistics are based upon rankings of treatment scores across rows. In this case the rows are pairs of images between two treatments: the predicted score for a randomly chosen site in the study area and the actual score from a photograph taken at that location. There are 30 rows (pairs of scores) for this study (n = 30) The treatments are the columns (m = 2). The rankings are summed and squared, to compute Kendall's W value Eq. (2). $(Rj)^2$ is the sum of the squares of the rankings for a column in computing the Kendall's W value [8, 9].

$$W = \frac{12\sum_{j=1}^{n} R_j^2 - 3\,m^2\,n\,(n+1)^2}{m^2\,n(n^2-1)} \tag{2}$$

HEALTH	environmental quality index
X1	perimeter of immediate vegetation
X2	perimeter of intermediate non-vegetation
X3	perimeter of distant vegetation
X4	area of intermediate vegetation
X6	area of distant non-vegetation
X7	area of pavement
X8	area of building
X9	area of vehicle
X10	area of humans
X11	area of smoke
X14	area of wildflowers in foreground
X15	area of utilities
X16	area of boats
X17	area of dead foreground vegetation
X19	area of wildlife
X30	open landscapes = X2 + X4 + (2 × (X3 + X6))
X31	closed landscapes = X2 + X4 + (2 × (X1 + X17))
X32	openness = X30 − X31
X34	mystery = X30 × X1 × X7/1140
X52	noosphericness = X7 + X8 + X9 + X15 + X16

Table 1. Variables for the environmental quality/Health index in Eq. 1.

Kendall's W value is a number ranging between 0 and 1. When W is near 0, there is no strong overall trend of agreement among the respondents. If W is near 1, then the responses could be regarded as close to unanimous in their agreement. The W test statistic approximates a Chi-square distribution with n−1 degrees of freedom (Eq. 3). If computed values for Chi-square (Eq. 3) are greater than significant values in a Chi-square table for n−1 degrees of freedom (in this case 29 = 30–1), then there is a high level of agreement/concordance—the predicted scores and the actual scores are in agreement.

$$X^2 = m\,(n-1)\,W \tag{3}$$

3. Results

The sample of images gathered in the investigation include forested lands (**Figure 3**), agricultural lands (**Figure 4**), residential environments (**Figure 5**) (known as urban savanna), downtown-like environments (**Figure 6**) (know as cliff detritus) and industrial sites (**Figure 7**) [12].

http://dx.doi.org/10.5772/intechopen.71908

Figure 3. Image of sample number 26 of a forested landscape in southern Michigan (visual score of 54.40120) — Copyright © 2011, Yuemin Jin, all rights reserved, used by permission.

Figure 4. Image of sample number 16 of a farmland landscape in southern Michigan (visual score of 59.12320) — Copyright © 2011, Yuemin Jin, all rights reserved, used by permission.

Figure 5. An image of a residential landscape (urban savanna with visual quality score of 61.58117), sample number 10 — Copyright © 2011, Yuemin Jin, all rights reserved, used by permission.

Table 2 presents the rankings of the images from the study. The predicted ranks are scores generated by developing a map of the study area. The actual scores are values taken and measured from a random site in the study area. Kendall's Concordance analysis revealed a Chi-square score of 51.8 (see Min for details) [8]. This value is larger than 49.58788 (a 99% confidence level ($p \leq 0.01$) for 29 degrees of freedom. Since 51.8 is larger than 49.58788, the predicted scores and the actual scores are in agreement at a 99% confidence level ($p < 0.01$). These results suggest that it is possible to construct an environmental/visual quality map of Southern Michigan that is relatively reliable (**Figure 8**).

Figure 6. An image of sample number 12 of a downtown environment (cliff detritus with a visual score of 69.28333) — Copyright © 2011, Yuemin Jin, all rights reserved, used by permission.

Figure 7. An image of sample number 22 of an industrial environmental (with a visual score of 86.56068) — Copyright © 2011, Yuemin Jin, all rights reserved, used by permission.

Property	Predicted Ranking	Images from Set Number 2	Actual Score	Set 2 Ranking
Industrial	3	55	107.80	1
Downtown	8	42	89.68	2
Industrial	3	52	83.12	3
Industrial	3	51	82.92	4
Commercial	13	31	82.12	5
Downtown	8	45	81.84	6
Downtown	8	41	77.91	7
Industrial	3	54	77.12	8
Commercial	13	33	76.05	9
Downtown	8	43	75.23	10
Commercial	13	32	74.15	11
Downtown	8	44	74.04	12
Residential	23	39	73.41	13
Commercial	13	35	72.92	14
Industrial	3	53	67.91	15
Commercial	13	34	66.76	16
Farmland	18	46	66.15	17
Residential	23	36	65.50	18
Residential	23	38	62.66	19
Farmland	18	47	62.13	20
Residential	23	37	61.68	21
Residential	23	40	61.64	22
Forested	28	58	55.1	23
Forested	28	59	55.09	24
Farmland	18	49	54.74	25
Forested	28	60	54.71	26
Forested	28	56	53.83	27
Forested	28	57	53.74	28
Farmland	18	48	53.27	29
Farmland	18	50	51.94	30

Table 2. The predicted rankings of sites across Michigan and the actual scores and ranking of images on those sites.

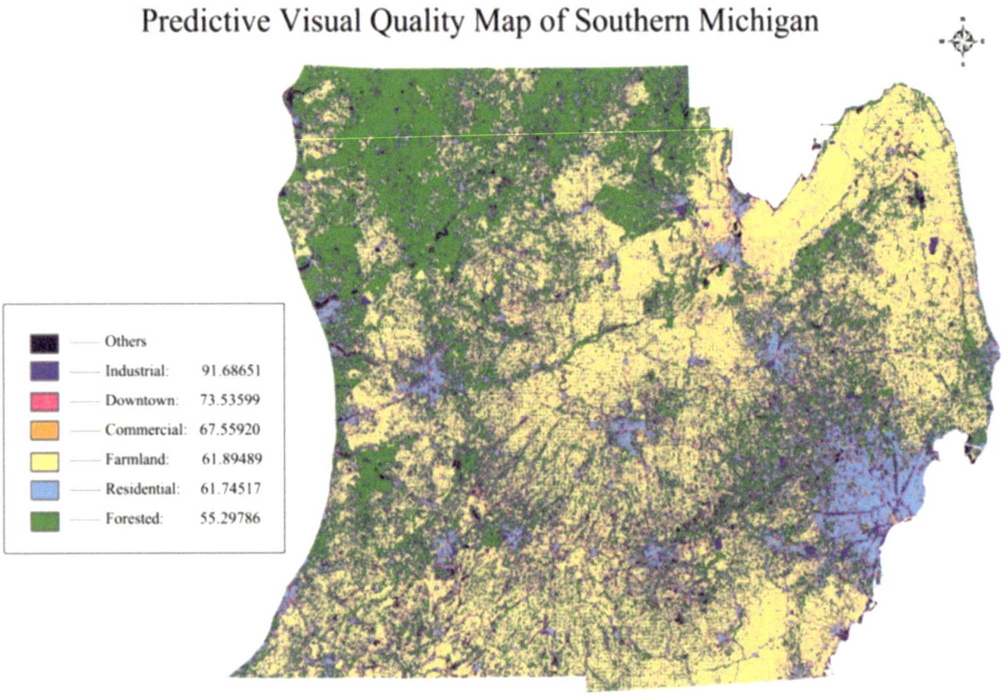

Figure 8. A map of the predicted environmental/visual quality of southern Michigan.

4. Discussion

4.1. Metropolitan Detroit environmental quality transformation

While maps as artifacts are interesting to inspect, we were interested in examining the environmental quality transformation of the environment in our study area, especially the Detroit metropolitan area. The Detroit metropolitan area is in the southeast portion of the study area (**Figure 7**). The area is comprised of numerous communities with the city of Detroit at its core. The area can be characterized as containing competing communities, distributed, and entrepreneurial. Using the middle-ages as a metaphor, the Detroit metropolitan area can be described as having a weak monarchy and ailing centralized authority (Detroit proper), with strong nobility and robust vassal states (the suburbs). Instead of one dominant business district, there are many competing commercial centers, industrial zones, and visions for urban organization [13–16]. The expansive metropolitan area is quite vibrant, enterprising, active, and engaging, at the cost of its core. While Detroit proper is suffering, the suburbs are thriving. Visitors to the Detroit metropolitan area are often surprised how normal, active, and busy the suburban communities are, because in the news, the perception is that the area is ailing. However, it is primarily the core that is failing and the rings around the core are thriving. It is in this setting that we desired to examine urban transformation.

http://dx.doi.org/10.5772/intechopen.71908

We applied our results in this study to a 2008 land-use map of the Detroit metropolitan area, supplied by SEMCOG (Southeast Michigan Council of Governments) [17]. The result was a map of the generalized environmental/visual quality for the Detroit metropolitan region

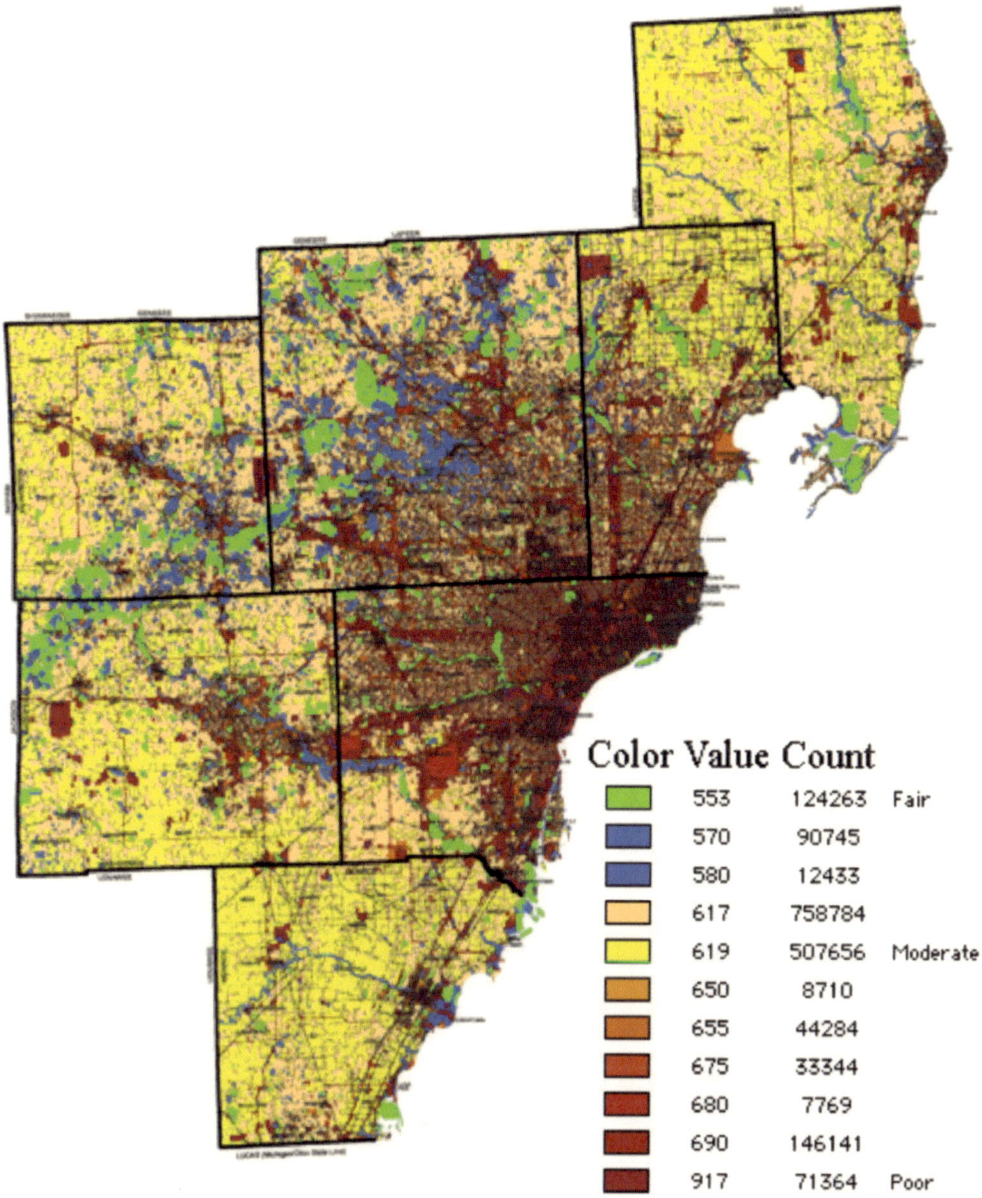

Figure 9. A map of the 2008 predicted environmental/visual quality of the seven county Detroit metropolitan area, divide each value by 10 to obtain the correct decimal reading.

(**Figure 9**). We then applied our results to a map from the 1800s containing pre-settlement land cover data [18]. This resulted in a map of the generalized and estimated environmental/visual quality from 200 years ago (**Figure 10**).

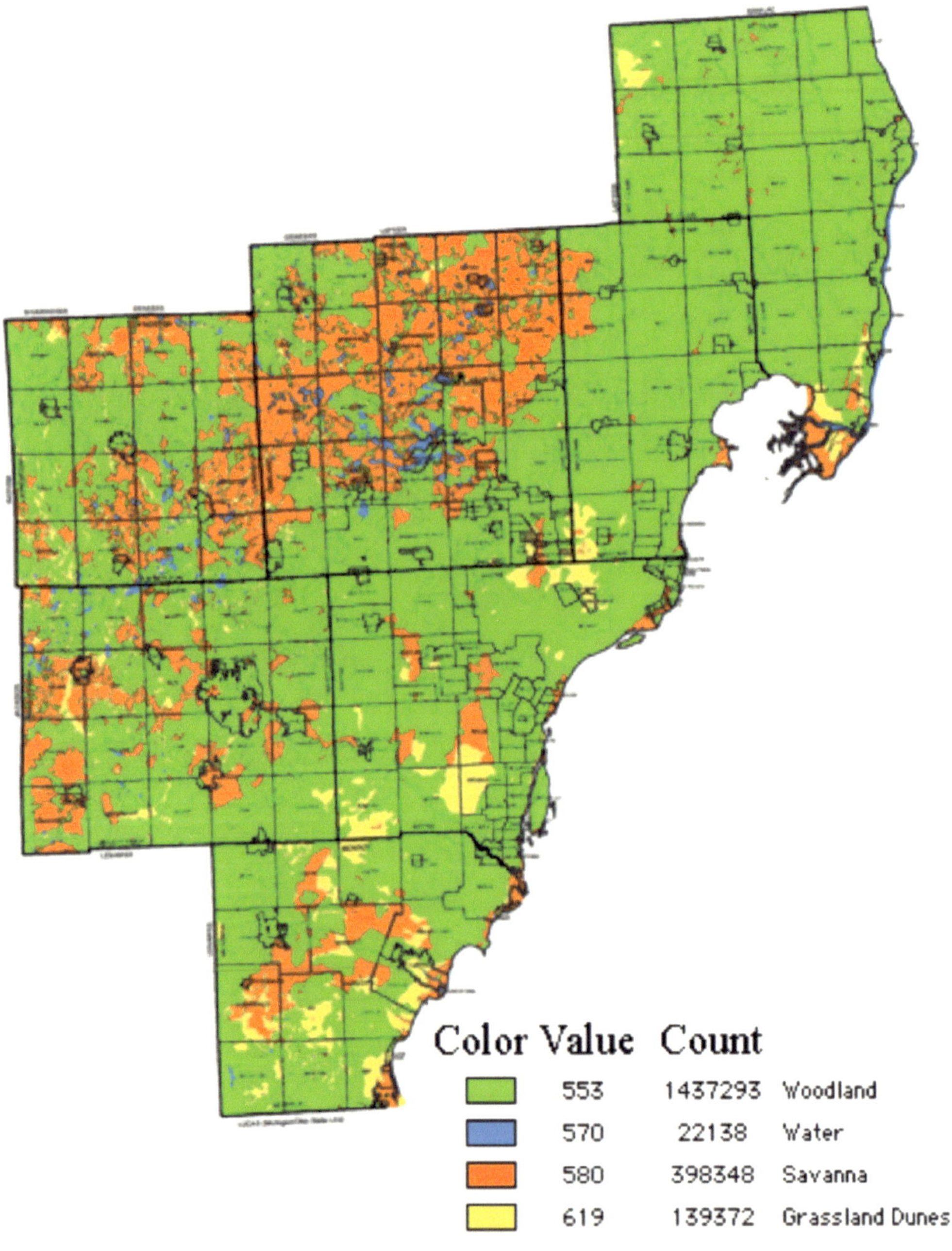

Figure 10. A map of the predicted environmental/visual quality of the seven county Detroit metropolitan area for the 1800s, divide each value by 10 to obtain the correct decimal reading.

The values (integers) in the legend for **Figures 9** and **10**, must be divided by 10 to obtain the environmental quality score (an artifact of the GIS software employed, where no floating point decimal was used). To interpret the scores, Burley notes that scores in the 30s indicate highly preferred environments. Scores in the 50s and 60s are often modestly preferred environments. Scores in the 70 are less preferred and scores near 100 are not preferred [19]. The 95% confidence interval for any scores is +5 points [20]. Thus it takes a separation of 10 points for any pairs of images to be notably different as perceived by respondents.

We subtracted the scores from **Figure 9** with the scores from **Figure 10**, producing **Figure 11**. This map (**Figure 11**) represents the expected change environmental/visual quality for the last 200 years. The light gray areas indicate the landscape where the greatest negative change occurred, meaning where the environmental/visual quality became worse. Some areas actually improved in visual quality (**Figure 12**). The improvement primarily took place on grassland and dunes that become vegetated with woody plants.

The largest transformations (as much as 30 points) came from landscapes that were forested to environments that became industrial areas, downtowns, and large multi-unit housing complexes, often comprising the cliff detritus. Most of the cliff detritus is centered around and in Detroit proper, representing an evolution of transformations as the urban area grew. Cliff detritus allows planners and designers to provide higher densities of human populations and activities. The idea is that creating density is less expensive, offering affordable solutions and preserving more countryside [21]. The problem is that cliff detritus, whether in Paris, France, San Francisco, California, or Detroit, Michigan is considered relatively marginal as a place to live when compared to the Loire River Valley, France, Marin County, California, or Charelvoix, Michigan. The environmental preference models employed in this study only reconfirm the obvious but often unstated. Packing people into less preferred cliff detritus environments may potentially be an inhuman planning and design choice.

Smaller transformations near zero or only modest changes (statistically insignificant transformations), can found in the change from woodland (55) to urban savanna (61) (note the change must be great than 10 points to be statistically notable). Much of the land transformation has been from woodland (55) to agriculture (61) to urban savanna (61). The change actually took place with the felling of trees and installation of agriculture. Transforming the agriculture to urban savanna produced predominantly no change. For us, this in an interesting observation as urban sprawl is often portrayed as an undesirable effect. Yet, urban savanna with its highly diverse and productive gardens are environmentally/visually acceptable, high in wildlife diversity, productive in biomass growth, diverse in vegetation composition, and extensive food production, meaning that the urban savanna has some positive attributes [12, 22]. The urban savanna represents a relatively new and evolving ecology, possibly being equal in value and contribution to greenways and other efforts to preserve natural areas in the urban fabric. We are not proposing the urban savanna replace greenways, rather we are suggesting that the urban savanna has been undervalued. In addition, the pre-settlement landscape of the study area did not consist of landscape with strong visually pleasing environments (scores in the 20s, 30, and 40s) such as those in the American mountain west. Therefore the measured changes from woodland to agriculture to urban savanna were not drastic.

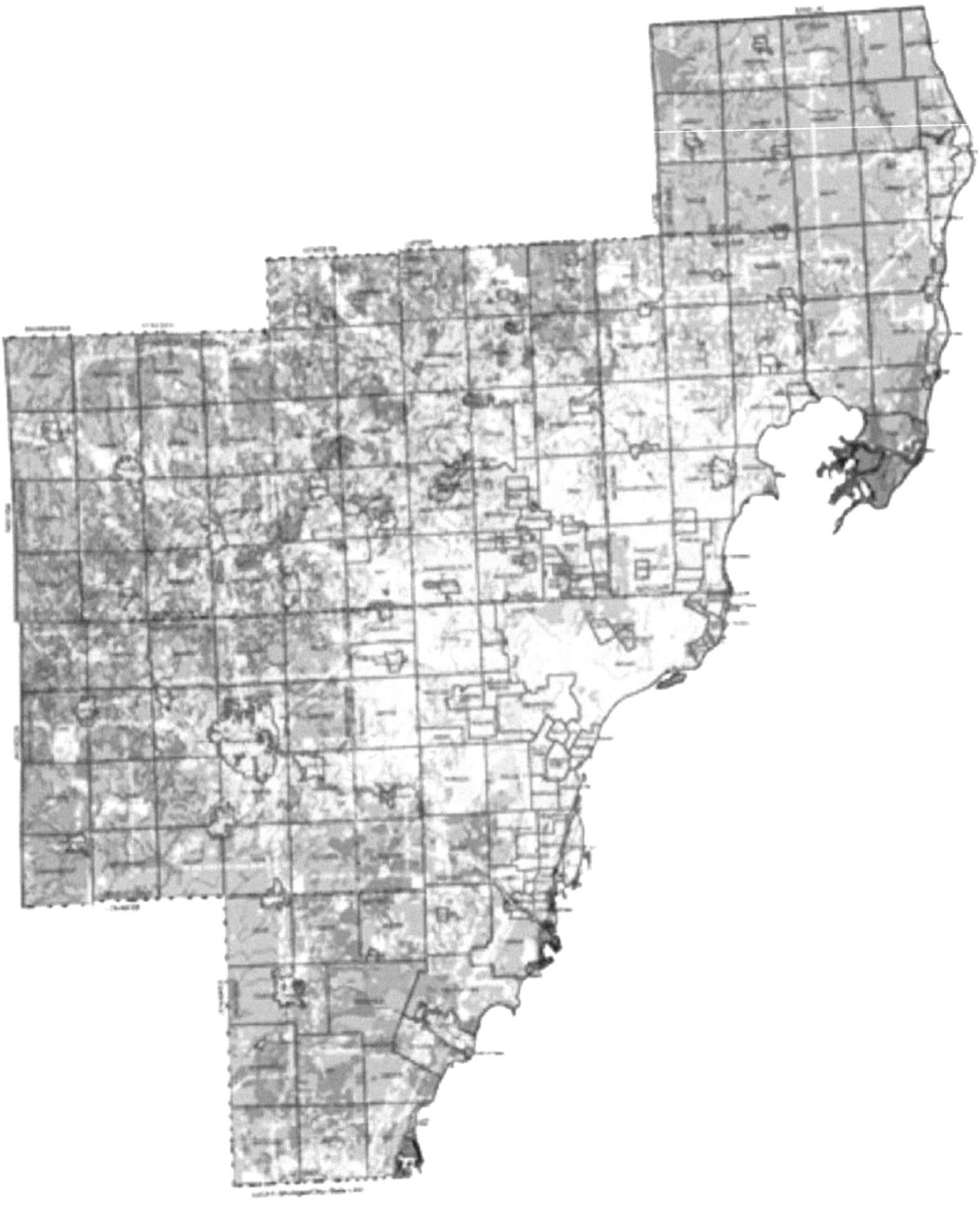

Figure 11. A map of change/transformation in environmental/visual quality over 200 years. The light gray areas indicate where the change tended towards less preferred environments. The mid-gray levels indicate no detectible change and the dark gray areas indicate a slight improvement in environmental/visual quality.

4.2. Comparison of visions

Recently planners, designers, citizens, and public employees have been embracing ideas first presented by Frank Lloyd Wright concerning the distributed and savanna-like character of

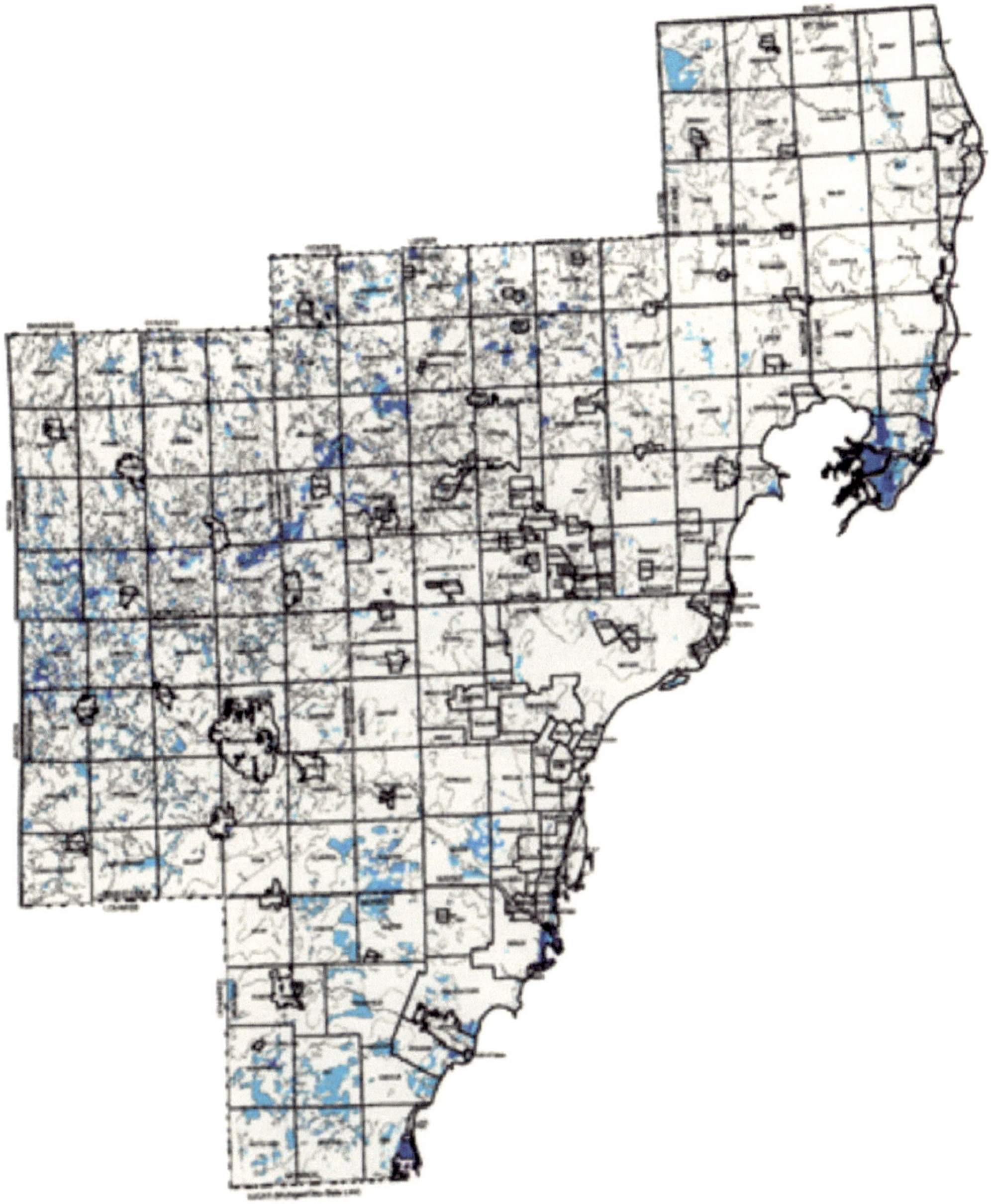

Figure 12. A map of change/transformation in environmental/visual quality over 200 years. The blue areas indicate zones with a slight improvement in environmental/visual quality.

the urban fabric [23, 24]. Long held notions about transforming landscape into extensive cliff detritus are being reconsidered. **Figures 13** and **14** are images from L Corbusier's project in Rezé-Nantes. **Figure 13** scores 73.4 and **Figure 14** scores over 110 (the upper limit for the regression line predicting environmental quality). These scores indicate the environment is much less preferred than much of the environment in Detroit. In addition, **Figures 15** and **16**

Figure 13. This is an image of the le Corbusier structure in Rezé-Nantes. This building is popular for artists and architect to live within — Copyright © 2012, Jon Bryan Burley, all rights reserved, used by permission.

Figure 14. This is an image of the ground floor area le Corbusier structure in Rezé-Nantes and illustrates the realities of urban structures — Copyright © 2012, Jon Bryan Burley, all rights reserved, used by permission.

Figure 15. A view from an urban dwelling across the shanghai towers — Copyright © 2017, Haoxuan Xu, all rights reserved, used by permission.

Figure 16. Another view of the shanghai urbanscape, even with the sea of urban trees, the image does not score well — Copyright © 2017, Haoxuan Xu, all rights reserved, used by permission.

are images from the tower urbanscape of Shanghai. **Figure 15** scores 78.41 and **Figure 16** scores 79.12, similar to a score for **Figure 13**. The expanse of towers is much less preferred than the urban savanna of metropolitan Detroit. The pastoral urban savanna is being given a renewed look. Our study provides some insight into why this fresh approach is being explored in the Detroit Metropolitan area. If these reinventions of the Detroit metropolitan area's cliff detritus are successful, we could imagine the environmental/visual quality scores of these areas to significantly improve (20 to 30 points) over the next generation. We encourage other investigators to employ this approach to study other areas of the world.

5. Conclusion

Predictive, respondent based models have been constructed to measure environmental and visual quality. This work is based upon over 50 years of research by investigators in the social, recreational, and planning and design disciplines/profession. The attributes of the landscape can be measured to form reliable maps of environmental/visual quality, providing a metric to assess landscapes, including urban landscapes. We were able to produce such a metric map for southern Michigan. Then, we applied our approach to study landscape transformation in the Detroit metropolitan area. We discovered that from the 1880s until 2008, much of the area had only modest change from woodland, to agriculture to urban savanna. Some small areas even improved. The predominant areas with degraded environments were in the large cliff detritus complex near and within Detroit. Much of this cliff detritus is being transformed again to resemble a pastoral urban savanna, similar to visions for the urban fabric as originally expressed by Frank Lloyd Wright. Numerous authorities, planners, and designers have developed visions concerning the creation of urban environments, from densely packed skyscrapers to pastoral living spaces. We believe our approach allows investigators to evaluate these visions and assess, measure, and quantify the environmental perceptions of these visions.

Author details

Yuemin Jin[1], Jon Burley[2]*, Patrica Machemer[2], Pat Crawford[2], Haoxuan Xu[2], Zhen Wu[3] and Luis Loures[4]

*Address all correspondence to: burleyj@msu.edu

1 Planning Design and Construction Division, Tongji University, Shanghai, China

2 Landscape Architecture, School of Planning, Design, and Construction, Michigan State University, East Lansing, MI, USA

3 Nanjing Forestry University, Nanjing, China

4 Polytechnic Institute of Portalegre, Portalegre, Portugal

References

[1] Mo F, Le Cléach G, Sales M, Deyoung G, Burley JB. Visual and environmental quality perception and preference in the People's Republic of China, France, and Portugal. International Journal of Energy and Environment. 2011;**4**(5):549-556

[2] Partin S, Burley JB, Schutzki R, Crawford P. Concordance between photographs and computer generated 3D models in a Michigan highway transportation setting. In: Buhmann E, Ervin S, Pietsch M, editors. Peer Reviewed Proceedings of Digital Landscape Architecture 2012 at Anhalt University of Applied Science. Berlin, Germany: Wichmann; 2012. pp. 482-471

[3] Burley JB, Deyoung G, Partin S, Rokos J. Reinventing Detroit: Grayfields—New metrics in evaluating urban environments. Challenges. 2011;**2**:45-54

[4] Liu C, Burley JB. Landscape evaluation criteria: dispersed expectations. Modern Landscape Architecture. In: Proceedings of the 6th WSEAS International Conference on Landscape Architecture (LA'13), Nanjing China, November 17-19, 2013. pp. 19-23

[5] Yilmaz R, Burley JB. Dissecting the Smyser index in visual quality. Modern Landscape Architecture. In: Proceedings of the 6th WSEAS International Conference on Landscape Architecture (LA'13), Nanjing China, November 17-19, 2013. pp. 32-36

[6] Lu D, Burley JB, Crawford P, Schutzki R, Loures L. Chapter 7: Quantitative methods in environmental and visual quality mapping and assessment: A Muskegon, Michigan watershed case study with urban planning implications. In: Burian J, editor. Advances in Spatial Planning. InTech; 2012. pp. 127-142

[7] Sbriglio J. Le Corbusier - L'Unite d Habitation de Marseille/the United Habitation in Marseilles: Et les Autres Unites D'habitation a Reze-les-Nantes, Berlin, … et Firminy/and the four Other Unite Blocks. Birkhauser; 2004 244 p

[8] Burley JB. Visual and ecological environmental quality model for transportation planning and design. Transportation Research Record. 1997;**1549**:54-60

[9] Daniel WW. Applied Nonparametric Statistics. Houghton Mifflin Company; 1978 560 p

[10] Kendall M. The problem of m rankings. The Annals of Mathematical Statistics. 1939; **10**(3):275-287

[11] Jin Y. Validating a Visual Quality Prediction Map of Southern Michigan. Master of Arts in Environmental Design. East Lansing, Michigan: Michigan State University; 2012

[12] Brady RF, Tobias T, Eagles PFJ, Ohrner R, Micak J, Veale B, Dorney RS. A typology for the urban ecosystem and its relationship to larger biogeographical landscape units. Urban Ecology. 1979;**4**:11-28

[13] LeDuff C. Detroit an American Autopsy. New York, New York, USA: Penguin Press HC; 2013

[14] Binelli M. Detroit City Is the Place to be: The Afterlife of an American Metropolis. New York, New York, USA: Metropolitan Books; 2012

[15] Martelle S. Detroit: A Biography. Chicago, Illinois, USA: Chicago Review Press; 2012

[16] Kersey P. Escape from Detroit: The Collapse of America's Black Metropolis. North Charleston, South Carolina, USA: CreateSpace Independent Publishing Platform; 2012

[17] Southeast Michigan Council of Governments. Generalized Land Use: Southeast Michigan. Detroit, Michigan, USA: SEMCOG; 2008. p. 2010

[18] Corner PJ, Albert DA, Wells HA, Hart BL, Price DL, Moore JR, Kashian DM, Schuen DW. Michigan's Native Landscape as Interpreted for the General Land Office Surveys. Lansing, Michigan, USA: Michigan Natural Features Inventory; 1816-1856. p. 1995

[19] Burley JB. The science of design: Green vegetation and flowering plants do make a difference: Quantifying visual quality. The Michigan Landscape. 2006;**49**(8):27-30

[20] Burley JB. A quantitative method to assess aesthetic/environmental quality for spatial surface mine planning and design. WSEAS Transactions on Environment and Development. 2006;**5**(2):524-529

[21] Burley J, Wang Y, Loures L. New ecologies: Emergence of the urban savanna and cliff detritus in a post post-modern era. In: Panagopoulos T, editor. New Models for Innovative Management and Urban Dynamics. Faro, Portugal: Universidade do Algarve; 2009. pp. 10-18

[22] Burley JB. The new ecology of urban savanna. The Michigan Landscape. 2008;**51**(10):41-47

[23] Gallagher J. Reimagining Detroit: Opportunities for Redefining an American City (Painted Turtle). Detroit, Michigan, USA: Wayne State University; 2010

[24] Gallagher J. Revolution Detroit: Strategies for Urban Reinvention (Painted Turtle). Detroit, Michigan, USA: Wayne State University; 2013

Do Degradation of Urban Greenery and Increasing Land Prices Often Come along with Urbanization?

Yu-hsin Tsai, Jhong-yun Guan and Yu-Hsin Huang

Additional information is available at the end of the chapter

http://dx.doi.org/10.5772/intechopen.71651

Abstract

In the wake of urbanization, driven by a variety of individual and socio-economic merits, human's basic residential needs and standard of living may be compromised in the urban areas, as the population agglomerates. However, the knowledge of the associations of urbanization with urban greenery and residential land prices is still in the pursuing process. This empirical research aims to contribute whether the degradation of essential living conditions is a trade-off for the pursued urban life. Hence, Taiwan is selected as the case to analyze the associated relations primarily between 1976 and 2016. The research methods involve descriptive statistics, the panel data analysis, and the cluster analysis. The panel data analysis demonstrates that degraded urban greenery and increasing residential land prices came along with the urbanization in Taiwan between 2001 and 2016. Policy implications include rethinking of the building coverage rate for renewed buildings for more plant-friendly ground, the adoption of building setback policy for more accessible mid-air mini-parks, and avoiding residential units as an investment commodity.

Keywords: urban greenery, land prices, urbanization

1. Introduction

The momentum of urbanization driven by a variety of forces is usually beyond the human capacity to change or detour. Individual's demand for moving to the urban areas for better job opportunities, standards of living, etc. is a primary force of urbanization. Some policies are also implemented to embrace such potential socio-economic benefits of urbanization as the compact city of sustainable urban form, and, to the extreme, the politically incorrect policy to contain human's pollution to their settlements. In turn, dealing with likely congestion issues caused by the urbanization is significant, which is particularly true since the global urban population is predicted to reach 69% in 2050 [1].

Rising housing costs, a likely issue emerging along with the urbanization, directly affect human's basic residential needs and standard of living. It is listed as one of the most crucial targets of the United Nation 2017's 17 sustainable development goals (SDGs), that is, ensuring adequate, affordable housing, inclusive settlements, and equality [2]. The city's green coverage, on the other hand, lays out the foundation of a sustainable urban built environment, decisively affecting such city's capacity as the mitigation and adaptation to climate change [3] and purification and maintenance of urban air quality [4]. Growing urbanization may drive some areas to thrive while others to shrink that may necessitate specifically customized spatial plans and strategies. Nonetheless, past research sheds limited light on the association between the urbanization and land/housing prices, and the research of the relationship between the urbanization and urban greenery is developing.

This research aims to contribute empirically to whether the degradation of essential living conditions is a trade-off for the pursued urban life in the wake of urbanization. The purposes of this research are twofold to analyze the impacts of urbanization on the urban green coverage and to analyze the association/impacts on land/housing prices. To fulfill these purposes, Taiwan is selected as the case to analyze the associated situations primarily between 1976 and 2016.

2. Urbanization, land/housing prices, and green coverage

This section provides some theoretical background on the urbanization, urban hierarchy that is highly associated with the urbanization, the associations of urbanization with land or housing prices, and the urban greenery.

2.1. Urban hierarchy, urbanization, and standard of living: inputs, outputs, and outcome

The urban hierarchy is associated with its service level of living environment and in turn the standard of living. The associated capacities of an urban area at certain urban hierarchy may include its financial sources, the capacities in administration and maintenance, the mandated implementation of spatial plans, etc., which constitute the input factors for formulating its built environment as the output. This input may be mandatory, or as a result of urbanization to cope with growing population or employment, that is the level of input and supply may also be the result of urban agglomeration. In return, the demand for public resources may also increase. The input resources and level of demand jointly decide the level of services for the public infrastructure of various kinds. Hence, the standard of living in the urban area usually is higher than that of the rural areas and can largely attribute to the planned result and economic agglomeration.

Nonetheless, the concentration of population is likely to cause such issues as congestion, regarding the quality of life, gridlock, the concentration of pollutants, urban heat island, less natural environment, and less affordable housing. Theoretically, the higher the degree of urbanization, the deteriorated these conditions could be. In particular, within an urban area, the conditions may be even worse in the highly urbanized sub-area, such urban core. Two of the most significant effects are degraded green open space and increasing land/housing cost, which decisively affects the fundamental needs for living and natural environment in the urban settings. However, little research has been devoted in these regards at the bourgeoning phase of urbanization research.

2.2. Urbanization and land/housing prices

The impact mechanism of urbanization on land/housing prices may go from the rural-urban migration to natural growth, together with urban plans and the resulted urban form, and to the effected land/housing demand and prices. Other than the natural change of population affecting a city's population, the migrations, either interregional or rural-urban, largely constitute the current rapid urbanization trend. As a result, both urban boundary and urban population grow significantly.

The urbanization may affect land and housing prices within a city or city region through a few routes. First, at the city region level, the housing and land demand increase as a result of population growth. City regions with different level of population growth as a result of the different degree of urbanization may experience different impact on housing/land prices. Second, the population size, a result of urbanization, of a city region may affect its land prices through the course of the overall needs for the land in the city region. Furthermore, density can represent the level of the demand or bid [5] for the land, given the land supply is limited, and the cap of the floor area is set by the government policy like in the urban plan and by the limits of the natural resources. Third, within a city region, the constituent sub-areas may also show different levels of urbanization, affecting the level of demand for the land and housing. The appearance of the degree of urbanization for a sub-area can be represented by density. Additionally, the degree of urbanization can be affected by the socio-economic conditions, accessibility, etc.

2.3. Urbanization and green coverage

The urbanized land grows in expanding urban boundary, converting green space in the urban area into built environment to house more population, or employment migration from the rural area. That is, land is urbanized as a response to the rural-urban immigration, and the reduction of green space comes as a price of urbanization. The reduction of green space in and around urban areas turns space into less natural, and the resulted negative impacts involve such loss of natural or semi-natural environment [6]. The green space loss will lead to the loss of low-temperature space [7], emerging urban heat island [8], deterioration of air quality, noise cancelation, beautification of urban built environment [4], outdoor open space for social, sporting, and active-life activities, the green infrastructure for flood mitigation [3], and disaster sanctuary, among others. The controversy is that as the urban population grows and expected to grow more, the green space would be replaced by the built environment on the one hand; while on the other hand, there could be more need for the green space to cope with the negative impacts due to the growing agglomeration of built environment, human activities such as employment and tourism, or population per se.

3. Research methods

The research methods involve descriptive statistics, the panel data analysis, and the cluster analysis. Two panel data analyses are developed to assess the associations of urbanization with green coverage and land prices. With the application of the cluster analysis, such constituent units as towns/districts are classified into various groups based on the degree of urbanization, green coverage, and land prices. The primary data sources include census data, employment, land prices, and satellite images. The software applied includes ArcGIS, Excel, SPSS, and Stata. Urbanization indexes adopted are composed of the Taiwan government's classification definitions and continuous variables to quantify the degree of urbanization. The normalized difference vegetation index (NDVI) [9, 10] is applied to extract the green land cover data from the Landsat satellite images [11].

4. Urbanization, green coverage, and land prices in Taiwan

This section provides the statistical background of urbanization, green coverage, and land prices in Taiwan, dating back as far as 1976, when the earliest data are available. The cross-analysis is also conducted to provide preliminary knowledge on the associations of the degree of urbanization with green coverage and land prices, respectively.

4.1. Urbanization in Taiwan: 1976–2016

This section of urbanization analysis of Taiwan adopts a definition of urbanization to develop an urbanization index to conduct a universal comparison with other countries; two major concerns are prioritized to serve this purpose: urban and rural areas can be differentiated, and the data for both areas are available. A variety of methods have been developed and applied to classify an area as an urban or rural area, based on which population living within is classified as urban or rural population. These methods for urban/rural classification include the direct dichotomy distinction from existing official documents or the distinction derived from the joint consideration of an array of aspects/variables, for example, population density, population size, and population growth rate. Among a few possible urban/rural definitions for the case of Taiwan, a place assigned with a mandatory urban plan is selected for defining an urbanized area and vice versa. This definition enjoys the pros of the existence of the needed data for this analysis, and furthermore, within the areas public infrastructure, well-designed spatial plans are required by law that leads to higher level of infrastructure in the urban area. Bear in mind that this polarized definition cannot tell the degree of urbanization, which is better for other purposes like the multivariate analysis in the following sections.

Figure 1 shows the urbanization trend in Taiwan between 1976 and 2016. During these four decades, the total population grew by some 42% (or an average of about one percentage point annually), from some 16 million to some 23 million. During the same period, however, urban population grew double the rate of population growth by some 84% (or an average of two percentage points annually). Urbanization rate grew from some 61% in 1976 to 80% in 2016, among the highly urbanized nations in the world.

Before showing the urbanization situation in the sub-areas in Taiwan, an urban hierarchy system is introduced first that can lay up the base for the comparison between the degree of actual urbanization and the urban hierarchy defined by the Taiwan government. In a latest urban hierarchy system officially defined in the latest nationwide regional plan [13], each town/district is designated accordingly to one of the four levels: core, sub-core, local core, and others (**Figure 2**). These four levels are defined by its development conditions, such as the population size, growth rate, and level of public infrastructure on the one hand and are applied for spatial planning regarding future goals and strategies regarding infrastructure's investment and the level of services, etc. [13]. This urban hierarchy system divides Taiwan main island into six city regions (or "region-level living cycles" as literally translated; **Figure 2**) [14]: Taipei City-New Taipei City-Keelung City-Yilan County (01 TNKY), Taoyuan City-Hsinchu County-Miaoli County (02 THM), Taichung City-Changhua-Nantou County (03 TCN), Yunlin County-Chiayi County (04 YCT), Kaohsiung City-Pingtung County (05 KP), and Hualien County-Taitung County (06 HT) [14].

Figure 3 displays the degree of urbanization in 2016 for each of the county/city in the main island of Taiwan; the most urbanized county/city is in the cores of North, Central, and Southern parts of Taiwan. The most urbanized areas with 90% of the population living in the urban areas are located in the cores of the urban hierarchy of the top three city regions, that is, 01 TNKY, 03 TCN, and 05 KP, plus Hsinchu City of 02 THM, and Chiayi City of 04 YCT.

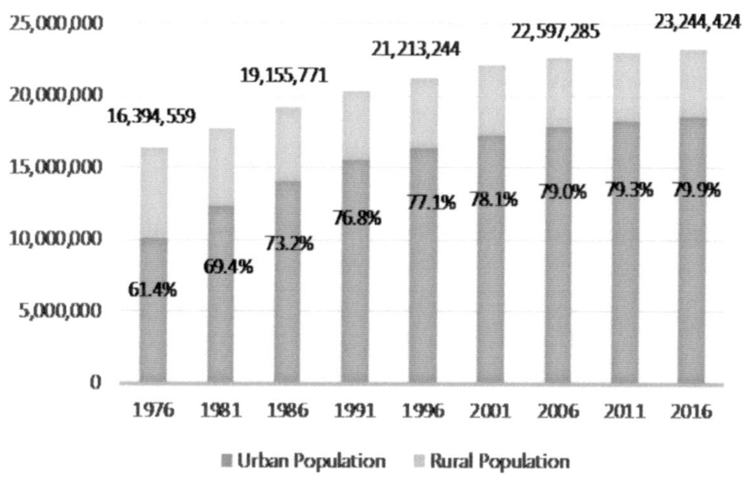

Figure 1. Urban population versus rural population in Taiwan (main island): 1976–2016. Data source: [12].

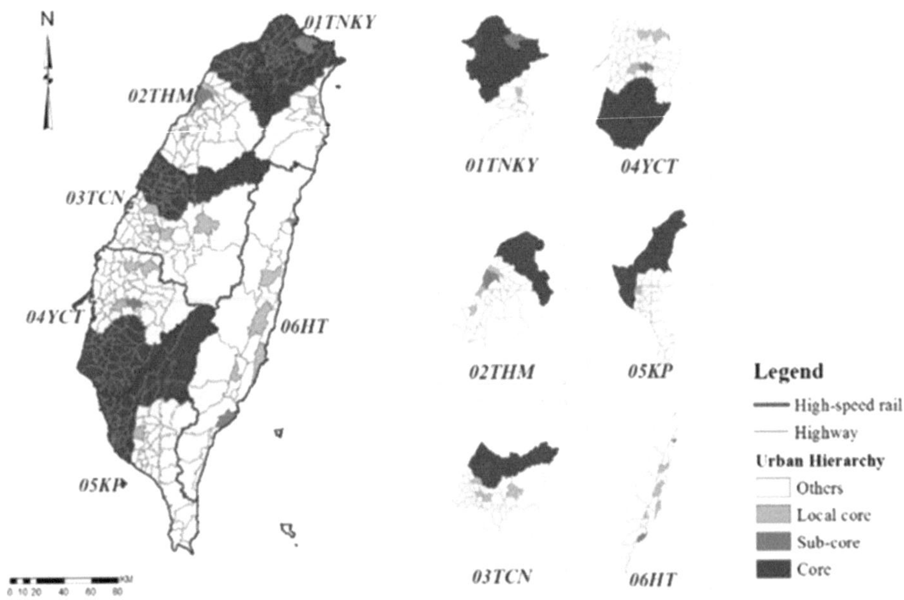

Figure 2. Urban hierarchy in Taiwan, 2013 by town/district. Data source: [13, 14].

4.2. Green coverage in Taiwan: 2001–2016

This section analyses the green coverage for Taiwan between 2001 and 2016, during which the free-quality satellite images are available for developing the green coverage data for this research. The main data source of the satellite image comes from the U.S. Geological Survey's (USGS) Landsat database [11]. The Landsat contains 30-meter resolution land coverage images appropriate for this research to develop NDVI that is often used to probe remote sensing data to tell whether the target area contains live green vegetation or not [9, 10]. Then, the green coverage data are derived for each 30-meter grip and aggregated for each town/district of Taiwan for the years 2001, 2006, 2011, and 2016. While taking advantage of the NDVI techniques in obtaining the green coverage information, it is important to note that there is a limit on accuracy rate due to a range of systematic reasons, including the degree of the original image resolution, clouds, and shade [5].

In 2001, the median green coverage rate of some 350 towns/districts across the island of Taiwan was 80%, with a range between 3.9 and 99.5% (**Table 1**). The green cover land surface has worsened between 2001 and 2016: annually, the median green land area has dropped by 0.07 km^2 and 0.12%. In general, the higher the urban hierarchy of a district/town, the lower is its green coverage level (**Figure 4**).

Figure 5 displays the annual percentage point change in green coverage rate between 2001 and 2016. Over these 15 years, some areas' green coverage rates reduced relatively high, alternatively indicating increasing built environment, which occurred in few spatial patterns. The first pattern occurred in the outer districts or suburban areas of a city region,

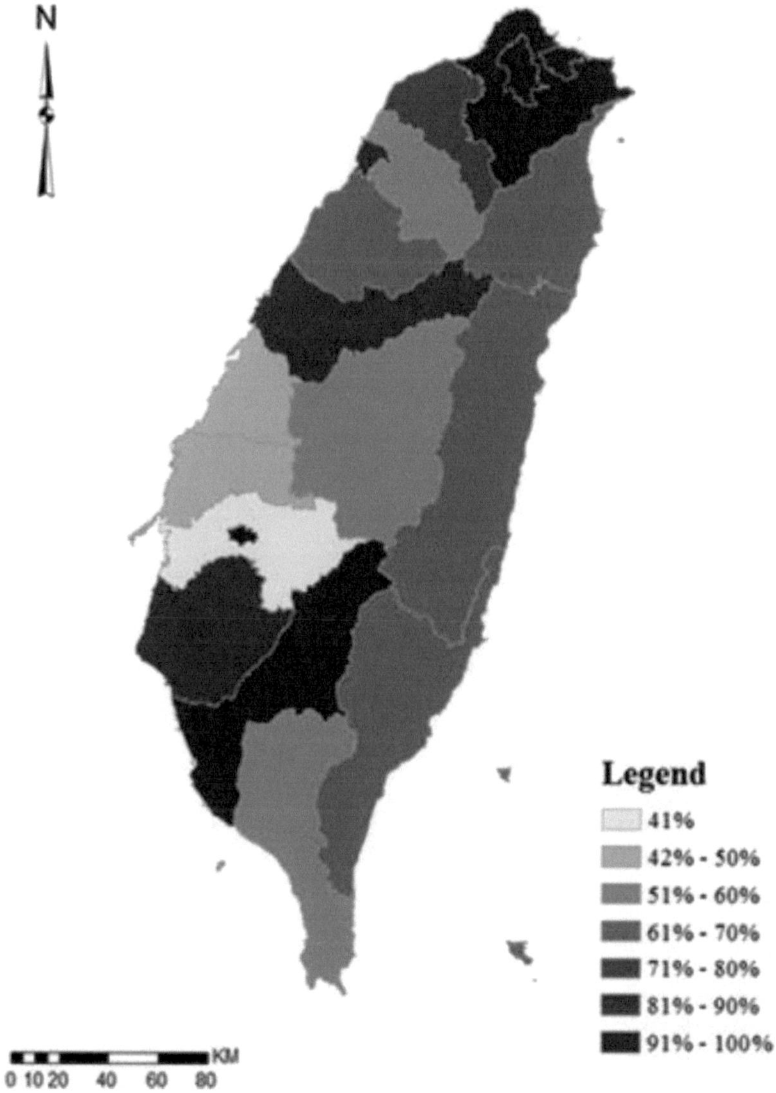

Figure 3. Percentage of urban population in Taiwan (main island), 2016, by county/city. Data source: [12].

where the most developed core districts were probably saturated for further development, such as in Taipei of the 01 TNKY city region; on the other hand, the outer districts with acceptable transportation accessibility afforded room to accommodate further development needs. The second pattern occurred in the sub-core or local core towns/districts, where there was still room for more development, such as in Keelung City and Yilan City of the 01 TNKY. The third pattern occurred in some rural towns possibly because of their unique local attractive factors for potential development, which might be boosted by such factor as improved intercity transportation, for instance, some towns in Yilan County of the 01 TNKY.

	Mean	Median	S.D.	Min.	Max.	N
Green coverage rate in 2001	80.0%	89.6%	23.5%	3.9%	99.5%	348
Annual average change in green coverage land area, (km^2) 2001–2016	−0.10	−0.07	0.15	−1.71	0.26	345
Annual average percentage point change in green coverage rate, (%) 2001–2016	−0.15%	−0.12%	0.15%	−0.69%	0.46%	345

Data source: Calculated from the Landsat [11] with NDVI.

Table 1. Green coverage in Taiwan, 2001–2016, by town/district.

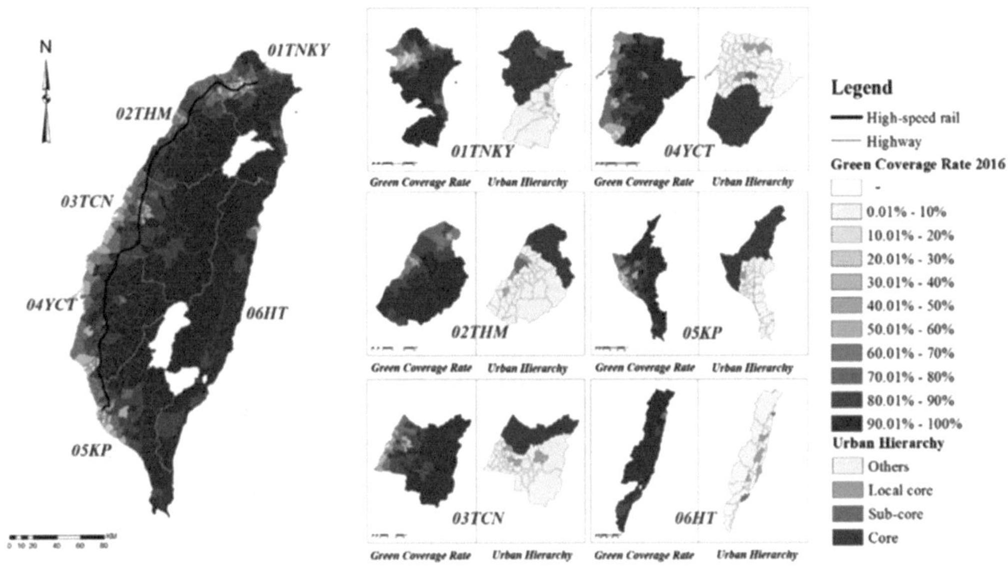

Figure 4. Green coverage rate in Taiwan, 2016, by town/district. Note: The blank areas are those suffering from large-scale landslides due to typhoons or heavy rain [15], which is excluded to avoid data distortion. Data source: Calculated from the Landsat [11] with NDVI.

Additionally, the cross-references between the change of green coverage rate and the urban hierarchy, representing the degree of urbanization, implicitly reveal some pros and cons of the Taiwan's four classes of urban hierarchy. On the one hand, some districts/towns at the top of the hierarchy tend to urbanize at a higher pace and vice versa. On the other hand, the four-class classification might not be delicate enough to distinguish the differences in the green coverage rate among the towns/districts of the same class, which drives the needs for quantitative variables of urbanization.

Figure 6 shows the annual average change of green land area, as opposed to **Figure 5**, providing an absolute scale of change as in where large reductions of green land area occurred between 2001 and 2016. The previous three spatial patterns based on the change in the green coverage rate of **Figure 5** still hold in general, but the first and third patterns stand out as large-scale reductions of green coverage seem to occur in the outer districts or the core and rural towns with the unique local attractiveness for development.

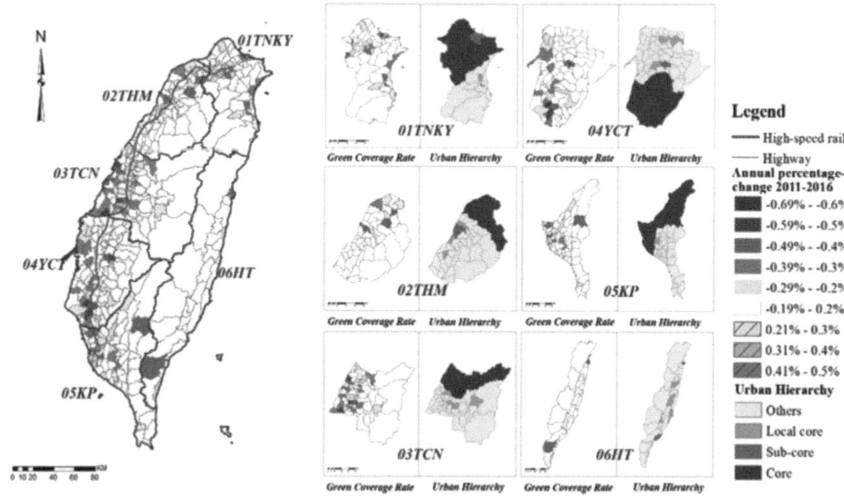

Figure 5. Annual average percentage point change in green coverage rate in Taiwan, 2001–2016, by town/district. Note: The white symbol ranges from −0.19 to 0.20% is designed to refer to none or marginal change. It is also designed to mitigate the possible error derived from NVDI. Data source: Calculated from the Landsat [11] with NDVI.

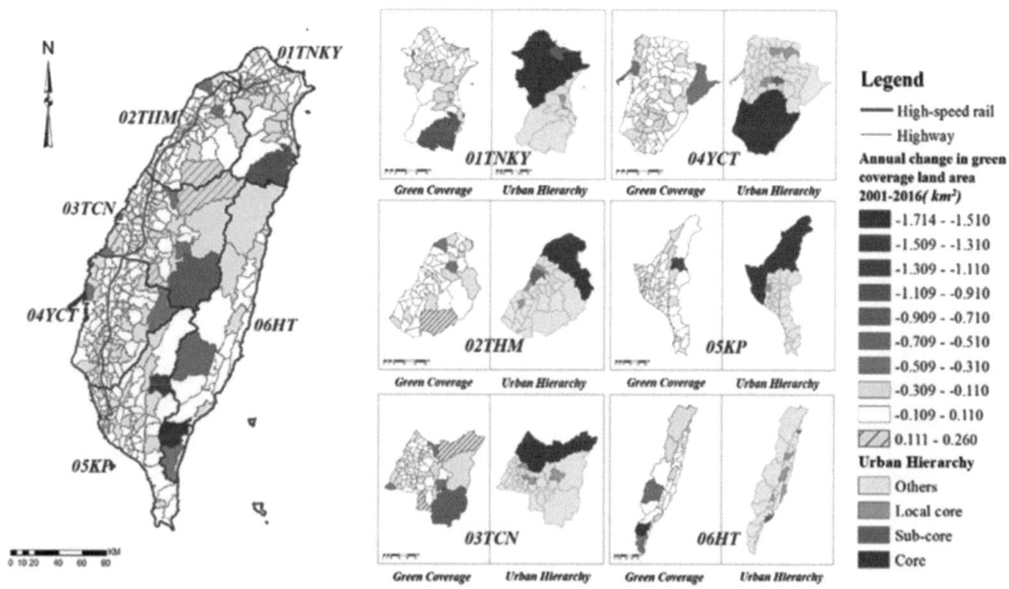

Figure 6. Annual average change in green coverage land area in Taiwan, 2001–2016, by town/district. Note: The white symbol ranges from −0.109 to 0.110 km^2 is designed to refer to none or marginal change. It is also designed to mitigate the possible error derived from NVDI. Data source: Calculated from the Landsat [11] with NDVI.

4.3. Land price index in Taiwan: 2001–2016

This section presents the residential land prices of urban areas in Taiwan for each town/district majorly for the period of 2001–2016 and also in 1996 for analysis purposes. The Urban Land Price Index (ULPI) for the residential areas [16], the only land or housing price index for this spatial scale available in Taiwan, is collected as a proxy of a housing index, since the land price is a significant factor in the housing price, so they are highly correlated. The ULPI is the average land price of a selected sub-area with the median land price in a town/district's urban areas. The ULPI is available after 1993 for all residential, commercial, and industrial uses. The residential median land price indexes are then applied for the cross-sectional analysis and trend analysis for between 2001 and 2016, almost covering the latest ascending part of the housing price cycle in Taiwan, peaked around 2014–2015 [17]. Additionally, the land price of 1996 is also incorporated, since it provides a price reference of the previous cycle.

Figures 7 and **8** show the residential median land prices of urban areas of each town/district for 2001 and 2016, respectively. The highest land price indexes across the country are located in Taipei City of the 01 TNKY for both years. Over these years, the highest land prices expanded overwhelmingly in the 01 TNKY and somewhat in 02 THM and 03 TCN. Also, some rural areas with great tourism attractiveness, including Kenting area in the 05 KP and Taitung in the 06 HT, became prosperous during this period, as reflected in their land price jump (**Figure 9**). It is worth noting that this land price index only reveals the median condition rather the extreme cases, such as luxury housing market, which skyrocketed overwhelmingly during the same period.

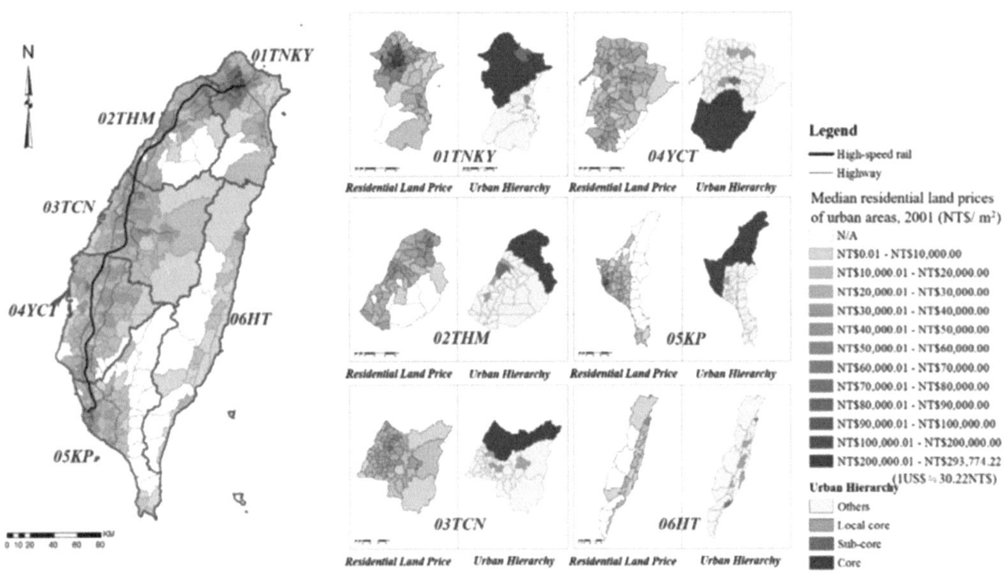

Figure 7. Residential median land price of urban areas in Taiwan, 2001, by town/district. Data source: [16]. The land prices have been adjusted by the consumer price index to the year 2001.

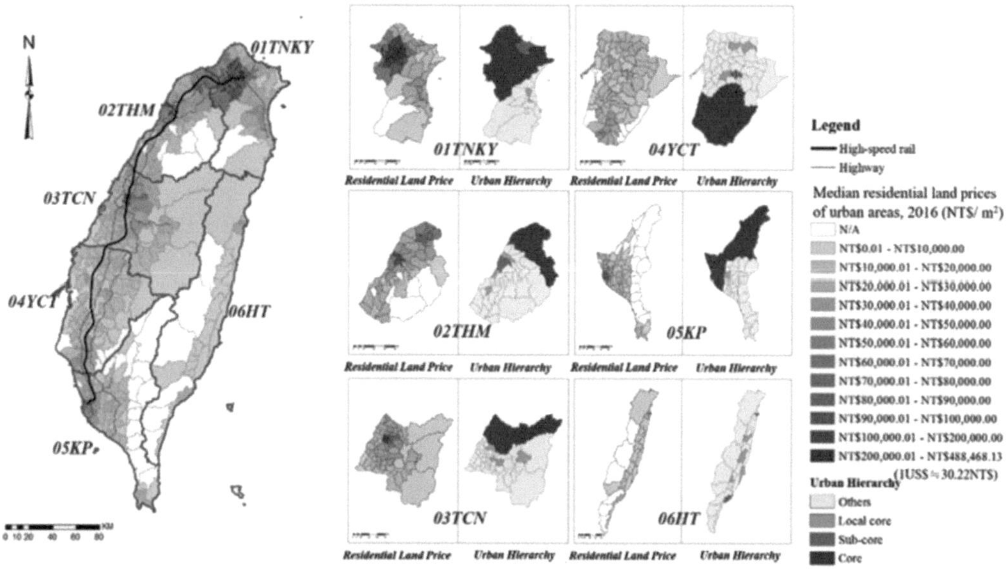

Figure 8. Residential median land price of urban areas in Taiwan, 2016, by town/district. Data source: [16]. The land prices have been adjusted by the consumer price index to the year 2001.

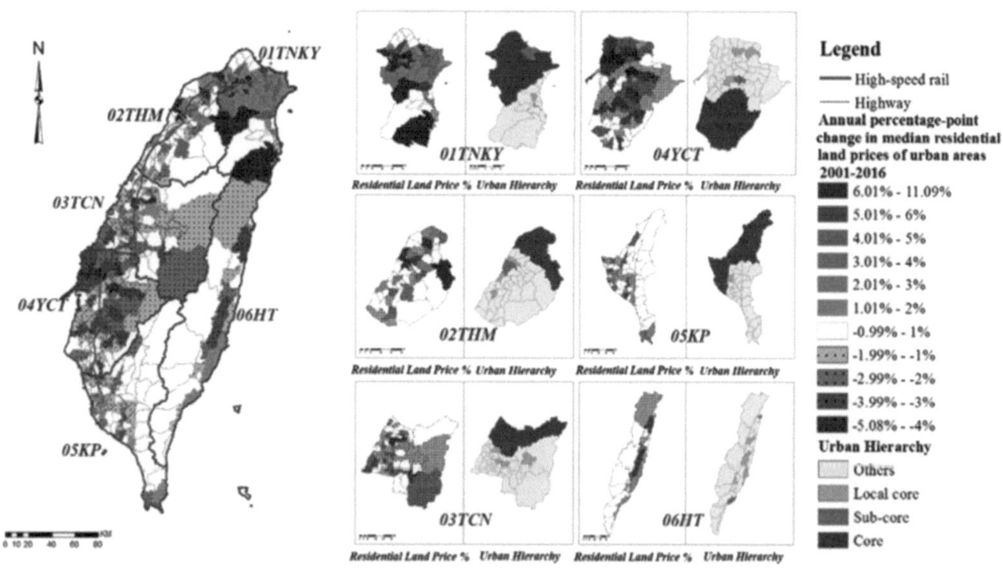

Figure 9. Annual percentage point change in residential median land price of urban areas in Taiwan, 2001–2016, by town/district. Data source: Calculated from Ref. [16]. The land prices have been adjusted by the consumer price index to the year 2001.

Table 2 shows the annual average percentage point change in the residential median land price for the towns/districts in Taiwan. First, surprisingly the land prices decreased by one percentage point between the previous (i.e., 1996) and current (i.e., 2016) near-peaks of the

Period	Mean (%)	Median (%)	S.D. (%)	Min. (%)	Max. (%)	N
1996–2001	−2.8	−3.0	3.9	−12.0	10.2	313
2001–2006	−2.9	−3.1	4.2	−13.7	14.3	313
2006–2011	−0.9	−1.4	4.0	−11.6	24.3	314
2011–2016	3.5	2.5	5.4	−18.0	25.5	318
1996–2016	−1.0	−1.7	2.6	−4.6	15.1	313

Data source: Calculated from Ref. [16]. The land prices have been adjusted by the consumer price index to the year 2001.

Table 2. Annual average percentage point change in residential median land price of urban areas in Taiwan, 1996–2016, by town/district.

residential price cycles. Second, opposite to the general impression that land prices increased between 2001 and 2016, they only increased between 2011 and 2016, which is probably because it is the median land prices rather than those housing whose prices jumped significantly such as the luxury housing and in the major city region like the 01 TNKY and hence grabbed the public attention.

4.4. Cross-analysis of green coverage and land price against urbanization

The differences in green coverage rate among the four classes of the urban hierarchy are statistically significant, and there is barely difference statistically across the city regions. The overall green coverage rate in Taiwan was 78% in 2016 (**Table 3**), the rest of which majorly consisting of the built environment and water body. The core and sub-core on average had the

	City region 01 TNKY (%)	City region 02THM (%)	City region 03 TCN (%)	City region 04 YCT (%)	City region 05 KP (%)	City region 06 HT (%)	Total[e] (%)
Core	62	78	66	76	60	N/A	67[c,d]
Sub-core	71	54	N/A	74	N/A	71	68[d]
Local Core	75	78	83	93	58	92	83
Others	90	93	83	87	89	93	88[a,b]
Total[f]	74	76	77	83	69	85	77.6

Data source: Calculated from the Landsat [11] with NDVI.
[a]The mean difference with the "Core" is significant at the 0.05 level in pairwise comparison in ANOVA.
[b]The mean difference with the "Sub-core" is significant at the 0.05 level in pairwise comparison in ANOVA.
[c]The mean difference with the "Local Core" is significant at the 0.10 level in pairwise comparison in ANOVA.
[d]The mean difference with the "Others" is significant at the 0.05 level in pairwise comparison in ANOVA.
[e]The mean differences across all urban hierarchy types are significant at the 0.05 level in ANOVA.
[f]The mean differences across all city regions are not significant at the 0.05 level in ANOVA.

Table 3. Two-way ANOVA green coverage rate by urban hierarchy/degree of urbanization and by city region in Taiwan, 2016.

	City region 01 TNKY	City region 02THM	City region 03 TCN	City region 04 YCT	City region 05 KP	City region 06 HT	Total[b]
Core	$4488	$1460	$1314	$653	$1104	N/A	$2,024[a]
Sub-core	$718	$2406	N/A	$963	N/A	$936	$1145
Local Core	$1825	$1968	$1060	$592	$788	$270	$1026
Others	$787	$785	$499	$389	$445	$237	$499
Total[c]	$1954	$1655	$958	$649	$779	$481	$1256

Data source: Calculated from Ref. [11]. The land prices have been adjusted by the consumer price index to the year 2001.
[a]The mean difference with the "Others" is significant at the 0.05 level in pairwise comparison in ANOVA.
[b]The mean differences across all the urban hierarchy types are significant at the 0.05 level in ANOVA.
[c]The mean differences across all the city regions are not significant at the 0.5 level in ANOVA.

Table 4. Two-way ANOVA median residential land price by urban hierarchy/degree of urbanization and by city region in Taiwan, 2016 (US$/M^2).

lowest green coverage rates of 67–68%, while the local cores' green coverage was 83%, only five percentage point short of the "others" class. City region-wise, the 05 KP had the lowest green coverage rate of only 69%, followed by the 01 TNKY, 02THM, and 03 TCN in the range of 74–77%.

There were statistically significant differences in the residential land prices across all the four classes of urban hierarchy as a whole in 2016, while most of the paired differences between classes were barely statically significant. The median residential land prices in the core towns/districts, on average, were twice as high as those in the sub-cores and local cores (**Table 4**). City region-wise, the highest land prices were located in the two northern city regions, that is, the 01 TNKY and 02 THM, doubling that of the 03 TCN and 05 KP. However, the overall differences across the city regions are not statistically significant.

5. Panel data regression: the relationships of urbanization with green coverage rate and land price

Two separate panel data regression analyses are conducted to assess the relationships of urbanization with the green coverage rate and the land price, respectively. This section goes beyond the previous bivariate analysis by adopting the multiple regression of panel data analysis to control the impacts of other possible factors. The panel data analysis is selected concerning the data structure of these empirical analyses, which contain data for some 350 towns/districts of Taiwan main island for the years 2001, 2006, 2011, and 2016. The town/district is selected as the analysis unit, since a larger spatial unit of county/city divides Taiwan main island into 19 units, providing less homogeneity of urbanization and an insufficient size of units for multiple regression. On the other hand, smaller spatial units do not provide data for all the required variables for the analysis.

In the two panel data analyses, the primary relationships to evaluate are how the degree of urbanization associates with the green coverage rate and the land price, respectively (Eqs. (1) and (2)). In both these two panel data models, two population-derived variables are adopted to quantify the degree of urbanization. One is the population size of a city region that quantifies the overall degree of urban agglomeration for the whole city region; the other is the population density of a town/district that quantifies the degree of urbanization within a sub-area of a city region. The regression models do not necessarily mean to prove the existence of a cause-effect relationship between the urbanization variables and the green coverage rate and the land price, but a correlation that proves the trend between two variables without a clear cause-effect explanation; the cause-effect relation may only exist given, at least, theoretical foundations. In these cases of the population size of city region, there may exist a correlation as a result of urbanization process, since it is also a result of urbanization process, like the green coverage rate and the land price; nonetheless, their theoretical cause-effect relationship is not clear.

Another definition of the city region for Taiwan, consisting of 36, is adopted over the 6 city region version, since it is more based on the daily activity rather than some political concern of the latter (**Table 5**). In both of the panel data regression models, region and county/city variables are incorporated as control variables to single out the net relationships of urbanization variables (**Table 5**). Curve estimations between the dependent and independent variables are first conducted before constructing the two panel data regressions to explore the most appropriate curve forms. In the panel data analysis, both fixed- and random-effects models are developed first, and then, the Hausman test is conducted to select the most appropriate model.

Variables	**Mean**	**S.D.**
Dependent variable		
Green coverage rate $_{t^*}$ (Town/District)	78.2%	24.6%
Inflation-adjusted land price $_t$ (USD/m^2; Base year: 2011)(Town/District)	US$1078	US$1497
Independent variable		
Degree of urbanization		
Population density $_t$ (Persons/km^2) (Town/District)	2874	5810
Population $_{t=2001}$ (City region, N. = 36)	1,610,953	1,953,901
Region, county/city		
Region (0/1)(North, Central, South, East, N. = 4)	N/A	N/A
County/city (0/1)(N. = 19)	N/A	N/A

No. of observations: 1396 (N. of towns/districts: 349; No. of time periods: 4 (2001, 2006, 2011, 2016).*t stands for time period.

Table 5. Descriptive statistics of the variables for the panel data regression models.

http://dx.doi.org/10.5772/intechopen.71651

5.1. Panel data regression of green coverage rate

Eq. (1) shows the panel data regression model of the green coverage rate:

$$\text{Green coverage rate}_t = a_0 + a_{Urban}{}^* X_{Urban,t} + \left(a_{Rgn} X_{Rgn} + a_{Cty} X_{Cty}\right) + e \tag{1}$$

where green coverage rate$_t$ is the green coverage rate of a town/district for the time period t; X_{Urban} is the array of the variables quantifying the degree of urbanization; X_{Rgn} and X_{Cty} are the arrays of region and county/city variables, respectively; a_0 is the interception, and a_{Urban}, a_{Rgne}, and a_{Cty} are coefficients, and *e* is the residual.

Both the fixed-effects and random-effects models of the panel data analysis are developed and then based on the significant p-value in the Hausman test (**Table 6**), the fixed-effects model

	Panel data regression, fixed-effects		**OLS regression model**	
Dependent variable	Green coverage rate $_t$ (Town)		Green coverage rate $_t$ (Town)	
Independent variable	**B** (Sig.)	**Std. Err.**	**B** (Sig.)	**Std. Err.**
Degree of urbanization				
Pop. density $_t$ (Town)	−2.69e-06(0.02)	1.14e-06	−0.000032(0.00)	7.18e-07
Pop. $_{t-2001}$ (City region)	−1.19e-07(0.00)	1.44e-08	−8.16e-08(0.00)	9.17e-09
Pop.^{2} $_{t-2001}$ (City region)	5.56e-15(0.00)	1.58e-15	1.11e-14(0.00)	1.94e-15
Region, county/city				
Hsinchu County			0.111(0.00)	0.020
Taoyuan County			0.094(0.00)	0.020
Taipei City			0.081(0.05)	0.041
New Taipei City			0.076(0.03)	0.034
Chiayi County			0.076(0.00)	0.016
Nantou County			0.056(0.01)	0.021
Yilan County			0.047(0.02)	0.021
Taichung City			0.038(0.01)	0.015
Hsinchu City			−0.103(0.01)	0.036
Tainan City			−0.025(0.04)	0.012
Constant	0.92 (0.00)	0.015	0.898(0.00)	0.007
Overall	Prob. > F = 0.00		Prob. > F = 0.00	
	R-sq.: Within = 0.15		Adj. R-sq.: 0.75	
	Between = 0.30			
	Overall = 0.30			
	Hausman Test: P-value = 0.00			

Table 6. Green coverage rate models: fixed-effects panel data regression model and OLS regression model.

is selected over the random-effects model. The fixed-effects model shows that both the population size of the city region and the population density of the town/district are statistically significant factors negatively associated with the green coverage rate. This finding indicates the higher the population size of a city region or the higher the population density of a town/district, the lower is its green coverage rate, ceteris paribus. The causal relationship between the population density of the degree of urbanization and the green coverage rate may occur primarily due to higher competition in high-density areas between the demand for green open space and other land uses, such as residential, commercials, and road and parking of the built environment.

The fixed-effects model, however, does not reveal the content of the fixed effects regarding the differences among counties/cities or regions. One solution is an OLS regression model, adopting binary regressors for spatial or administrative units (**Tables 5** and **6**). An OLS

	Panel data regression, random-effects	
Dependent variable	ln(Inflation-adjusted land price $_t$ (USD/m^2; Base year:2001)(Town)	
Independent variable	**B** (Sig.)	**Std. err.**
Degree of urbanization		
ln (Pop. density $_t$ (Town))	0.31(0.00)	0.020
Pop. $_{t=2001}$ (City region)	3.29e-08(0.00)	6.44e-09
Region, county/city		
Taipei City	0.48(0.00)	0.06
Hsinchu City	0.29(0.00)	0.08
Hsinchu County	0.28(0.00)	0.05
Yilan County	0.24(0.00)	0.05
Miaoli County	0.21(0.00)	0.04
Taoyuan County	0.18(0.00)	0.04
New Taipei City	0.16(0.00)	0.05
Nantou County	0.13(0.01)	0.05
Yunlin County	0.11(0.00)	0.03
Taichung City	0.07(0.00)	0.03
Constant	3.22(0.00)	0.05
Overall	Prob. > F = 0.00	
	R-sq.: Within = 0.04	
	Between = 0.83	
	Overall = 0.76	
	Hausman Test: P-value = 0.15	

Table 7. Land price random-effects panel data regression model.

regression model shows that there exist fixed differences in the green coverage rates among counties/cities (**Table 6**). Some counties/cities in the Northern Taiwan such as Hsinchu County, Taoyuan City, Taipei City, and New Taipei City had higher green coverage rates than the average county/city. In contrast, Hsinchu City and Tainan City had lower green coverage rates.

5.2. Panel data regression of land price

Eq. (2) shows the panel data regression model of the land price:

$$\ln\left(\text{Inflation-adjusted land price}_t\right) = a_0 + a_{Urban}{}^{*} \ln\left(X_{Urban,t}\right) + \left(a_{Rgn} X_{Rgn} + a_{Cty} X_{Cty}\right) + e \quad (2)$$

where the inflation-adjusted land price $_t$ of a town/district is the median land price adjusted by the consumer price index to year 2001; X_{Urban} is the array of variables quantifying the degree of urbanization; X_{Rgn} and X_{Cty} are the arrays of region and county/city variables, respectively (**Table 5**); a_0 is interception, and a_{Urban}, a_{Rgn}, and a_{Cty} are coefficients, and e is the residual.

Based on the curve estimation where power relationship has the highest R-squared value, log-log model is adopted for between inflation-adjusted land price and population density in the panel data regression. Both fixed-effects and random-effects models are developed, and then, the random-effects model is selected, since the p-value in the Hausman test was insignificant (**Table 7**). The random-effects panel data regression shows that the higher the degree of urbanization, the higher is its median land price, as supported by the statistically significant positive coefficients of the population density of the town/district, as well as the population of the city region. The high density of a town/district may cause high land prices due to the high demand for residences, which is also demonstrated in the bid-rent theory [18] that the high rent corresponds with the high building height or population density [19]. In addition, the median land prices of the towns/districts of some cities are higher than the others such as Taipei City, Hsinchu City and County, Yilan County, etc.

6. Cluster analysis of towns/districts, by green coverage and land price

This section is intended to classify the towns/districts in Taiwan based on their level of greenery and land price to address possible challenges or opportunities in the wake of urbanization.

The characteristics of towns/districts selected for the cluster analysis are the level of green coverage rate, residential land price, and their change patterns; the method is the two-step cluster analysis. The green coverage rate and residential land price are two important aspects of the standard of living that are highly associated with the level of urbanization. Adopting these two aspects to classify the towns/districts is intended to reveal the types of quality of life under different levels of urbanization. Additionally, not only the level of green coverage rate and residential land price but also their change patterns are adopted to provide both the latest levels but also changing fashions, respectively. Then, regarding the method of cluster analysis, the two-step cluster analysis and two traditional cluster methods hierarchical and k-means are all tried. The final selection of the method is the two step due to its results comply with expectation and theoretically sensible and reasonable to interpret.

Based on the levels and change patterns of greenery and residential land price, the towns/districts in Taiwan are classified into four clusters: (1) Gray expensive urban core; (2) green moderately priced suburbs; (3) green lowly priced suburbs; and (4) green rural areas (**Table 8**).

Cluster	**N (%)**	**Mean population density, 2016 (Persons/km²)**	**S.D.**	**Range (Persons/km²)**
1. Gray expensive urban core	29 (9%)	18,894	7279	3455–39,203
2. Green moderately priced suburbs	45 (14%)	3085	2999	19–12,476
3. Green lowly-priced suburbs	95 (30%)	2006	2278	90–14,071
4. Green rural areas	143 (46%)	639	1114	8–10,635
Total	312 (100%)	3105	5881	9–39,203

F-test: 331.607, P-value: 0.000.

Table 8. ANOVA of town/District's population density, by cluster, 2016 Taiwan.

Cluster	**Green coverage rate, 2016[1]**		**Residential median land price, 2016[2] (USD/m²)**		**Annual percentage point change of green coverage rate, 2001–2016[1]**		**Annual percentage point change of residential median land price, 2001–2016[2]**	
	Mean (%)	**S.D. (%)**	**Mean**	**S.D.**	**Mean (%)**	**S.D. (%)**	**Mean (%)**	**S.D. (%)**
1. Gray expensive urban core	15	10.2	US $5016	US $4515	−0.09	0.13	2.1	3.68
2. Green moderately-priced suburbs	75	16.5	US $2349	US $1831	−0.16	0.12	4.8	2.61
3. Green lowly-priced suburbs	75	15.1	US $733	US $410	−0.31	0.12	−1.4	1.60
4. Green rural areas	90	7.4	US $525	US $330	−0.06	0.08	−1.4	1.48
Total	76	24.0	US $1269	US $62,577	−0.15	0.15	−0.17	3.03

[1]Data source: Calculated from the Landsat [11] with NDVI.
[2]Median urban land price of residential district inflation-adjusted to year 2011 [16].

Table 9. Cluster distribution and descriptive statistics of towns/districts in Taiwan, by green coverage rate, land price, and their growth patterns.

The first cluster is coined as the gray expensive urban core, since most of them were located in the urban core with the lowest green coverage rate of a mean of 15%, and the highest median residential land price (**Table 9**, **Figure 10**). This cluster is the most urbanized class, composed of less than 10% of all districts/towns with valid data, most of which were the most urbanized core constituents of the most developed city regions of the 01 TNKY and 05 KP, as well as some in the core of 03-TCN. In the last 15 years of 2001–2016, the green coverage rate of this cluster descended, and the median land prices were on the rise with an annual mean of 2.1 percentage points, only second to the second cluster of green moderately priced suburbs detailed below.

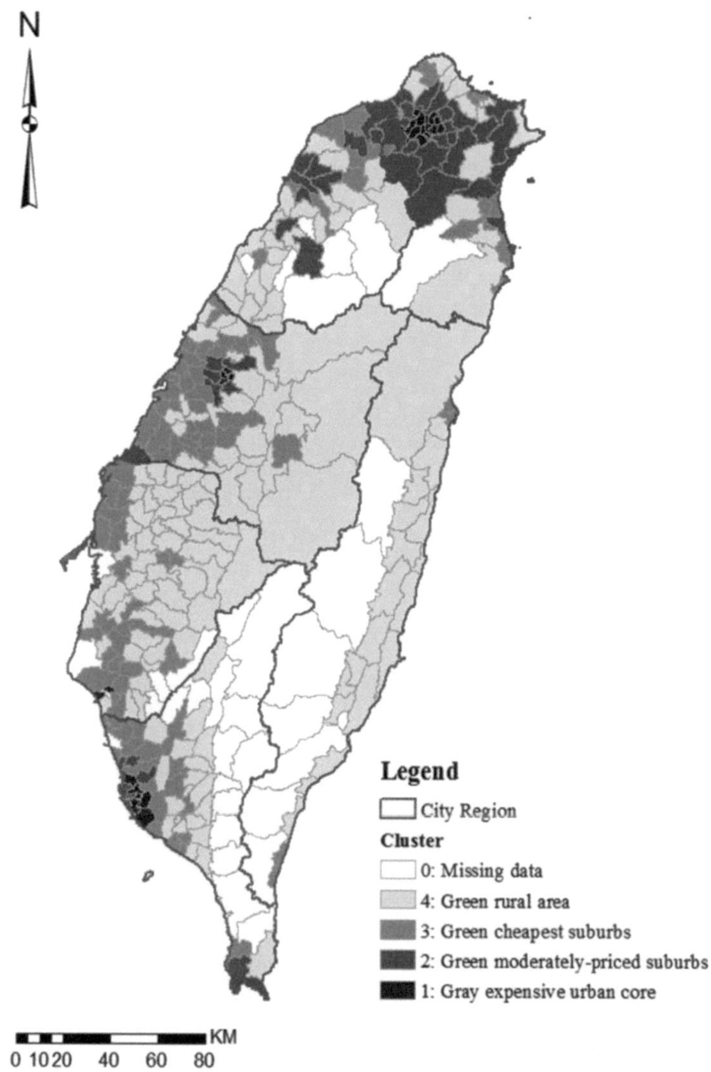

Figure 10. Cluster distribution towns/districts in Taiwan, by green coverage rate, land price, and their growth patterns.

The other three clusters all had higher green cover rates of 75% or more and lower residential land prices than the gray expensive urban core, indicating better natural environment and lower housing cost, but lower accessibility to the urban core (**Table 9**, **Figure 10**). The second cluster coined as the green moderately priced suburbs mostly contains districts/towns adjacent to the cluster gray expensive urban core, including some second-tier urban cores of the city region. Its median residential land price was only some half of the cluster gray expensive urban core, but its annual growth percentage point was the highest at about 5%. These inner parts of city regions of the cluster gray expensive urban core and/or cluster green moderately priced suburbs were the only urban areas with rising median residential land prices. Then, the cluster green lowly priced suburbs mostly lied further out from the urban core with significantly lower residential median land prices, but its residential prices decreased between 2001 and 2016. The rest is the rural areas unsurprisingly with the highest green coverage and lowest residential median land prices.

Overall, these four clusters comply with the degree of urbanization to some degree, the higher the degree of urbanization, the higher the residential land price and the lower the green coverage rate. The mean population density of the cluster gray expensive urban core was as high as about 19,000 persons per square km in 2016, followed by a large gap, by the second tier of the two clusters of green suburbs, and the green rural cluster (**Table 8**).

7. Conclusions and policy implications

Similar to many other fast urbanizing countries, Taiwan's urbanization grew considerably between 1976 and 2016 from 61 to 80%. Meanwhile, the population grew abruptly by 42%, which made sustainable urbanization an even more crucial issue, since the total urban population was enlarged at an even higher pace. This research focuses on the social and environmental aspects of sustainable urbanization regarding housing affordability and natural living quality in the urban areas, respectively. The panel data analysis demonstrates the negative associations between the green coverage rate and the degree of urbanization represented by the population density of the sub-area of a city region-town/district and the population size of the city region. Moreover, urbanization's relationship with the median land price is positive. In summary, degraded urban greenery and increasing residential land prices came along with the urbanization in Taiwan between 2001 and 2016.

Degraded living quality regarding urban greenery and housing affordability seems to come as the cons of urbanization, largely driven by the rural emigrants seeking for the better level of service in urban areas. Arguably, in the sense of "user pays," the local scale impacts of human development on the natural environment may be reduced to some degree for the other non-human stakeholders by confining human developments in urbanized areas, and there is still needs to resolve the degraded standard of living in urban areas. Other than a whole bunch of potential resolutions of raising the level of urban greenery by planting wherever possible, horizontally or vertically, on the ground or roof, two specific types of urban planning tools are emphasized here that may essentially affect accessible green space on the ground or in the buildings. The first one is to delicately reconsider building coverage rates for various land uses to allow the most possible natural plant-friendly ground, when the buildings are renewed; this will lay up the fundamental permeable settings in the urban area. The second is to consider the building setback policy to provide gardens located in the mid- or low-floors of a building, which are highly accessible to building users or even pedestrians. This type of mid-air gardens may also avoid strong sunlight blocked by the building itself. Due to their vertical locations in the building, as opposed to the roof garden, it can be more accessible and function better as a community mini-park.

Rising residential land/housing prices is probably a common, tough issue for many cities or counties. In the case of Taiwan, there was lack of cures at the peaks of the last two housing market cycles. There has even been some ignorable portion of vacant housing units in the greater Taipei area of the 01 TNKY city region, but the least possible expectation during the peak was for the housing prices to decline, much of which was to blame the situations and policy that allow housing units to become an investment commodity.

The low green coverage rate and high land price in the highest urbanized urban core may cancel out the benefits of urban agglomeration as in the high service level of public infrastructure. Whether the low mean green coverage rate of 15% of the gray expensive urban core cluster leads to unhealthy, unpleasant, inefficient, or unbalanced urban settings may not seem as initiative as it looks or even sensible to the general public. However, this condition may be planners' responsibility to think ahead to consider if this level has already passed the acceptable threshold. This greenery issue may also be further complicated by the concerns over when to adopt green cover land area as a whole or per capita green coverage when assessing various aspects. For instance, when considering the conversion of carbon dioxide produced by residents into oxygen needed for each resident, the index of per capita green coverage may make more sense. In contrast, the overall green coverage may serve to evaluate the natural landscape or streetscape in the urban settings.

Finally, based on the cluster analysis, the green moderately priced suburbs and green lowly priced suburbs had better greenery but lower residential prices. Governments may employ policies to leverage higher level of infrastructure in these towns/districts or even to establish another urban core to formulate a polycentric urban form. A rule of thumb for the government is to take advantage of the urban agglomeration by providing better infrastructure in needed urban areas and smartly drive residents to move to the places with less congestion issue. On the other hand, for the individuals or households, the suggestion would be to reconsider if some greener districts or cities with cheaper housing may serve their needs; however, they have never popped up on their residential wish list.

Funding

Ministry of Science and Technology, Taiwan Government. MOST 105-2621-M-004-003-MY3.

Author details

Yu-hsin Tsai*, Jhong-yun Guan and Yu-Hsin Huang

*Address all correspondence to: yuhsintsai@gmail.com

Department of Land Economics, National Chengchi University, Taipei, Taiwan

References

[1] United Nations, Department of Economic and Social Affairs, Population Division. World Urbanization Prospects: The 2014 Revision, CD-ROM Edition [Internet]. 2014 [Updated: 2014]. Available from: https://esa.un.org/unpd/wup/CD-ROM/ [Accessed: August 20, 2017]

[2] United Nations, editor. The Sustainable Development Goals Report 2017. New York, US: United Nations; 2017. p. 64

[3] Demuzere M, Orru K, Heidrich O, Olazabal E, Geneletti D, Orru H, Bhave AG, Mittal N, Feliu E, Faehnle M. Mitigating and adapting to climate change: Multi-functional and multi-scale assessment of green urban infrastructure. Journal of Environmental Management. 2014;**146**:107-115

[4] Davis RG, Barbosa O, Fuller RA, Tratalos J, Burke N, Lewis D. City-wide relationships between green spaces, urban land use and topography. Urban Ecosystems. 2008;**11**: 269-287

[5] Xie YC, Sha ZY, Yu M. Remote sensing imagery in vegetation mapping: A review. Journal of Plant Ecology. 2008;**1**(1):9-23

[6] Ngom R, Gosselin P, Blais C. Reduction of disparities in access to green spaces: Their geographic insertion and recreational functions matter. Applied Geography. 2016;**66**:35-51

[7] Yanga J, Suna J, Geb QS, Li XM. Assessing the impacts of urbanization-associated green space on urban land surface temperature: A case study of Dalian, China. Urban Forestry & Urban Greening. 2017;**22**:1-10

[8] Gallo KP, Tarpley JD, McNab AL, Karl TR. Assessment of urban heat islands: A satellite perspective. Atmospheric Research. 1995;**37**(1-3):37-43

[9] Rouse JW, Haas RH, Scheel JA, Deering DW. Monitoring vegetation systems in the great plains with ERTS. In: 3rd Earth Resource Technology Satellite (ERTS) Symposium Proceedings. 1974;**1**:48-62

[10] Nouri H, Beecham S, Anderson S, Nagler P. High spatial resolution WorldView-2 imagery for mapping NDVI and its relationship to temporal urban landscape evapotranspiration factors. Remote Sensing. 2014;**6**(1):580-602. DOI: 10.3390/rs6010580

[11] U.S. Geological Survey, U.S. Department of the Interior. Landsat Missions [Internet]. https://landsat.usgs.gov/ [Updated: 6/22/2017]. [Accessed: 15-07-2017]

[12] National Development Council, Taiwan Government. Compilation of Urban and Regional Development Statistics [Internet]. 2016 [Updated: 2016]. Available from: http://statistic.ngis.org.tw/index.aspx?topic=4 [Accessed: 25-07-2017]

[13] Ministry of the Interior, Taiwan Government. Nationwide Regional Plan. Taipei, Taiwan: Ministry of the Interior, Taiwan Government; 2013. p. 137

[14] Legislative Yuan, Taiwan Government. National Spatial Development Strategic Plan. Legislative Yuan, Taiwan Government: Taipei, Taiwan; 2010. p. 147

[15] Council of Agriculture, Taiwan Government. Debris Flow Disaster Prevention Information [Internet]. June 3, 2017 [Updated: June 3, 2017]. Available from: http://246eng.swcb.gov.tw/ [Accessed: July 31, 2017]

[16] Ministry of the Interior, Taiwan Government. Land Price Indexes [Internet]. July 2017 [Updated: July 2017]. Available from: https://www.land.moi.gov.tw/chhtml/hotnewsall.asp?cid=102 [Accessed: August 2017]

[17] Ministry of the Interior, Taiwan Government. Information Platform for Real Estate [Internet]. 2017 [Updated: 2017]. Available from: http://pip.moi.gov.tw/V2/E/SCRE0106.aspx [Accessed: June 20, 2017]

[18] Alonso W. Location and Land Use. Cambridge, Massachusetts: Harvard University Press; 1964

[19] Brueckner JK. Chapter 20 the structure of urban equilibria: A unified treatment of the muth-mills model. Handbook of Regional and Urban Economics. 1987;**2**:821-845

Chapter 12

The Corridor Island: A New Space to Redesign the Landscape of Tenerife

Miguel Ángel Mejías Vera

Additional information is available at the end of the chapter

http://dx.doi.org/10.5772/intechopen.72190

Abstract

The urbanization processes in the island of Tenerife in the last 50 years have left a very fragmented landscape. Due to the growing and constant process of construction-infrastructures-construction, the abiotic and biotic space is being reduced but still more the productive space of the primary sector. The reaction to the threat, especially the "natural" space—abiotic and biotic—was protection. The result is that half of the island has some protection figure, but the other half does not. To this "unprotected" half, I call the risk island and concentrate the stress or tension, because it is where economic activities and habitat are developed and combined. For its management, in spite of existing an island territorial planning, it is the municipalities that have the competences in urban planning, always under the norms of superior administrative rank. Therefore, there are 31 different models, one per municipality. We have to focus on the corridor island understood as the unsealed soil. This is the area where we should intervene, redesigning the new spaces, to connect organically and in balance the two islands.

Keywords: urban sprawl, soil sealing, urban planning, risk island, corridor island, geo-design, landscape

1. Introduction

Tenerife is divided into two distinct areas: the protected island and the unprotected island. I could even say, the static island and the dynamic island. The first, where time stops, the pace becomes slower, where it hardly changes, the island preferred, the island imagined, the island visited. The second is the opposite. It is the dynamic island, inhabited, transformed, but at the same time the least valued, desired or visited.

This general, complex hypothesis, where many components and variables intersect, starts with a partial hypothesis that centers this article: The Island unprotected is the one that has a higher risk for the conservation of its landscape. This section has a wide and fractionated space corridor that connects both "islands". Here, it will be the biggest change. Therefore, it is where we must act redesigning it in the key of landscaping.

The whole island is actually regulated territorially and urbanistically through the instruments of the planning system of the Canary Islands that has recently been modified in Law 14/2014 [1]. It is made up of territorial and environmental planning, which includes general and sectoral planning guidelines, and island planning and urban planning of municipal scope. In spite of this, I consider and enhance the epithet of the unprotected island.

here are two key concepts in this phase of the research: one is the urban sprawl, and the other is soil sealing. The first one, urban sprawl or suburban sprawl [2–7], is characterized by spaces composed of low-density [8], unstructured, and amorphous, which generate problems related to the increase in the temporal and spatial distances between the family residence and the different destination centers (work, education, leisure, etc.). The dispersed space entails, among other aspects, dependence on the private vehicle, and with it, it generates spaces of exclusion [9]. They imply a high cost for the construction, management, and maintenance of infrastructures and services, usually of a public nature. Richard Ingersoll uses the term sprawl scape [10] or landscape of dispersion, where urbanization processes have built a landscape, in which "oil stains" multiply around highways and roundabouts [11].

According to the European Environment Agency (EEA), "Urban sprawl is commonly used to describe physically expanding urban areas" and "have described sprawl as the physical pattern of low-density expansion of large urban areas, under market conditions, mainly into the surrounding agricultural areas" [12]. "Urban sprawl is visually perceptible" [6]. "A landscape suffers from urban sprawl if is permeated by urban development or solitary building" [6]. On the other hand, it is the concept of "soil sealing" that is defined as the covering of the ground by a waterproof material and is considered as one of the main causes of degradation of the soil in the European Union (EU) (European Commission). The european environment agency has defined soil sealing as "the covering of soil due to urbanization and infrastructure construction, such that soil is no longer able to perform the range of functions associated with it. Soil sealing is not adverse *per se*, rather it is the irreversibility in practical terms of sealing the soil and the consequent loss of soil functions" [13]. "Soil sealing" is one of the main factors of the studies on the risks of "land degradation (LD)" next to the salinity of the soil or the erosion. Soil sealing is associated with the fragmentation of landscapes or the risks of fires [14].

Previously, in 2002, the EEA [15] published the results of a meeting in Copenhagen in 2001, entitled "Proceedings of the Technical Workshop on Indicators for Soil Sealing" which were raised at the European level of multiple proposals for indicators to measure the soil sealing, change in the land use, motorization for the measurement of the soil loss, etc. This was the result of projects developed in different members.

The concept of soil sealing is currently redefined and also the way of how to tackle them from different techniques. It has gone from being a merely physical analysis to a complex analysis and global, that affects whole ecosystems. In this sense, recently, a review of the concept of soil sealing has been published, trying to unify the methodological criteria applied to the concept of scale [16].

The analysis, spatial entity, and complexity are underway from models of raster data with satellite and aerial images of a very high resolution (VHR) [17].

According to the document of work published by the European Commission in 2012 (from the study commissioned by the European Commission to Prokop [18] and other studies and data provided by a group of experts from the member states which advised in 2011 to the commission services), the surface total of soil sealing in 2006 is estimated to 100,000 km^2 or to 2.3% of the territory of the EU, with a half of 200 m^2 by citizen. On the island of Tenerife, I think that the percentage exceeds the European average. To an insular, limited, and finite space, it is a great concern.

If the model of low density, both built urban land and the dispersed space in rustic soil, requires more connectors in the form of infrastructure, the level of "soil sealing" will increase dangerously. If the solution is to search for compactness, I believe that it will generate more destruction than construction. It would have to find other ways of integration, especially in the peri-urban areas, and in relation to the development and defense of the agricultural productive space, as developed in the Agrarian Park of Llobregat, Barcelona, Spain [19], or many actions of awareness and education in the peri-urban agricultural landscape, as developed the village of El Tablero on the island of Tenerife [20]. These initiatives, among others, mark the trend to follow in these spaces. The connectivity between the urban and the rest of the territory must be connected through corridors characterized by primary production, integrated dispersed buildings and camouflaged with the environment and natural spaces, even if they are not protected (Landscape "Island of Tenerife" award 2010 by the Council of Tenerife and the European Network of Local and Regional Authorities for the implementation of the European Landscape Convention RECEP ENELC [21]. http://laserasdeeltablero.blogspot.com.es/).

Now, create a more sustainable city model? According to experts, after analyzing and comparing different urban systems, the model that conforms to the principle of urban efficiency and livability is the city compact in its morphology, complex in their organization, efficient metabolically, and cohesive socially [22].

In terms general it dispersion urban in the island of Tenerife tends to be a problem important to form nuclei and cities compact, which not makes but increase the difficulty in it planning territorial and in it management urban. It is for this reason that we need to make a diagnosis of the scope of the insular dispersion and georeferenced spaces that tend to be the concentration or the compactness of the soil. Their progress clearly limited biodiversity, because scattered space environments tend to abandon early activities, mainly agricultural activities, to become in the waiting space. Often the changes happen over very long periods of time. This process implies that landscapes deteriorate because they do not have a specific function.

To demonstrate these hypotheses, it is necessary to first size up the problem at the island level and build the necessary reference frameworks: island, risk island, and corridor island. They are in this article, my goals.

1.1. Island, risk island, and corridor island

General system theory understands interconnected space. It produces connections that affect the whole. When we work on a local problem, we often try to look for boundaries, which often cut the system. It is necessary to extract the organizational patterns in the complexity of the system, but is limited, because these are possibly dependent on others that form or even that is formed in our limits and that is transferred to the environment. When we work the whole of a finite space, at least from the terrestrial point of view, as it is an island, we come closer to systemic conclusions, but certainly, not even the islands are saved to other global interconnections that break or bridge the limits terrestrial.

Working on the concept of compactness, dispersion, connectivity, or green corridor on an island-wide scale is a challenge and involves an important effort in selecting, editing, and combining data, often scattered and without uniform criteria. If it is achieved, the overall analysis capacity and its diagnoses reinforce territorial planning and the measures to be taken into account.

The concepts of island, risk island (Ri), and corridor island (Ci) are mere abstractions that allow me to dissect large sets and subject them to different pressures. The island cushions the processes of pressure among its abiotic, biotic, and cultural spaces. If we reduce the total insular space, to the space that really supports the tension of the changes, all its pressure indicators will increase. Their results would allow us to evaluate the risk more. Pressure-reducing valves are the corridors that connect the substantial organs of equilibrium (abiotic and biotic) with the green mosaic of the risk island—agricultural space, livestock, abiotic and unprotected biotic, urban green, etc. (Ci).

How is deconstruction formed? It is a sequence of systematic reductions from a whole, the island. We exclude spatially those less compromised components and those that have no pressure, either because they are protected by some figure or because they are already part of the waterproof space. The first exclusion is made up of spaces declared and protected by the current law as protected natural areas. In most of them with values of abiotic and biotic character. These spaces are the vital organs of diversity and biodiversity. They have no discussion. The spatial result of this reduction is the risk island. Often colloquially, politically and mediated, 50% of the island is protected, to which I have always expressed a second phrase in an interrogative key, and the other 50%, what? If we consider that the island is made up of multiple landscapes, I understand the landscape as a social product, that is until the abiotic spaces have been used by man in different phases of history, and accepting their consideration of protection, my concern is in the form of planning and ordering the rest of that 50%.

Given the data I provide in this work, the concern is more than evident. It is urgent to make decisions in other directions that bring us closer to the protected vital organs. The unprotected and pressed envelope is a threat to itself, but also to vital organs and therefore to the island corpus. Why does this happen? There are several causes, as always, economic, social, political, cultural, I would not know which is more influential, is not my object here, but what has caused, is a spatial change in land uses that as a result leaves us waterproof space–construction + infrastructures–very important, a reduced agricultural space and a space urban and projected disproportionate.

From this conclusion and spatial deconstruction, which pierces the risk island we have the corridor island, the island permeable, connector and balancer. Corridor island is where new renovation strategies should be designed. These must be conceived from the concept of landscape. With this, I do not say that you do not have to act in the excluded part of the corridor island, on the contrary, it must act in all areas, but you have to do it in truth and with a collective look, at all, of the Island.

http://dx.doi.org/10.5772/intechopen.72190

2. Objectives

This work has two objectives:

1. Design and construct new units of spatial references, perhaps a unit of landscape.
2. Measure territorial problems at scales more realistic, as are the urban sprawl and the soil sealing.

3. Methodology

Corredor Island is a process of deconstruction. It is the result of successive spatial reductions. The level of deconstruction is marked by the number of components involved in the process. It will be more intense and detailed as we incorporate more components.

In short, the island (I) is reduced, extracting protected natural areas (PNA). Its result is the risk island (Ri). If we introduce the pressure or tension components (T), the built urban space (Uab), the built rural space (Rab), the non-urban road network (Nur), we obtain the first model of Corredor Island (Ci).

$$I - PNA = Ri - T\left[U\{(Uab) + (Rab) + (Nur)\}\right] = Ci \tag{1}$$

The methodological process is developed through the three moments or levels of information [23], at the time of problem definition and data selection (analysis and synthesis), inventory level; at the time of simplification and interpretation of results (combinations and relationships), level of treatment; at the moment of communication of the results, level of communication (**Scheme 1**).

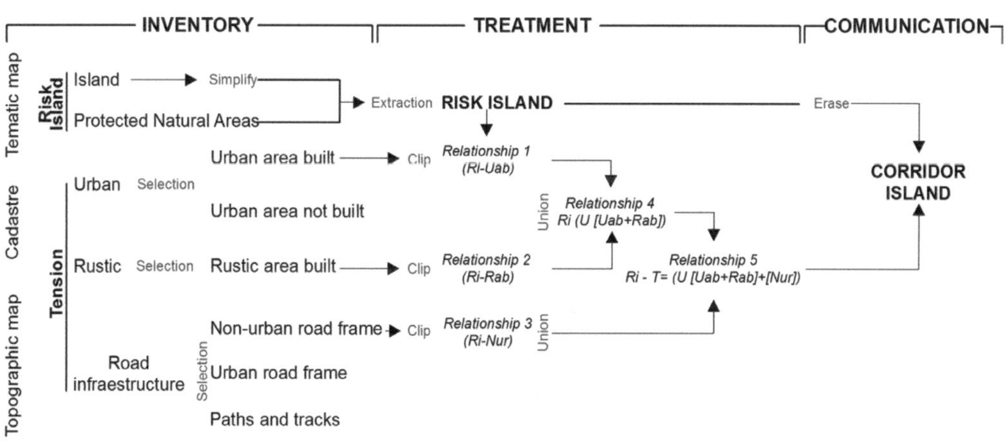

Scheme 1. Methodology.

For its development we use three different and basic cartographic sources: thematic cartography, that by means of the selection and treatment of the components of insular delimitation and of protected natural spaces we construct the risk island. This map is provided by the regional company GRAFCAN S.A. Cadastral and topographical cartography are the sources from which we extract and treat the components for the construction of the existing tension within the risk island. The cadastral provided by the General Directorate of Cadastre of Spain and the topographic ones by the company GRAFCAN S.A.

In order to identify the constructed parts (understood as spaces destined to houses, industries, commerce … or to constructions destined to the agricultural use), in the urban space as in the non-urban it has been necessary to exclude elements registered cadastrally. The result of this exclusion makes it possible to define constructed soil surfaces and to establish patterns. Exclusions differ according to the rustic or urban nature. In both cases, the excluded elements are classified into three groups: (1) orchards and plots; (2) water infrastructures; and (3) sports areas.

The non-urban road frame (Nur) is composed of the sum of motorways, roads, tracks and urban structural way. Urban routes (Ur) are formed by the fabric of streets of the villages. The roads and trails frame (Pat) consists of soft, non-sealed and permeable routes.

In this article we will focus in the first group: them frames road not urban, i.e., the network general and structural of the island of Tenerife paved. Will be in a second phase of this research where complete the synthesis of them way urban and the edition of them roads and the paths, although, priori, these latest not contribute to them processes of sealed, although if to them of dispersion.

3.1. Inventory level

3.1.1. Components of the risk island (Ri)

The Island (I) and the protected natural areas (PNA) are simple data layers, which has intervened. In the first dissolving the separation administrative and eliminating those polygons referred to islets and rocks that surround the perimeter coastal. As a result, remains only the part "continental" of the island. The objective is the simplification. Keep those elements multiplies the number of polygons and not provide data substantial to the target, as are very small, not are inhabited or processed by any activity. In the second (PNA), has intervened, only and for the same reason, in the islets and rocks of the coastal perimeter.

3.1.2. Components of the tension (T)

Its pressure or tension (T) on the risk island is in the process of sealing of soil [15], or in the waterproofing and destruction of the space biotic and productive primary. The variables influencing this process are: the built space and infrastructure. This problem is even more acute if there is an expansive and dispersed, model which creates a heterogeneous and irregular landscape [24], apparently interesting from the perspective of landscape, but enormously complex and difficult order [25].

The space built. It is the most complex and important stress variable. The cadastral information is divided in urban nature and rustic nature. To do this, it relies on legal ground–urban or rural class division. The patterns of organization of the space built in floor urban and flooring rustic are very significant and explain them models of settlement.

The urban area built (Uab) and urban area not built (Uanb) and rustic area built (Rab) are three variables of synthesis built starting from the data space and alphanumeric of Cadastre. Its construction was performed using multiple selections. This process is detailed in the headings of criteria.

Road structure. It is the other variable of pressure or tension (T), which possibly expands and channeled the built space. The analysis is constructed from synthesize, on the one hand, the wefts non-urban road (Nur); for another, the way urban (Ur); and third, the paths and roads frame (Pat). This synthesis is the result of the process of extraction and fusion of the road with polygonal geometry of the different sheets of the topographic map of the island scale 1:5000.

3.2. Level of treatment

3.2.1. Relationship level 1

Risk island (Ri). It is generated by the combination of the island components (I) and protected natural areas (PNA). The operation space made is its removal.

$$I - PNA = Ri \tag{2}$$

Tension (T). It is built by multiple relationships between the components (level 2, 3 and 4).

$$T\left[U\left\{(Uab) + (Rab) + (Nur)\right\}\right] \tag{3}$$

3.2.2. Relationship level 2

Risk island (Ri) with the urban area built (Uab)

$$Ri - Uab \tag{4}$$

Risk island (Ri) with the rustic area built (Rab)

$$Ri - Rab \tag{5}$$

Risk island (Ri) with the non-urban road frame (Nur)

$$Ri - Nur \tag{6}$$

3.2.3. *Relationship level 3*

Risk island (Ri) and the sum of the urban built environment and rustic (*U [Uab + Rab]*).

$$Ri - (U\,[Uab + Rab]) \tag{7}$$

3.2.4. *Relationship level 4*

Risk island (Ri) and the soil sealing component. (*T*) = *U* [*Uab* + *Rab*] + [*Nur*]

$$Ri - (T) \;=\; U\,[Uab + Rab] + [Nur] \tag{8}$$

3.3. Level of communication

It is the synthesis of the previous levels.

Corridor island (Ci) is obtained from the elimination in the risk island (Ri) all the space occupied by the tension (T).

$$Ci \;=\; Ri - T \tag{9}$$

4. Results

4.1. Inventory level

4.1.1. *Components of the risk island (Ri)*

4.1.1.1. *Island (I)*

The island of Tenerife has a surface area of 2034.7 km^2. It is the largest of the Canary Islands. Administratively it is divided into 31 municipalities. The surface range of municipalities is 201 km^2. The smallest municipality is the Puerto de la Cruz with 8.8 km^2, and the largest, La Orotava, with 210 km^2. In this municipality is located the Teide National Park (ISTAC from data of the National Institute of Statistics -INE-). The superficial media by municipality is 65 km^2 (**Table 1**). Population, Tenerife, is also the largest, with 42.3%, representing 889,936 residents in the year 2014 (**Table 2**).

The pressure that exerts the number of inhabitants in the island is important (**Table 2**), but if to these you add the population floating, i.e., the number of tourists that overnight in the island, the pressure is multiplies. This fact affects the distribution of the settlement and its management

Frequency	31
Minimum Area	8.8 km^2
Maximum Area	210 km^2
Accumulated Sum	2034 km^2
Average	65.6 km^2

Table 1. Statistics of surface by municipality.

	Inhabitants	%	km^2	Inhab/km^2
Canary islands	2104.815	100	7446.95	282.64
Tenerife	889.936	42.3	2034.38	437.44
Gran Canaria	851.157	40.4	1560.10	545.57
Lanzarote	141.940	6.7	845.94	167.78
Fuerteventura	106.930	5.1	1659.74	64.42
La Palma	83.456	4	708.32	117.82
La Gomera	20.721	1	369.76	56.03
El Hierro	10.675	0.5	268.71	39.72

Table 2. Population statistics of the Canary islands (2014).

model. The resident population is organized in concentrated urban settlements and scattered settlements, on the other hand, the tourist population, mainly choose tourist places (**Table 3**). Paradoxically, these tourist places are more efficient in the consumption of soil and more oriented toward a polycentric structure and high density. As urban backbone, are walks, streets or avenues where green space has an occupation and outstanding value, which next to the gardens and pools of accommodation infrastructures comprise scenically rich from the point of view of the landscape. This fact, at least admits a reflection: these spaces are attractive for travelers, but also for the residents, there is an ideal of paradise, then, why our model of organization of the settlement is so distant from that model?

	N° tourists	N° overnight stays	Average length of stay (days/tourist)	Occupancy rates
Tenerife	5148.453	39,540.119	7.68	67.14%

Table 3. Statistics of tourism receptive of the island of Tenerife (2014) (source: Area of employment, economic development, trade and foreign policy. Council of Tenerife).

4.1.1.2. Protected natural areas (PNA)

From 1987 to 2016 the legal history of protected areas in the Canaries is written as a succession of changes brought about by human pressure on the fragile and rich natural ecosystem of the Islands and adaptations to different national and regional laws of territorial organization. Part of a need for delimitation and conservation promoted in 1987 to a relativization of this delimitation or conservation, while as objective one, the draft law on the land of the Canaries (2016), promulgates that "management of the Canary Network Protected Natural Areas must meet the objectives of conservation, socio-economic development and sustainable use." Objective three states that the "socioeconomic development of the populations settled in protected areas, especially in rural parks and protected landscapes, will have a special consideration in the planning of them." It follows that the current delimitation (**Figure 1**) may change over the next few years. Possibly, the restrictions that the protected island, as defined in this work, exerts on the rest, the risk island, tend to break and modify those limits and uses. Clearly, the problem we present in this work is currently having a great impact and it is key to provide figures of this tension and with urgency.

The current result is that the island of Tenerife goes from having in 1987 a total of 32 protected natural spaces to 43, at present and its distribution by categories increases by 3–8. This represents 1098 km^2 (**Figure 1**).

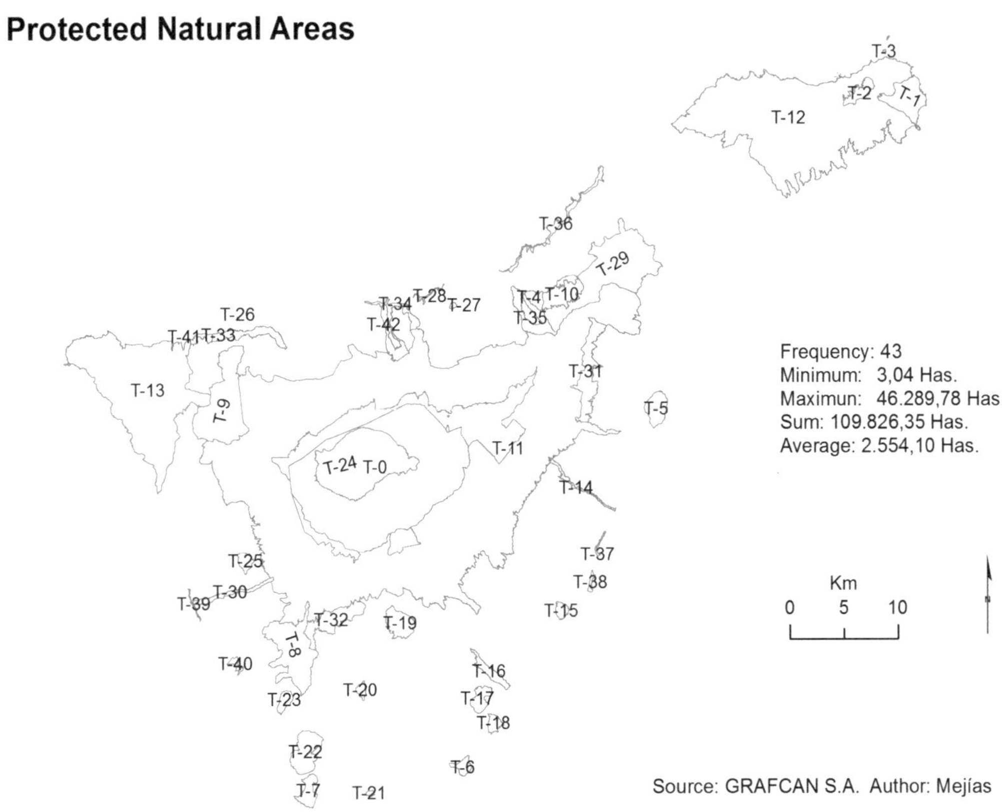

Figure 1. Distribution and description of protected natural areas (2016).

4.1.2. *Components of the tension (T)*

4.1.2.1. *The space built*

- Urban area builds (Uab). The land registry organizes the information on different items. Spatially, fragmented each cadastral reference into as many parts as divisions, including volumetric differences. In this way, the island of Tenerife deals space built in urban land between 140,954 cadastral references, and these in turn are divided in such a way that the number of polygons associated with cadastral references goes to 490,915. The sum total of the space built in urban land is from 34.11 km^2. The spatial distribution of the cadastral references is quite heterogeneous, because you can concentrate on an only polygon or reach distributed in 694. These ends respond to two models different. The first refers to the single-family constructions, the last responds to fragmented hotel establishments. But these are the minimum and maximum extremes. It is common to find, as it indicates the average–3.5–poor division. This fact indicates clearly the model of dispersion and of low density. The surface average built urban space is 242 m^2. The percentage of soil sealing is 1.67% (**Figure 2**).

- Urban area not built (Uanb). Excluded for not being built urban space adds 100 km^2, three times more than the space strengthened by building. The cadastral reference number is 146,344 and just has partitions, 1.3 divisions by reference. The average surface of each cadastral reference is 687 m^2. No doubt are plots very large and homogeneous to the not be subdivided (**Figure 2**).

- Rustic area builds (Rab). There are 6.75 km^2 of space built on rustic land. If count the rustic not built but valued, where incorporated fields for soccer, golf, tennis, or floors considered as courtyards or next area to building (which tend to be orchards in production), etc., the footprint is 16.85 km^2, nearly three times built in rural and half of the urban built space. If we add both figures rustic, the space in risk of construction is of 23.6 km^2. If we speak of "soil sealing," the data must be of 6.75 km^2. The number of built-up cadastral references is 43.475 and the average area of is 155 m^2. These dimensions are not buildings oriented farming or animal husbandry activities linked to the productive rustic features. If talk of the subdivisions within each of them references can specify that the media in of 2.16 divisions what explains them 94,061 polygons. The percentage of soil sealing is of 0.33% (**Figure 3**).

4.1.2.2. *Road structure*

The terrestrial communications network is complex. It's formed by motorways, dual carriageways, different roads and road urban. These make up the most important soil sealing. But they related other tissues consisting of roads and paths. Together we talk about 20,200 km. Are that support more than 4,556,240,521 vehicles in the year 2014, distributed in them 1157.50 km of roads not urban, what supposed an IMD average of 10.784 vehicles/day (Daily average intensity (IMD) is the most commonly used to characterize any magnitude. It is defined as the total number of vehicles spanning a section in a year divided by 365 days).

- Non-urban road (Nur). The surface that seals is of 17.80 km^2. Motorways sealed 1.80 km^2 in its few 113 km. The roads of different categories along with the rest of roads that make up the main structure of the nuclei of population total 16 km^2, divided in its impressive 6055 km. If I

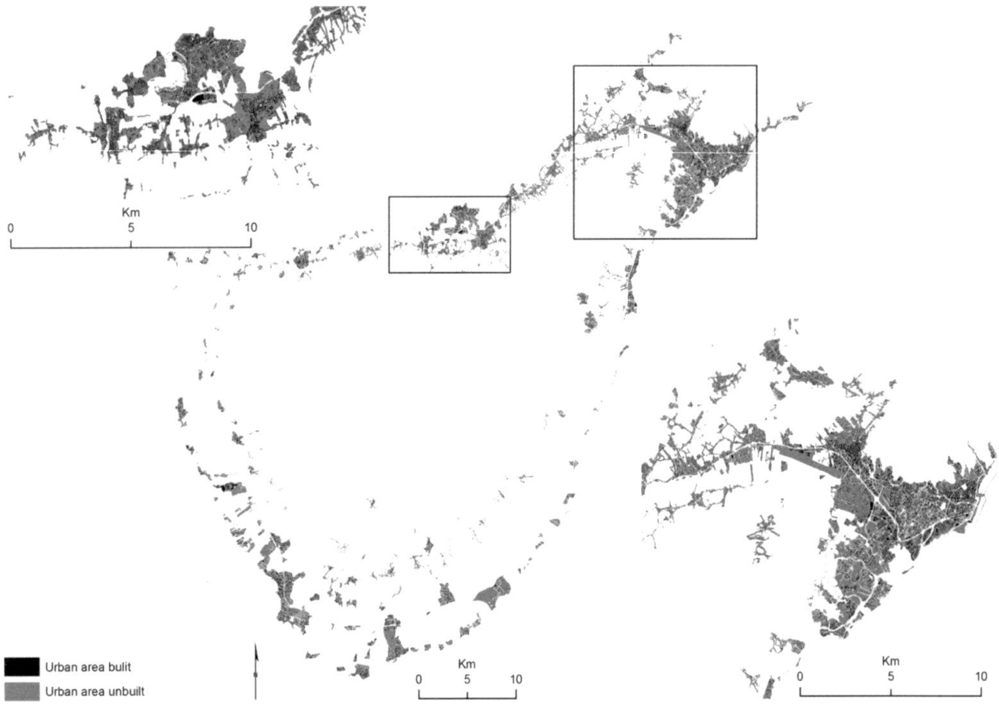

Figure 2. Distribution of urban area built (Uab) and unbuilt (Uanb).

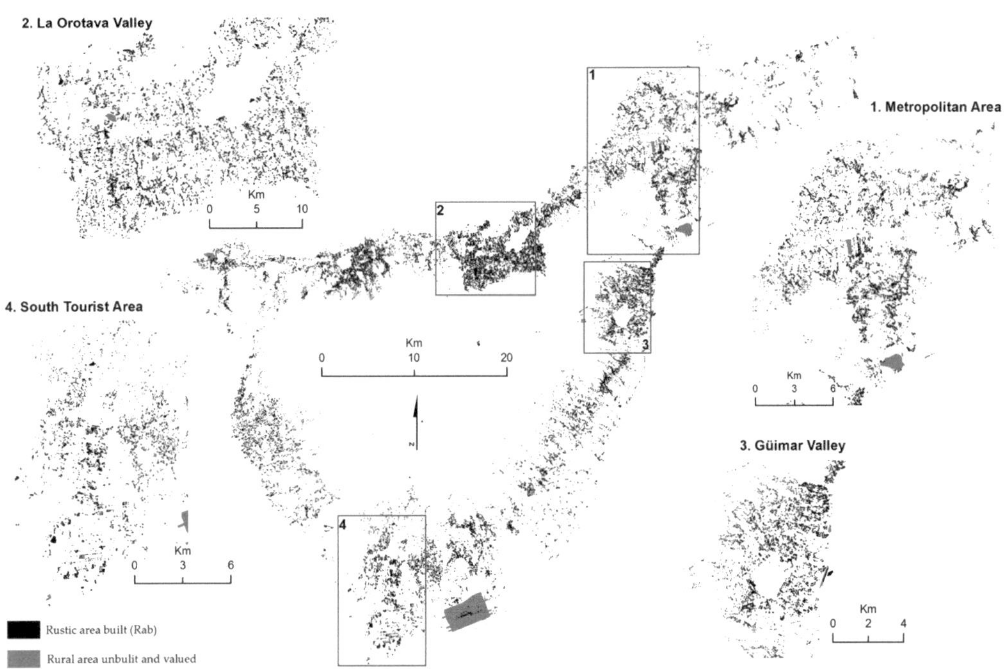

Figure 3. Distribution of the rustic area built (*Rab*) and unbuilt and valued.

relate it to the population every inhabitant would be 144 km of track. The percentage of sealing of soil is of 0.87% (**Figure 4**).

- Paths and roads (Par). They do not enter into the analysis of sealed, but those that are close to concentrated nuclei or even to the scattered pieces, are tending to become in developing that part of the seal. These routes are which channeled the housing development irregular on the island of Tenerife. This structure shows the agrarian paleo-landscape of the island of Tenerife, symbol of its identity [26]. The accumulated length is 14,172 km. Its density throughout the island is 6.5 km/km^2.

4.1.2.3. Synthesis inventory level

The analysis of the inventoried components already shows, at insular level, the intensity of the fragmentation (**Table 4**) and the important percentage of soil sealing. (**Table 5**).

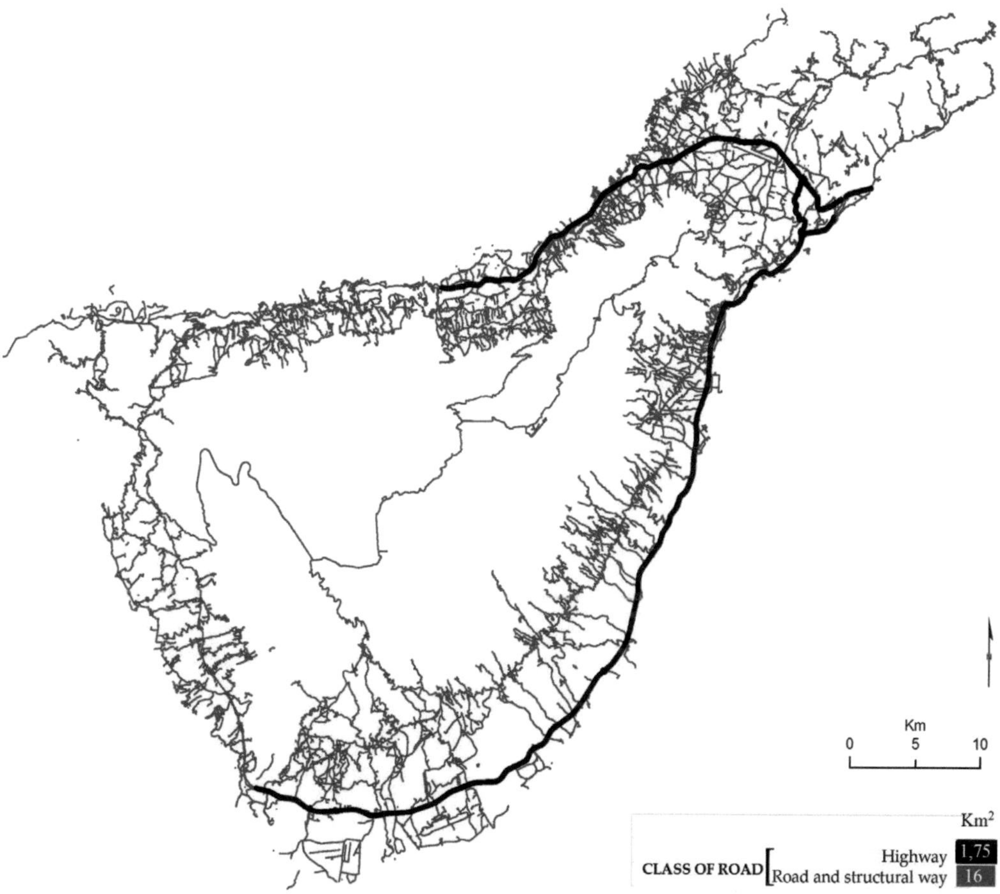

Figure 4. Distribution of the road structure.

Component	Code	Area	Length	Average (m^2)	Cadastral reference	Cadastral	Average polygon/ cadastral reference
		(km^2)	(km)		(frequency)	polygon	
Urban area built	*Uab*	34.11		242	140,954	490,915	3.48
Urban area not built	*Uanb*	100		687	146,344	196,287	1.34
Rustic area built	*Rab*	6.75		155	43,475	94,061	2.16
Rustic area not built	*Ranb*	16.85		1110	12,509	15,180	1.21
Non-urban road	*Nur*	17.8	6055				
Path and road	*Par*		14,172				
Total		**175.51**	**20,227**		**343,282**	**796,443**	**2.32**

Table 4. Synthesis of the fragmentation of the built space (inventory level).

Component	Code	Area	Soil sealing	Average soil sealing EU (%)
		(km^2)	(%)	
Urban area built	*Uab*	34.11	1.67	
Rustic area built	*Rab*	6.75	0.33	
Non-urban road	*Nur*	17.8	0.87	
Total		**58.66**	**2.87**	**2.3**

Table 5. Synthesis of soil sealing by components (inventory level).

4.2. Level of treatment

4.2.1. Relationships level 1

4.2.1.1. Risk island (Ri)

The risk island can be considered as a complex risk landscape unit. Here are determined by the processes of growth and restrictions on human activities. The projections of the territorial and urban planning are also concentrated in this space, hence the importance of these results and their magnification. If we calculate the pressure data, population, building and road frame on this new reduced spatial reference and real, and the results are triggered upward.

If we reduce the risk of transformation space 1046 km^2 and interact with the components that generate pressure diagnosis is more critical (**Figure 5**).

4.2.2. Relationships level 2

Tension (*T*). Combinations or risk island-generating components of strain relations express the following results:

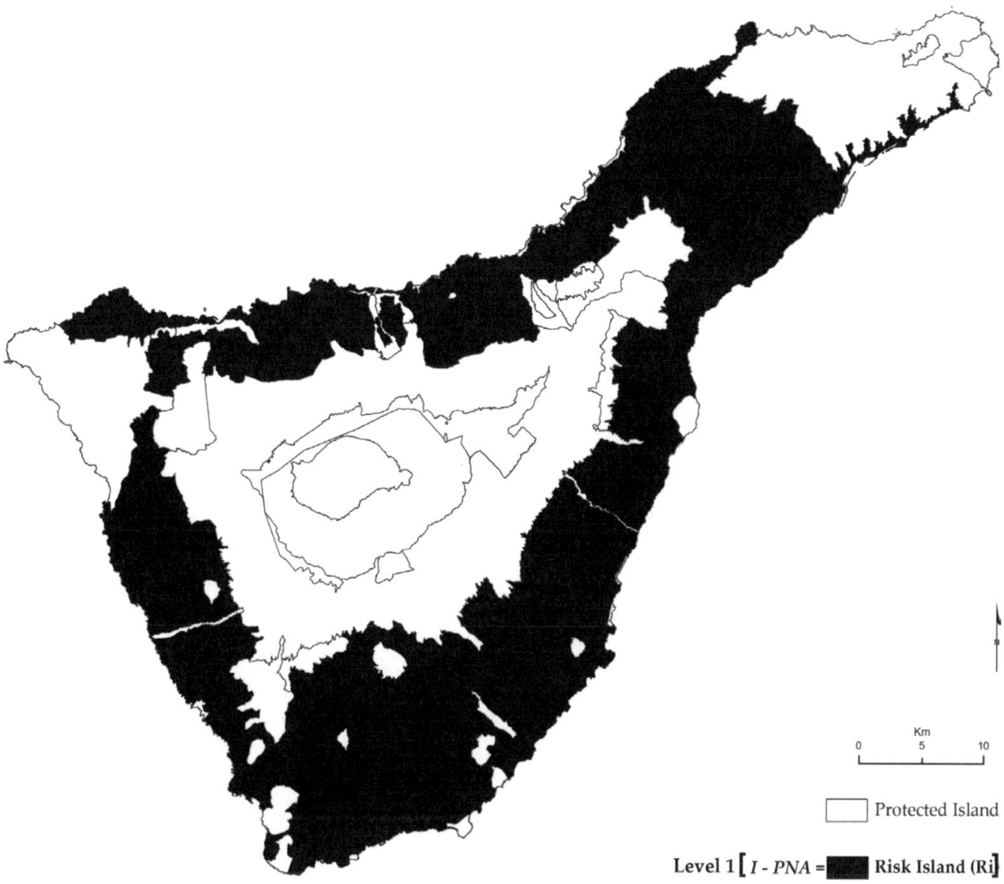

Figure 5. Risk island (*Ri*).

- Relationship of the risk island (*Ri*) with the urban area built (*Uab*)

The 100% the urban constructed land lies within the risk island. The sum of soil sealing is 34.11 km^2. This data represents the 3.26% of the risk island. If we compare it with the data from the EU of 2.3% of the total soil sealing, we are already talking of a tension in a single component. If we associate it with the absolute population is 38 m^2 per capita.

- Relationship of the risk island (*Ri*) with the rustic area built (*Rab*)

Risk island. The space built in rustic area is of 6.75 km^2. Within the risk Island are 6.36 km^2, representing 94.65%. The number of cadastral references is 40,133 and its media division–frequency–is 2.2 polygons/ reference. The average area of each reference is 82.27 m^2. The sealing of soil of this component is of 0.60% of the risk island. His relationship with the absolute population is 7.14 m^2 per capita.

Protected island. On the other hand, out of the risk island and within any of the categories of space natural protected exist 3373 references, the 5.35% of the constructions in rustic. The average surface of each reference is 67.03 m^2 and the number of divisions of each reference is 1.77.

- Relationship of the risk island (*Ri*) with the non-urban road frame (*Nur*)

Within the risk island is the highways total (100%) and 88.75% roads and structural way. In total 16 km^2. Each inhabitant of non-urban road is 18 m^2. The percentage of sealing of soil is of 1.52%.

4.2.3. Relationships level 3

- Relationship between the risk island (*Ri*) and the sum of the urban built environment and rustic (**Figure 6**). *U [Uab + Rab]*

The result is a built space partitioned by the island-risk, with a development of space built on rural land very intense, dispersed and fragmented, which in some cases (Orotava Valley, Güímar Valley or two rural parks–Anaga and Teno-) exceeds the risk island. Soil sealing by building is 40.83 km^2: the cadastral reference number is 181,087. The soil sealing represents 3.86% and its relationship with the population builds up 45.14 m^2 per capita.

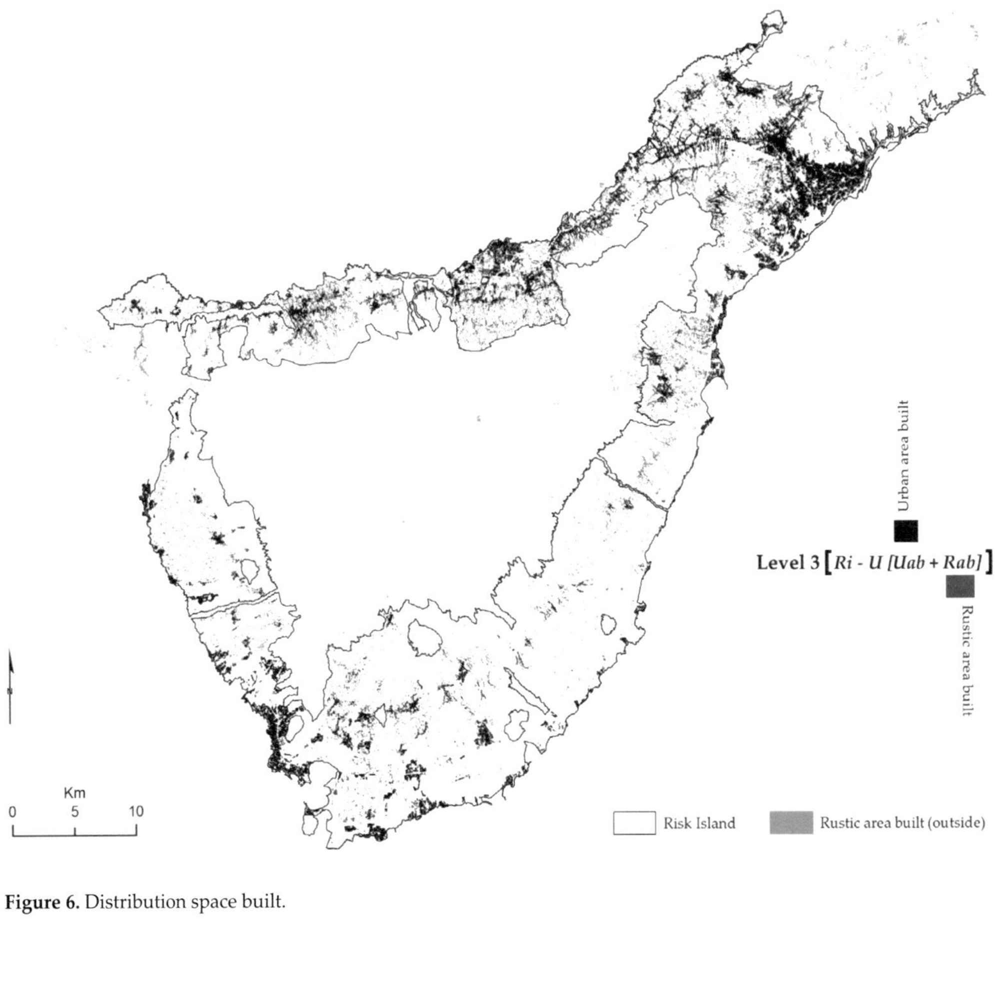

Figure 6. Distribution space built.

http://dx.doi.org/10.5772/intechopen.72190

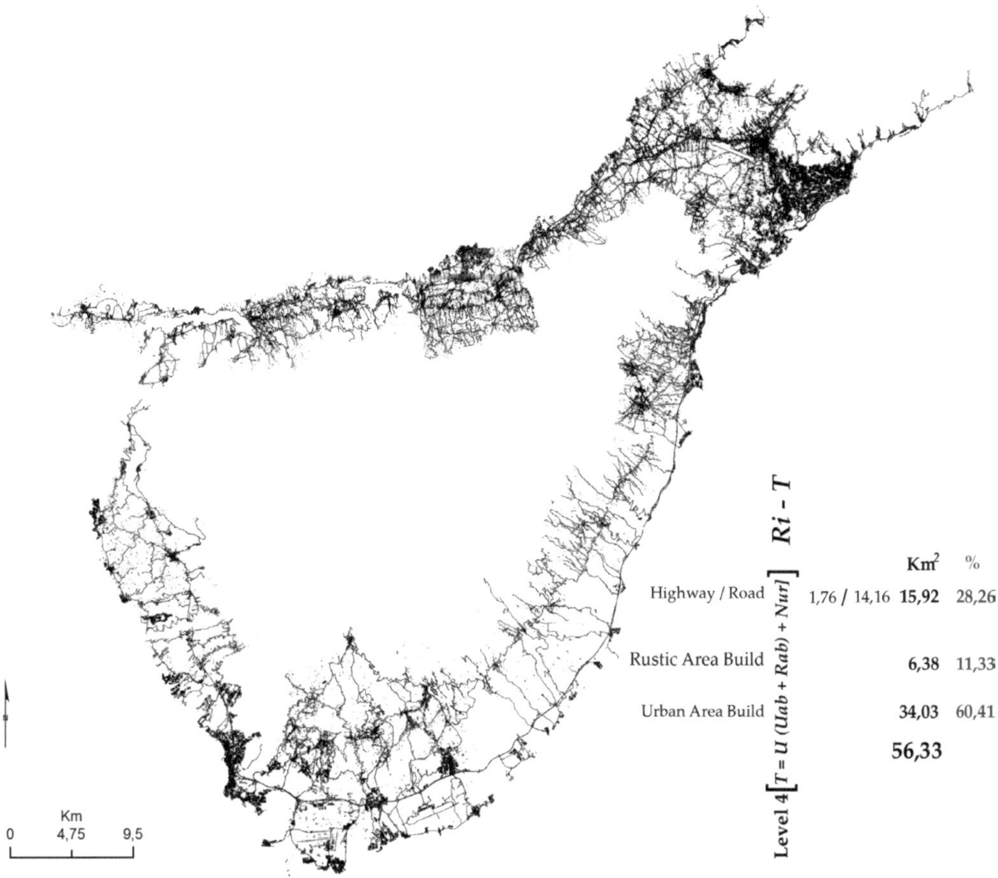

Figure 7. Distribution of the tension (pressure).

Component	Code	Area soil sealing (km²)	Average (m²)	Soil sealing (%)	m² per capita	Average soil sealing EU (%)
Urban area build	*Uab*	34.11	242	3.26	38.00	
Rustic area build	*Rab*	6.36	82.27	0.60	7.14	
Non-urban road	*Nur*	16		1.52	18.00	
Total		**56.33**		**5.38**	**63.29**	**2.3**

Table 6. Synthesis of soil sealing per component (treatment level).

4.2.4. Relationships level 4

- Relationship between the risk island (*Ri*) and the soil sealing component (**Figure 7**). $T = U[Uab + Rab] + [Nur]$

Tension on the risk island spatially focuses on urban construction and the intense network of roads and structural way, add 85% of the tension between the two. But at the same time the intensity of road network provides access to promote dispersion in non-urban land (11.33%). In any way the tension that generates the soil sealing in the risk island of Tenerife is of 56.33 km^2. This is 63.29 km^2 per capita. The percentage of sealing of soil is of 5.38% (**Table 6**).

4.3. Level of communication

Corridor island (*Ci*) (**Figure 8**).

The end result of this procedure leads to the corridor island. The remaining space on which rests the new tension. Our space unit of risk part with 1046 km^2. The value of the tension, represented by the soil sealing, is of 56.41 km^2. The corridor island features of 990.47 km^2. Those corridors are fragmented in 2200 polygons. These range from between 10 m^2 and 31 km^2.

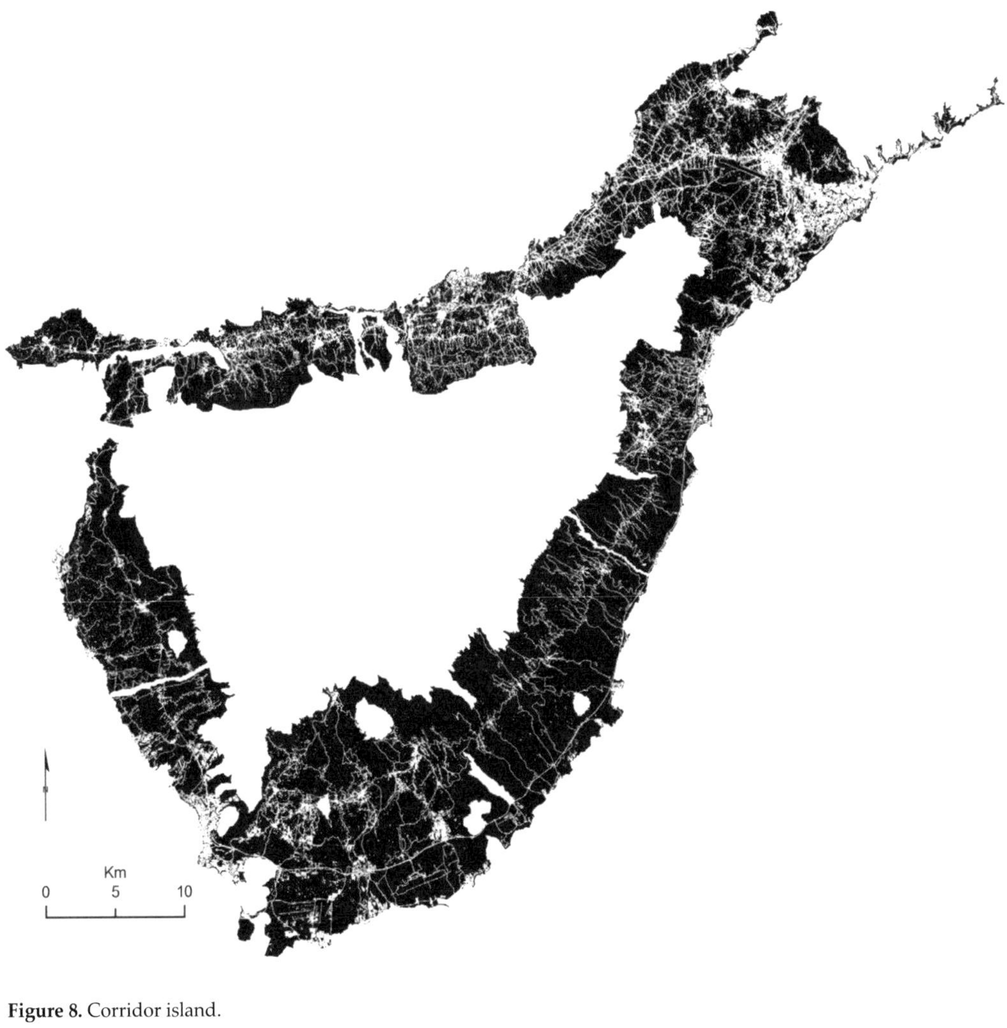

Figure 8. Corridor island.

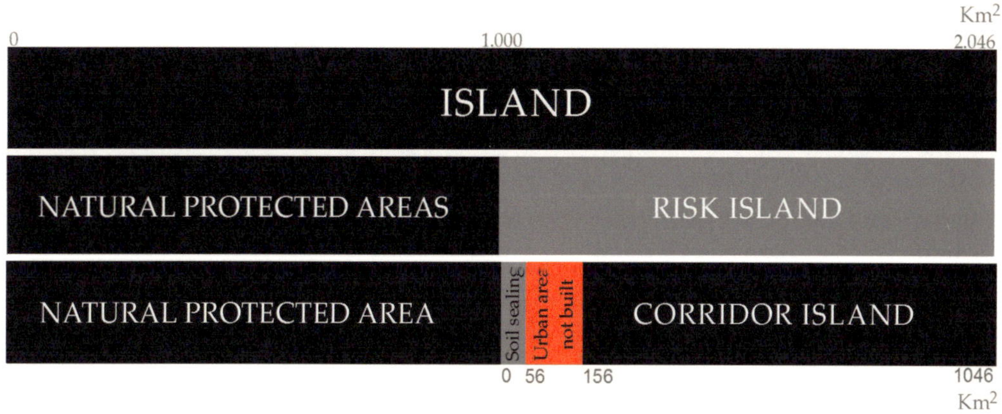

Scheme 2. Proportional distribution of the types of island.

The average is 0.45 km². This result demonstrates the importance of fragmentation. Also get the base space about which defines the complex strategies of intervention in this space (**Scheme 2**).

The key is now on characterizing the space corridor. This is divided into two main areas; on the one hand the cultural field represented by agricultural space and on the other the biotic field, both in contact, and the fruit of the activity or inactivity of the first. But also both in contact with the components that make up the stress, as defined in this article. What is the tendency of the polygons inter-road defined, how deals with territorial planning and the urban sprawl and the non-urban land management. These are issues that will deal of resolve in a next work starting from the delimitation of these units of reference obtained.

5. Conclusions

The island of Tenerife has a gross population density (ratio between absolute population and population) of 444.3 inhabitants/km², but a real population density, if we focus on its risk space is 850.79 inhabitants/km². This figure if we extrapolate it worldwide, Tenerife would occupy the position number 14 of the ranking. To this must be added the 14,100 daily tourists that coexist in the island, this supposes an increase of density in the risk island until the 864.28 Inhabitants/km².

In the island of Tenerife there are 43 protected areas between 3 and 46,000 hectares. Together they are 109,000 hectares. They are equivalent to 50% of the island territory. Almost 100% of the population residing or visiting the Island do not live in protected areas, that is, they are distributed within the other 50% of the total island. In spite of this, at the moment, it is being rethought from the present regional government, to modify the unique objective of conservation and delimitation of the protected spaces, incorporating and prioritizing the socioeconomic development for the resident population in those spaces.

The analysis of the space built in urban and rural soil diagnoses patterns of low density expansion models. The average surfaces of the built polygons are smaller in the rustic soil 155 m^2 than in the urban soil, 242 m^2. On the other hand, the reserve area that represents the unconstrained urban land triples that consolidated by the building. These data and their relationship with the insular surface indicate a percentage of soil sealing of 2%. According to the pressure levels determined by the study of soil sealing for the EU [17], Tenerife would be in the middle level only with this component.

The analysis of the space occupied by the non-urban road network indicates an important density, as well as the risk of unpaved roads and paths in the continuity of the dispersion. Only the asphalted road is counted as a component of soil sealing and this occupies 0.87% of the island.

If we add the values of the built-up space and the asphalted roadway, the total sealing result, on an island scale, is 2.87% of its territory. This result would be located in the medium-high border of the scale proposed by Prokop [18].

The risk island is a new spatial reference that should be used as a unit of analysis. It is more realistic because it is the transformable space. It must be the new space to design and apply landscape policies. On the island of Tenerife, it accounts for 50% of its territory.

The pressure or stress experienced by the risk island is evidenced by the occupation of the built urban space with respect to the sealing of the soil, as it is 1% higher than the EU average and would enter the high pressure level according to the Prokop scale [18].

The space built on rustic soil is more significant, representing 0.6% of the sealed floor and average surface area of 82 m^2 make them the clear dispersion and low density model. In the risk island is 95% of the total of this type of buildings; the remaining 5% is inside the protected island.

The non-urban and asphalted road network is almost entirely within the risk island, 100% of motorways and 89% of the main roads. They are 16 km^2 to which we would have to add the urban plot not calculated in this work.

In the absence of accounting for this data, currently under development, the total pressure on the unprotected island is 5.8%. This percentage would place it in the fourth place of the European ranking, only behind Malta, Holland and Belgium. Undoubtedly, this classification at the state level does not reflect regional particularities, but the status of insularity confers a specific status to the pressure exerted on a finite space. It would be interesting to analyze this problem in other EU insular spaces to compare urban growth models and their soil sealing coverage.

The rest of the island, the corridor, is a pressed, highly fragmented island. It is an island threatened by the building expansion through the network of roads not paved, that the planning instruments in force have not been able to curb. It is a space where the urban surface not built and even the space according to the classification of soil in force would duplicate the urban space and therefore the increase of pressure and sealing of soil of the corridor island.

The rest of the island, the corridor island, is a depressed island, very fragmented. It is an Island threatened by the building expansion through the network of unpaved roads, which the planning instruments in force have not been able to stop. It is a space where the undeveloped urban area, and even the planned urban development, could duplicate the urban space, and with it, the demand for infrastructures that would only increase the sealing of the ground.

Author details

Miguel Ángel Mejías Vera

Address all correspondence to: mmejias@ull.edu.es

Department of Geography and History, Faculty of Humanities, University of La Laguna, Spain

References

[1] Law 14/2014 on December 26, harmonization and simplification in the field of protection of the territory and natural resources. Available from: https://www.boe.es/boe/dias/2015/02/06/pdfs/BOE-A-2015-1116.pdf

[2] Hasse J, Lathrop R. Land resource impact indicators of urban sprawl. Applied Geography. 2003;**23**:159-175. DOI: 10.1016/j.apgeog.2003.08.002

[3] O'Meara M. Los límites de la ciudad. Cómo frenar la dispersión urbana. Cuadernos Worldwatch. Bakeaz. Centro de Documentación y Estudios para la Paz. Gobierno Vasco. Bilbao, Spain; 2004

[4] Sudhira H, Ramachandra T, Jagadish K. Urban sprawl: Metrics, dynamics and modelling using GIS. International Journal of Applied Earth Observation and Geoinformation. 2004;**5**:29-39. DOI: 10.1016/j.apgeog.2009.08.001

[5] García MA, Muñiz I. ¿Policentrismo o dispersión? Una aproximación desde la nueva economía urbana. Investigaciones regionales. 2007;**11**:25-43. Available from: http://www.redalyc.org/articulo.oa?id=28901102

[6] Jaeger J, Bertiller R, Schwick C, Kienast F. Suitability criteria for measures of urban sprawl. Ecological Indicators. 2010;**10**:397-406. DOI: 10.1016/j.ecolind.2009.07.007

[7] Bhatta B, Saraswati S, Bandyopadhyay D. Urban sprawl measurement from remote sensing data. Applied Geography. 2010;**2010**(30):731-740. DOI: 10.1016/j.apgeog.2010.02.002

[8] Tarroja A. La dimensión social del paisaje. In: Busquets J, Cortina A, editors. Gestión del paisaje. Manual de protección, gestión y ordenación del paisaje; Ariel, Barcelona. Spain; 2009. pp. 239-251

[9] García LM, Smith N, Mejías MA. Gentrification, displacement and tourism in Santa Cruz de Tenerife. Urban Geography. 2007;**28**(2):276-298 http://dx.doi.org/10.2747/0272-3638.28.3.276

[10] Ingersoll R. Sprawlscape: il paesaggio como redenzione. In De Rossi A, Durbiano G, Governa F editors. Lineen el paesaggio Esplorazioni nei territorio della trasformazione. UTET, Torino. 1999

[11] Muñoz F. Paisajes metropolitanos. In: Busquets J, Cortina A, editors. Gestión del paisaje. Manual de protección, gestión y ordenación del paisaje; Ariel, Barcelona. Spain; 2009. pp. 61-75

[12] European Environment Agency. Urban Sprawl in Europe. The Ignored Challenge. Report. N°10/2006; EEA and Joint Research Center: Copenhagen, Denmark. ISBN: 92-9167-887-2

[13] European Commission. Soil. Available from: http://ec.europa.eu/environment/soil/sealing_guidelines.htm

[14] Salvati L, Zetti M. The environmental "risky" region: Identifying land degradation processes through integration of socio-economic and ecological indicators in a multivariate regionalization model. Environmental Management. 2009;**44**:888-898. DOI: 10.1007/s00267-009-9378-5

[15] Turner S. Proceedings of the Technical Workshop on Indicators for Soil Sealing. Technical Report 80. Office for Official Publications of the European Communities, Luxembourg. 2002. 62 p. ISBN: 92-9167-497-4

[16] García JM, Pérez ME, García, MP. Revisión del concepto de sellado de suelos y propuesta de tipología urbana. Anales de Geografía. 2014; **34**:87-103. Available from: http://dx.doi.org/10.5209/rev_AGUC.2014.v34.n1.45193

[17] Plugiese L, Scarpetta S. An object based analysis applied to very high resolution remote sensing data for the change detection of soil sealing at urban scale. In: Bassis et al, editors. Recent Advances of Neural Network Models and Applications, Smart Innovation, Systems and Technologies 26. Springer International Publishing: Switzerland. pp. 155-162; 2014. DOI: 10.1007/978-3-319-04129-2_16

[18] Prokop G, Jobstmann H, Schönbauer A. Overview of best practices for limiting soil sealing or mitigating its effects in EU-27. In: Report on best practices for limiting soil sealing and mitigating its effects. Technical Report - 2011 – 050. European Commission. 2011. 228 p. DOI: 10.2779/15146

[19] Sabaté J, editors. Patrimonio y proyecto territorial. Colonias, Sèquia de Manresa y Delta del Llobregat. Espai Blau. Diputació Barcelona: Barcelona, Spain. 2004. 274 p

[20] Mejías MA, Alonso AB, Vera JR, Rodríguez D. El Tablero ¿village ou quartier? Vers la consolidation d'une stratégie paysagère. In: Colloque International Paysages de la Vie Quotidienne. Perpignan-Girona. France: Spain; March 2011. Available from: http://www.paysage-developpementdurable.fr/IMG/pdf/colloquepdd_programme_definitif_fr.pdf

[21] Berengo C, Di Maio S. We Are the Landscape. Understanding the European Landscape Convention. Florence, Italy: Giunti Progetti Educativi, RECEP ENELC; 2009

[22] Agencia de Ecología Urbana de Barcelona. Sistema de indicadores y condicionantes para ciudades grandes y medianas. AL21. Red de Redes de Desarrollo Local Sostenible. Barcelona, Spain. 2002

[23] Bertin J. La gráfica y el tratamiento gráfico de la información. Madrid: Taurus; 1987. p. 310

[24] Gallego J, Escribano P, Christensen S. Comparability of landscape diversity indicators in the European Union. In: From Land Cover to Landscape Diversity in the European Union. 2000. pp. 84-97. Available from: http://europa.eu.int/comm/agriculture/publi/landscape/index.htm

[25] Mejías MA. ¿Cómo medir el fenómeno urban sprawl a través de indicadores paisajísticos? Aplicación a la isla de Tenerife. Boletín de la Asociación de Geógrafos Españoles. 2013;**62**:49-73. Available from: http://www.age-geografia.es/ojs/index.php/bage/article/viewFile/1569/1489

[26] Mejías MA. Geographic information systems for the design and management of tourist routes. Landscape as strategic value in new tourism and land policy. In: Díaz FM, editors. Competitive Strategies and Policies for Tourism Destinations. Quality, Innovation and Promotion. Nova Science Publishers: New York, USA. 2010, 171-198. ISBN: 978-1-60876-475-4

Chapter 13

Land and Infrastructure Development in Peri-Urban Areas: Case Study of Gomti Nagar, Lucknow, India

Kana Ram Godha

Additional information is available at the end of the chapter

http://dx.doi.org/10.5772/intechopen.73525

Abstract

The objectives of the paper is to determine the status of land and infrastructure in the peri-urban areas of Lucknow Municipal Corporation (LMC). Lucknow, being the capital of most populated state of the country, has lots of migration in search of jobs and better education. The growth of Lucknow peri-urban areas has propelled by the economic growth and demographic pressure. The growth of peri-urban areas has of the city has to sustain over a period of time. However, the predominance of growth form over the period of time has significant harmful impacts. It has thwarted mass transit development, separated rich and poor, caused unnecessary travel, consumed fragile land, and generated excessive public expenditures. The paper has tried to pursue three key objectives of the development of land and infrastructure in peri-urban areas of the Lucknow city. The first was to elaborate the main conceptual and theoretical debates concerned with peri-urban areas. Second, the paper considered planned development of land and infrastructure as an option for sustainable development of peri-urban areas. Third, how planned development of land and infrastructure would help other Indian cities in providing better services to their citizens.

Keywords: economic growth, demographic pressure, sustainable development, transit development, peri-urban infrastructure

1. Peri-urban development–an introduction

The peri-urban area is an area of mixed rural and urban population and land uses. It begins at the point where agricultural land uses appears near the city and extends up to the point where villages have distinct urban land uses or where some persons, at least from the village community commute to the city daily for work or other purposes [7]. The term is used to describe the built up area just outside the corporation limits of the city [11]. The term fringe suggests a borderline case between the rural and urban and actually lies on the periphery of urban areas, surrounding it and distinguishing it from the truly rural countryside [6].

The peri-urban development has arguments for and against it. However, it is evident that it alters the urban spatial structure to bring about negative externalities that affect the functioning of the city [2]. This implies that causes are to be treated to lessen the negative impacts. The growth of an urban area are being propelled by the economic growth and demographic pressure. The sociocultural characteristics of the context and preferences exhibited, shape growth into growth patterns. In the wake of rapid expansion of metropolitan cities, the local governments can barely cope up with only operation and maintenance of the existing services within their jurisdiction. While serviced urban land may indeed be far below the requirement of the urban area, the processes of speculative purchase and land conversion from agriculture to nonagricultural use continue unabated in the peri-urban areas. However, the form this growth takes and its after effects are contributed to the institutional inefficiencies and land management mechanisms within their legal framework. However, the benefits include access to a wider choice of neighborhoods with required levels of stimulus and rest, access to more natural environments, affordable land market and subsidized services [3]. Within the neighborhoods, congestion levels are lower and security is higher.

Critics of peri-urban areas are to maintain that the predominance of growth form over the period of time has had significant harmful impacts. It has thwarted mass transit development, separated rich and poor, caused unnecessary travel, consumed fragile land, and generated excessive public expenditures. On the other side, there are those who believe that citizens are getting what they want: single-family homes on large plots, safe communities with good school systems, and metropolitan locations far from the pace and problems of urban populations.

2. An empirical case study of Gomti Nagar peri-urban areas of Lucknow

This case study is concerned with Gomti Nagar peri-urban area of Lucknow Municipal Corporation (LMC). Lucknow is rapidly growing metropolis and capital city of Uttar Pradesh state of India. The city is famous for its Nawabi culture and traditional values. Gomti Nagar is located in north-east part of Lucknow Municipal Corporation. It consists of both residential and commercial settlements. Gomti Nagar is one of the largest and most preferred commercial destinations in the city along with hectic residential development [1].

The city of Lucknow is gradually wearing a look of modern city with its swank commercial properties, sprawling shopping malls, and modern residential real estate and metro rail. The city has attracted a large number of migrants from smaller towns and villages of state in search of education and employment and higher-order services; therefore, it is estimated that there is an additional population of 100,000 in every 5 years [9], which needs to be incorporated in additional areas in the city. The agricultural lands in close proximity of

http://dx.doi.org/10.5772/intechopen.73525

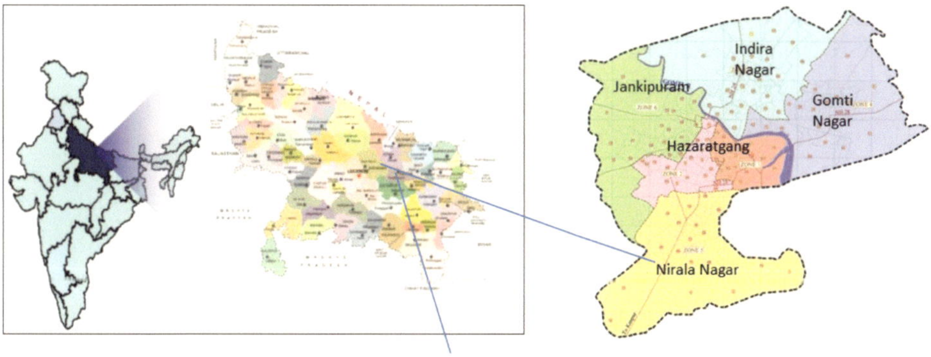

Figure 1. Location map of Gomti Nagar and Lucknow Municipal Corporation.

the built up areas of the city are slowly being converted to urban use. However, some of the conversion is a part of the planned growth; the rest of it is haphazard and is being carried out by private developers, property agents, land speculators, and individual owners (**Figure 1**).

The speculation of urban land by private developers for vested interested is leading to haphazard, unplanned, and illegal development in city. This is also resulting into a reduction in supply of serviced land for the public sector. The increasing population and aspirations of the peoples in new peri-urban areas like Nirala Nagar, Indira Nagar, Jankipuram, and Gomti Nagar are coming. Out of these peri-urban areas, Gomti Nagar is coming up as one of the most preferred commercial destinations in the city along with hectic residential development. This paper would elaborate the process of land development and its impact on peri-urban land and infrastructure in the Gomti Nagar of Lucknow city [10].

3. Land and infrastructure development in Gomti Nagar: peri-urban area of Lucknow

A Gomti Nagar peri-urban area is situated on the north-east side of the city in trans-Gomti river area, consisting of both residential and business settlements. It is situated on the banks of river Gomti, which flows through Lucknow, hence its name. It is one of the largest and upcoming areas of Lucknow. Gomti Nagar is an organized settlement developed by Lucknow Development Authority (LDA), taken up by the LDA in 1980s. LDA works as an autonomous body to fulfill the growing demand of housing in the State. It is home to high-end premium residential projects, malls, IT parks, commercial property, plots, business centers, multiplexes, clubs, banks, food courts, entertainment centers, and finance institutions.

Gomti Nagar is now developing as the commercial and IT hub of mainland Lucknow. Till now, the central Lucknow consisting of Hazratganj, Vidhan Sabha Marg, and Ashok Marg was the main commercial area. Realty of Gomti Nagar is growing at very fast. It has the software technology park, Reserve Bank of India building, NABARD, and many other important offices and commercial centers. It boasts of some important national and state level headquarters of SONY, Reliance retail, etc. It also has a developed industrial area with several factories and industrial units. The Sahara city is also situated in Gomti Nagar.

As earlier stated that the development of Gomti Nagar is divided into three phases:

1. Gomti Nagar—Phase-I
2. Gomti Nagar—Phase-II and
3. Gomti Nagar extension.

Phase-I and II have good occupancy rates, Gomti Nagar extension is still under construction (**Table 2**).

For development of colony, primary survey of the land was carried out by the LDA and Uttar Pradesh Housing Board (UPHB) has prepared plan and sent Government of Uttar Pradesh with a request for acquisition of land, after having been passed the state government issued a notification for preliminary intention to acquire land around 1080 ha (approx.) [8] (**Table 1**).

The Lower Income Groups (LIG) and Economic Weaker Section (EWS) houses are mostly located in Vinay Khand and in 1995 onwards, people started living and high income groups (HIG) colonies are mostly in Vipul Khand of Gomti Nagar [4]. The basic infrastructure and construction was carried out by the LDA and finally handed over to LMC for further maintenance (**Table 3**).

The infrastructure facilities provided by the Government agencies are adequate and able to sustain with the future growth of the area. The layout of the colony is cosmopolitan—with

S. No.	Name of colony	Gomti Nagar
01	Land area	1080 ha (approx.)/10.8 km^2
02	Project cost	Rs. 900 crore (approx.)
03	Location	Situated in trans-Gomti river area and along Faizabad road

Source: Lucknow Development Authority [5].

Table 1. Details of Gomti Nagar peri-urban area project, Lucknow.

Position	EWS	LIG	MIG-I	MIG-II	HIG	Total
Total taken up in Phase-I and Phase-II	5245	4686	5838	6597	4925	27,291
Completed	5245	4686	5838	6628	4908	27,305
Allotted	5010	4557	5691	6322	4450	26,030
Possession	5010	4526	5604	6162	4218	25,520

Source: Lucknow Development Authority [5].

Table 2. Details of plotted development for different income groups Gomti Nagar Phase I and Phase II.

http://dx.doi.org/10.5772/intechopen.73525

Name of colony	Total scheme area (sq.m.)	Total no. of houses	Total built-up area (sq. m.)	Road area (sq.m.)	Open space (sq.m.)	Other amenities area (sq.m.)	Assessment of infrastructure facilities	Other facility
Gomti Nagar	1,080,0000	27,305	3,800,000 (60%)	940,000 (17%)	900,000 (12%)	600,000 (11%)	Water supply, electricity and sewerage line available, internal roads in good condition	Street light, parks and community center

Source: Lucknow Development Authority [5].

Table 3. Infrastructure status in Gomti Nagar peri-urban areas.

roads running parallel and at the right angles to each other and services placed at vantage points. All basic amenities, like, shopping centers, schools/colleges, parks, hospitals and dispensaries, community centers, petrol pumps, bus stop, and also is stadium provided by the Government agencies.

All these development activities of peri-urban land development and infrastructure development by public sector agencies are carried out with respect to the existing policies of Lucknow Development Authority. However, access to land is made through the acquisition of land act, which has major drawbacks of procedural delay and encroachment on land by other interest groups.

3.1. Comparison of land and infrastructure of Gomti Nagar peri-urban area with other part of the cities

Gomti Nagar area witnessed a growth in property prices as development of Lucknow city started spreading outside its limits. The area being near to city limits experienced growth and soon captured the rising property prices. With the current property prices in the city rocketing toward sky, the land rate in the Chowk area is high as it is the center of the informal commercial market and the price varies 6000–9000 per sq. ft. While the land rate are higher in Hazaratganj than the Chowk area because it has better infrastructure and comparatively new development than the Chowk so the land price over here are 7000–9000 per sq. ft. on main road area as there are virtually no place for further development of the rate of land, which are available are very high. On the other hand, in case of new area development like in Indira Nagar, Nirala Nagar, and Gomti Nagar, land prices are comparatively lower than the city and these are affordable. In Gomti Nagar area, the land prices are as high as 6000 per sq.ft., while in the Nirala Nagar it is up to 5000 per sq.ft. In the new peri-urban areas, the infrastructure is better than in comparison to old areas, more open spaces and better living environment, and comparatively affordable price and this also shows that the old city center areas are getting more and more congested due to various reasons, and hence, people are preferably moving outside the city center toward the periphery, which has livable environment and also affordable to mass.

3.2. Comparison between open vs. built spaces

The graph number shows that Chowk area has highest built-up in comparison to other areas as this one is older development and quite congested than the Hazaratganj and other new areas, the open space is also very less and that is why this area is very much suffocating than the other areas. The other amenities and facilities are also not good in old areas such as road and basic infrastructures and this may be the reason why people who can afford are moving toward outer periphery along with migrating population. The data shows that all three peri-urban colonies namely Indira Nagar, Nirala Nagar, and the Gomti Nagar are newly developed areas basically after late 1970s and in 1980s in response to the congestion in the central city area, population increase and increasing aspirations of people. These three areas have wider road, better infrastructure, and housing with medium and low density built-up (**Figure 2**).

3.3. Comparison of problems in Gomti Nagar peri-urban area with other part of the cities

The residents of Gomti Nagar and other part of cities rated the problems in the area and majority of the peoples rated traffic as the major problem in the old city areas followed by

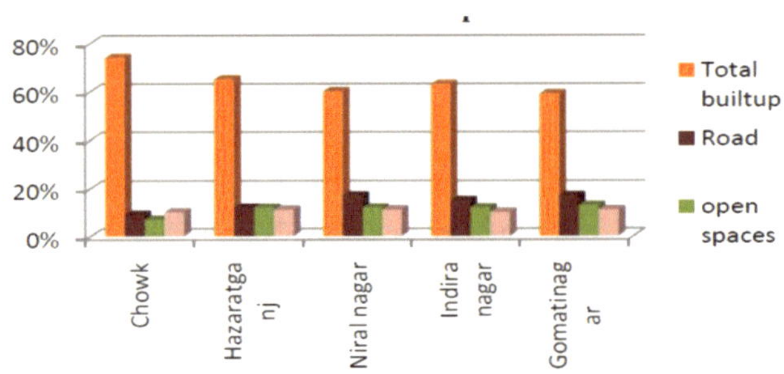

Figure 2. Comparison between open vs. built spaces of Gomti Nagar with other areas of city. Source: Lucknow Development Authority and filed visit.

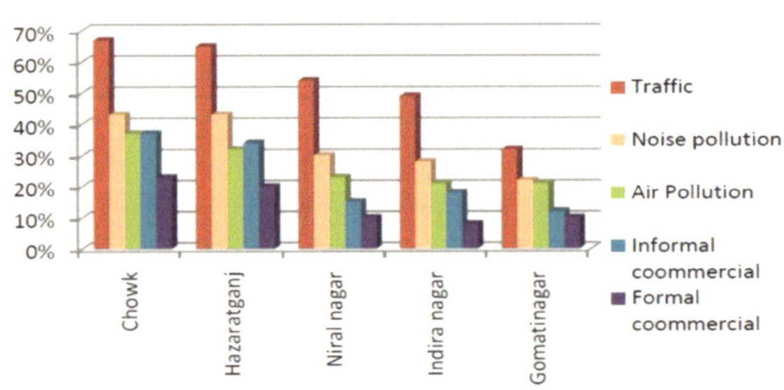

Figure 3. Comparison of problems in Gomti Nagar peri-urban area with other parts of the cities. Source: Lucknow Development Authority and filed visit.

http://dx.doi.org/10.5772/intechopen.73525

noise pollution and air pollution, which create congestion graphs. Residents in the Chowk area had rated traffic congestion as a major problem because the roads are very narrow and the informal activities are spread across roads. This creates lots of hassle and traffic jams for long hours, and therefore, lots of noise pollution and air pollution in the area. Hazaratganj has also face the same problems because of lots of formal and informal activities and conversion of residential units into office spaces and for commercial purposes. This also creates lot traffic jams, congestion, noise, and air pollution. While in Nirala Nagar, Indira Nagar and Gomti Nagar area this problem is very less because these area are new development and away from the central hub of the city and roads are also wide so has less traffic jams, noise and air pollution (**Figure 3**).

The Gomti Nagar peri-urban area has very less problems as roads are wider and area is well planned and newly developed.

4. Status of infrastructure in Gomti Nagar peri-urban area

4.1. Water supply

The River Gomti continues to be the main source of water supply to city, though a number of tube-wells have been bored to exploit ground water. Water supply in Gomti Nagar is through the household's water supply with metered connection, and water supplied to people is 3 h in the morning and 2 h in the evening. The houses sentiments which do not have household's level, municipal water supply have their own bore wells for water.

4.2. Sewerage and solid waste management

In Gomti Nagar, most of the households have septic tanks and few of them are connected with the sewer line. The door to door solid waste collection system is exist in Gomti Nagar peri-urban area. However, in some areas, usually waste from the households is thrown on the streets. Some households have begun to contact private collectors to deposit the waste into the nearby depots.

4.3. Roads and public transportation

The road network is wide and efficient to cater optimal traffic in Gomti Nagar. This was done in a planned manner keeping mind of future growth of the city. The roads of Gomti Nagar, Jankipuram, Aliganj, Indira Nagar, Nirala Nagar, and Sahara city are wide. The internal roads are 10 m wide and are enough to cater to traffic volume generated peak time in these areas. The infrastructure development activities in Gomti Nagar peri-urban areas has carried out by the public sector agencies like Lucknow Development Authority. Access to the land is made under the land acquisition act. The objectives of development of housing for HIG and MIG and land development process through private agencies are to earn maximum profit and developed the infrastructure for urban poor who cannot afford housing and other facilities.

Overall, in Gomti Nagar peri-urban area, the process of peri-urban land and infrastructure development is controlled by the public and private sector agencies.

5. Issues in development of land and infrastructure in Gomti Nagar peri-urban area

It is observed that Gomti Nagar peri-urban area has planned developments. However, still LDA is facing issues of development. These has its negative effects along with its positive benefits and its impacts can be seen in different heads like physical, legal, organizational, and financial issues and these are:

5.1. Land speculation in Gomti Nagar area

Gomti Nagar peri-urban area of Lucknow is one of the fastest growing peri-urban areas of the city. However, issue of land speculation is one of the major issue as real estate developers enters into the market the land rate in Gomti Nagar area getting highest in the city. It is being unaffordable for lower income and EWS people of the cities. The land rates are reaching the sky and caused problems of land speculation in Gomti Nagar areas.

5.2. Development in low lying area of trans-Gomti river

The Gomti Nagar is getting developed in the trans-Gomti river area, which is very low lying and marshy land. However, it is as risky for development as of the risk of floods from the river and other hazardous diseases such as water borne diseases can occur due to the dirty water quality. Development in low lying area is also a costly affair in Gomti Nagar area.

5.3. Restricted development permission in Gomti Nagar

In the Gomti Nagar area, permissible FSI is one which is very less as this area has the potential to absorb extra population of the city, development activities should be encouraged within this area, by giving benefits of extra FSI and Transfer of Development Rights.

5.4. Less developed physical infrastructure

Despite this area has planned development but still there is a lack of inadequate physical infrastructure facilities like water supply system, sewerage, and solid waste management etc.

Though, Gomti Nagar peri-urban area is developed by Lucknow Development Authority in a planned manner but still there are many milestones, which should be unturned for maximum benefits of the development and future growth of the area. The peri-urban must absorb the growing population and their aspirations. The model of Gomti Nagar peri-urban development could be a role model for other cities if this area would addressed the issues in a proper away. In larger context, there is a need of careful slight attention from development and implementation authority to control haphazard development in peri-urban area so that Gomti Nagar will be a living role model of peri-urban development to other Indian cities.

6. Way forwards and learning to other Indian cities

Gomti Nagar peri-urban area of Lucknow has come as one of the most planned development in Lucknow city with boasts of modern and Hi-tech residential and commercial complexes. The city is growing very rapidly and its boundary limit is changing. The villages located at surrounding often come under the influence of urbanization and pressure depending on the direction and intensity of growth of the city. The growth of a city is triggered by growth of the population by natural processes and indirectly by the ability of the city to offer economic potential in its context through the influx of migrant population. This outcome manifests itself in space and evolves over time to produce a growth pattern in the form of peri-urban development to fulfill the growing aspirations and this growth pattern along with the evolution of the spatial structure of the city gives clues on the efficiency of the city structure in fulfilling its functions. Much of the urban growth is taking place along transport corridors, and unrestrained by municipal jurisdictions; the distinction between urban and rural is getting blurred. As there is no place in the central city, the growth may decline but in the peripheries, there will be new growth because these areas are the new hopes for city. New peri-urban areas like Jankipuram, Nirala Nagar, Indira Nagar, Aliganj, and Gomti Nagar area etc. are coming very fast along the various trunk roads of the city and therefore they are very well connected with the central part of the city.

The Gomti Nagar area of Lucknow city is the modern development by Lucknow Development Authority where infrastructural facilities along with other facilities are good and development of this area has happened in planned manner, and it will become a role model for the other peri-urban land development in the city. For the cities of similar size like Kanpur, Allahabad, Meerut, and Moradabad, there is a need of an effective local government and sensible legal framework for the operation.

The presence of a legal mechanism, which facilitates action, autonomous central institution, and a regulated land market, and all these will make a planned development of the city without emergence of sprawl. This will ensure efficient utilization and equitable distribution of resources that enhance economic growth and secure environmental sustainability. In the recent past, the LDA has been proactive in plan preparation and implementation. Not only the development plan proposals have been projected for implementation but also the servicing of land for residential development have been phased out in the plan document. As the organization is able to involve stakeholders of various kinds, its intentions to facilitate development come with the minimal delays and encumbrances. The city has intended to achieve a more efficient urban form, support, and promote vitality of the urban economy, integrate urban poor into the urban fabric by providing them place to stay and live, plan for a sustainable urban environment in terms of better infrastructure, sanitation, open spaces, wide roads, and living places etc. and for this, the authority has started implementing plan for peri-urban